Atlantis Studies in Variational Geometry

Volume 3

More information about this series at http://www.atlantis-press.com

Gennadi Sardanashvily

Noether's Theorems

Applications in Mechanics and Field Theory

Gennadi Sardanashvily
Moscow State University
Moscow
Russia

ISSN 2214-0700 ISSN 2214-0719 (electronic)
Atlantis Studies in Variational Geometry
ISBN 978-94-6239-170-3 ISBN 978-94-6239-171-0 (eBook)
DOI 10.2991/978-94-6239-171-0

Library of Congress Control Number: 2016932506

Printed on acid-free paper

To my wife
Aida Karamysheva
Professor, molecule biologist

Preface

Noether's first and second theorems are formulated in a very general setting of reducible degenerate Grassmann-graded Lagrangian theory of even and odd variables on graded bundles.

Lagrangian theory generally is characterized by a hierarchy of nontrivial Noether and higher-stage Noether identities and the corresponding gauge and higher-stage gauge symmetries which characterize the degeneracy of a Lagrangian system. By analogy with Noether identities of differential operators, they are described in the homology terms. In these terms, Noether's inverse and direct second theorems associate to the Koszul–Tate graded chain complex of Noether and higher-stage Noether identities the gauge cochain sequence whose ascent gauge operator provides gauge and higher-stage gauge symmetries of Grassmann-graded Lagrangian theory. If these symmetries are algebraically closed, an ascent gauge operator is generalized to a nilpotent BRST operator which brings a gauge cochain sequence into a BRST complex and provides the BRST extension of original Lagrangian theory.

In the present book, the calculus of variations and Lagrangian formalism are phrased in algebraic terms of a variational bicomplex on an infinite order jet manifold that enables one to extend this formalism to Grassmann-graded Lagrangian systems of even and odd variables on graded bundles. Cohomology of a graded variational bicomplex provides the global solutions of the direct and inverse problems of the calculus of variations.

In this framework, Noether's direct first theorem is formulated as a straightforward corollary of the global variational formula. It associates to any Lagrangian symmetry the conserved symmetry current whose total differential vanishes on-shell. Proved in a very general setting, so-called Noether's third theorem states that a conserved symmetry current along any gauge symmetry is reduced to a superpotential, i.e., it is a total differential on-shell. This also is the case of covariant Hamiltonian formalism on smooth fibre bundles seen as the particular Lagrangian one on phase Legendre bundles.

Lagrangian formalism on smooth fibre bundles and graded bundles provides the comprehensive formulations both of classical field theory and nonrelativistic mechanics.

Non-autonomous nonrelativistic mechanics is adequately formulated as particular Lagrangian and Hamiltonian theory on a configuration bundle over the time axis. Conserved symmetry currents of Noether's first theorem in mechanics are integrals of motion, but the converse need not be true. In Hamiltonian mechanics, Noether's inverse first theorem states that all integrals of motion come from symmetries. In particular, this is the case of energy functions with respect to different reference frames.

The book presents a number of physically relevant models: superintegrable Hamiltonian systems, the global Kepler problem, Yang–Mills gauge theory on principal bundles, SUSY gauge theory, gauge gravitation theory on natural bundles, topological Chern–Simons field theory and topological BF theory, exemplifying a reducible degenerate Lagrangian system.

Our book addresses to a wide audience of theoreticians, mathematical physicists and mathematicians. With respect to mathematical prerequisites, the reader is expected to be familiar with the basics of differential geometry of fibre bundles. We have tried to give the necessary mathematical background, thus making our exposition self-contained. For the sake of convenience of the reader, a number of relevant mathematical topics are compiled in appendixes.

Moscow
October 2015

Gennadi Sardanashvily

Contents

Introduction

Noether's theorems are well known to treat symmetries of Lagrangian systems. Noether's first theorem associates to a Lagrangian symmetry the conserved symmetry current whose total differential vanishes on-shell. The second ones provide the correspondence between Noether identities and gauge symmetries of a Lagrangian system. We refer the reader to the brilliant book of Yvette Kosmann-Schwarzbach [84] for the history and references on this subject.

Our book aims to present Noether's theorems in a very general setting of reducible degenerate Grassmann-graded Lagrangian theory of even and odd variables on graded bundles.

We however start with even Lagrangian formalism on smooth fibre bundles (Chap. 2), and focus ourselves especially on first-order Lagrangian theory (Chap. 3) because the most physically relevant models (nonrelativistic mechanics, gauge field theory, gravitation theory, etc.) are of this type.

Lagrangian theory of even (commutative) variables on an n-dimensional smooth manifold X conventionally is formulated in terms of fibre bundles over X and jet manifolds of their sections [15, 53, 108, 133, 143], and it lies in the framework of general technique of nonlinear differential equations and operators [25, 53, 85]. This formulation is based on the categorial equivalence of projective $C^\infty(X)$-modules of finite rank and vector bundles over X in accordance with the classical Serre–Swan theorem, generalized to noncompact manifolds (Theorem A.10).

The calculus of variations and Lagrangian formalism on a fibre bundle $Y \to X$ can be adequately formulated in algebraic terms of the variational bicomplex (1.21) of differential forms on an infinite order jet manifold $J^\infty Y$ of sections of $Y \to X$ [3, 15, 56, 61, 108, 133, 143]. In this framework, finite order Lagrangians and Euler–Lagrange operator are defined as elements (1.32)–(1.33) of this bicomplex (Sect. 1.3). The cohomology of the variational bicomplex (Sect. 1.2) provides a solution of the global inverse problem of the calculus of variations (Theorems 1.15–1.16), and states the global variational formula (1.36) for Lagrangians and Euler–Lagrange operators (Theorem 1.17).

In these terms, Noether's first theorem (Theorem 2.7) and Noether's direct second theorem (Theorem 2.6) are straightforward corollaries of the global decomposition (1.36).

Noether's first theorem associates to any symmetry of a Lagrangian L (Definition 2.7) the conserved symmetry current (2.21) whose total differential vanishes on-shell (Sect. 2.3). One can show that a conserved symmetry current itself is a total differential on-shell if it is associated to a gauge symmetry (Theorem 2.8).

Treating gauge symmetries of Lagrangian theory, one is traditionally based on an example of Yang–Mills gauge theory of principal connections on the principal bundle $P \to X$ (8.3) where gauge symmetries are vertical principal automorphisms of this bundle (Sect. 8.2). They are represented by global sections of the associated group bundle (8.43) and, thus, look like symmetries depending on parameter functions. This notion of gauge symmetries is generalized to Lagrangian theory on an arbitrary fibre bundle $Y \to X$ (Definition 2.3) and on graded bundles (Definition 7.3).

Given a gauge symmetry of a Lagrangian system on fibre bundles, Noether's direct second theorem (Theorem 2.6) states that its Euler–Lagrange operator obeys the corresponding Noether identities (2.17) (Sect. 2.2). A problem is that any Euler–Lagrange operator satisfies Noether identities, which therefore must be separated into the trivial and nontrivial ones. These Noether identities can obey first-stage Noether identities, which in turn are subject to the second-stage ones and so on. Thus, there is a hierarchy of nontrivial Noether and higher-stage Noether identities which characterizes the degeneracy of a Lagrangian theory (Sect. 7.1). A Lagrangian system is called degenerate if it admits nontrivial Noether identities and reducible if there exist nontrivial higher-stage Noether identities. We follow the general analysis of Noether and higher-stage Noether identities of differential operators on fibre bundles when trivial and nontrivial Noether identities are described by boundaries and cycles of a certain chain complex [61, 123]. This description involves Grassmann-graded objects.

In a general setting, we therefore consider Grassmann-graded Lagrangian systems of even and odd variables (Chap. 6).

Different geometric models of odd variables either on graded manifolds or supermanifolds are discussed [28, 30, 46, 107, 134]. Both graded manifolds and supermanifolds are phrased in terms of sheaves of graded commutative algebras [7, 61, 131]. However, graded manifolds are characterized by sheaves on smooth manifolds, while supermanifolds are constructed by gluing of sheaves on supervector spaces. We follow the Serre–Swan theorem for graded manifolds (Theorem 6.2) [11, 61]. It states that, if a graded commutative $C^\infty(X)$-ring is generated by a projective $C^\infty(X)$-module of finite rank, it is isomorphic to a ring of graded functions on a graded manifold whose body is a smooth manifold X. Accordingly, we describe odd variables on a smooth manifold X in terms of graded bundles over X [61, 134].

Let us recall that a graded manifold is a local-ringed space (Definition C.1), characterized by a smooth body manifold Z and some structure sheaf $\mathfrak{A}$ of Grassmann algebras on Z [7, 61, 134]. Its sections form a graded commutative

$C^{\infty}(Z)$-ring $\mathcal{A}$ of graded functions on a graded manifold $(Z, \mathfrak{A})$. It is called the structure ring of $(Z, \mathfrak{A})$. The differential calculus on a graded manifold is defined as the Chevalley–Eilenberg differential calculus over its structure ring (Sect. 6.2). By virtue of the well-known Batchelor theorem (Theorem 6.1), there exists a vector bundle $E \to Z$ with a typical fibre V such that the structure sheaf $\mathfrak{A}$ of $(Z, \mathfrak{A})$ is isomorphic to a sheaf $\mathfrak{A}_E$ of germs of sections of the exterior bundle $\wedge E^*$ of the dual E^* of E whose typical fibre is the Grassmann algebra $\wedge V^*$ [7, 14]. This Batchelor's isomorphism is not canonical. In applications, it however is fixed from the beginning. Therefore, we restrict our consideration to graded manifolds $(Z, \mathfrak{A}_E)$, termed the simple graded manifolds, modelled over vector bundles $E \to Z$ (Definition 6.4).

Let us note that a manifold Z itself can be treated as a trivial simple graded manifold (Z, C^{∞}_Z) modelled over a trivial fibre bundle $Z \times \mathbb{R} \to Z$ whose structure ring of graded functions is reduced to a commutative ring $C^{\infty}(Z)$ of smooth real functions on Z (Remark 6.1). Accordingly, a configuration fibre bundle $Y \to X$ in Lagrangian theory of even variables can be regarded as the graded bundle (6.29) of trivial graded manifolds (Remark 6.4). It follows that, in a general setting, one can define a configuration space of Grassmann-graded Lagrangian theory of even and odd variables as being the graded bundle $(X, Y, \mathfrak{A}_F)$ (6.32) over a trivial graded manifold (X, C^{∞}_X) modelled over the smooth composite bundle $F \to Y \to X$ (6.33) (Sect. 6.5). If $Y \to X$ is a vector bundle, this is the particular case of graded vector bundles in [73, 107] whose base is a trivial graded manifold.

By analogy with the calculus of variations and Lagrangian theory on smooth fibre bundles, Grassmann-graded Lagrangian theory on a graded bundle $(X, Y, \mathfrak{A}_F)$ (Sect. 6.5) comprehensively is phrased in terms of the Grassmann-graded variational bicomplex (6.54) of graded exterior forms on a graded infinite order jet manifold $(J^{\infty}Y, \mathfrak{A}_{J^{\infty}F})$ [6, 11, 12, 61, 134, 137]. Graded Lagrangians and Euler–Lagrange operator are defined as elements (6.69) and (6.70) of this graded bicomplex. The cohomology of a Grassmann-graded variational bicomplex (Theorems 6.9–6.10) provides a solution of the global inverse problem of the calculus of variations (Theorem 6.11), and states the global variational formula (6.73) for graded Lagrangians and Euler–Lagrange operators (Theorem 6.13).

In these terms, Noether's first theorem is formulated in a very general setting as a straightforward corollary of the global variational formula (6.73) (Sect. 6.6). It associates to any supersymmetry of a graded Lagrangian L (Definition 6.9) the conserved supersymmetry current (6.84) whose graded total differential vanishes on-shell (Theorem 6.20). One can show that a conserved supersymmetry current along a gauge supersymmetry is a graded total differential on-shell (Theorem 7.11).

Given a gauge supersymmetry of a graded Lagrangian, Noether's direct second theorem (Theorem 7.10) states that an Euler–Lagrange operator obeys the corresponding Noether identities. As was mentioned above, a problem is that any Euler–Lagrange operator satisfies Noether identities, which therefore must be separated into the trivial and nontrivial ones, and that there is a hierarchy of Noether and higher-stage Noether identities.

We follow the general analysis of Noether identities and higher-stage Noether identities of differential operators on fibre bundles (Appendix D). If a certain homology regularity condition (Definition 7.2) holds, one can associate to a Grassmann-graded Lagrangian system the exact Koszul–Tate chain complex (7.28) possessing the boundary Koszul–Tate operator (7.26) whose nilpotentness is equivalent to all complete nontrivial Noether identities (7.12) and higher-stage Noether identities (7.29) (Theorem 7.5) [11, 12, 61, 134, 137].

It should be noted that the notion of higher-stage Noether identities has come from that of reducible constraints. The Koszul–Tate complex of Noether identities has been invented similarly to that of constraints under the regularity condition that Noether identities are locally separated into the independent and dependent ones [6, 44]. This condition is relevant for constraints, defined by a finite set of functions which the inverse mapping theorem is applied to. However, Noether identities of differential operators, unlike constraints, are differential equations (Appendix D). They are given by an infinite set of functions on a Fréchet manifold of infinite order jets where the inverse mapping theorem fails to be valid. Therefore, the regularity condition for the Koszul–Tate complex of constraints is replaced with the above mentioned homology regularity condition.

Noether's inverse second theorem formulated in homology terms (Theorem 7.9) associates to the Koszul–Tate chain complex (7.28) the cochain sequence (7.38) with the ascent operator (7.39), called the gauge operator, whose components are complete nontrivial gauge and higher-stage gauge supersymmetries of Grassmann-graded Lagrangian theory [12, 61, 134, 137].

The gauge operator unlike the Koszul–Tate one is not nilpotent, unless gauge symmetries are Abelian (Remark 7.12). Therefore, an intrinsic definition of nontrivial gauge and higher-stage gauge symmetries meets difficulties. Another problem is that gauge symmetries need not form an algebra [48, 60, 63]. However, we can say that gauge symmetries are algebraically closed in a sense if the gauge operator admits a nilpotent extension, termed the BRST (Becchi–Rouet–Stora–Tyitin) operator (Sect. 7.5). If the BRST operator exists, the above mentioned cochain sequence is brought into the BRST complex. The Koszul–Tate and BRST complexes provide a BRST extension of original Lagrangian theory by Grassmann-graded ghosts and Noether antifields.

This extension is a preliminary step towards the BV (Batalin–Vilkovisky) quantization of reducible degenerate Lagrangian theories [6, 13, 44, 50, 63].

For the purpose of applications in field theory and mechanics, Chap. 3 of the book addresses first-order Lagrangian and covariant Hamiltonian theories on fibre bundles since the most of relevant field models is of this type. We restrict our consideration to classical symmetries represented by the projectable vector fields u (3.24) on a configuration bundle $Y \to X$ (Sect. 3.3). Since the corresponding symmetry current (3.29) is linear in a vector field u, one usually deals with the following types of symmetries.

(i) If u is a vertical vector field, the corresponding symmetry current is the Noether one (3.32). This is the case of so-called internal symmetries

(ii) Let τ be a vector field on X. Then its lift $\gamma\tau$ (B.27) onto Y (Definition B.2) provides the corresponding energy-momentum current (3.33) (Definition 3.3). We usually have either the horizontal lift $\Gamma\tau$ (B.56) by means of a connection Γ on $Y \to X$, e.g., in Yang–Mills gauge theory (Sect. 8.4), or the functorial lift $\widetilde{\tau}$ (Definition B.28) which is an infinitesimal general covariant transformation. This is the case of gauge gravitation theory (Sect. 10.4).

Applied to field theory, the familiar symplectic Hamiltonian technique takes a form of instantaneous Hamiltonian formalism on an infinite-dimensional phase space, where canonical variables are field functions at each instant of time [65]. The true Hamiltonian counterpart of first-order Lagrangian theory on a fibre bundle $Y \to X$ is covariant Hamiltonian formalism, where canonical momenta p_i^μ correspond to jets y_μ^i of field variables y^i with respect to all base coordinates x^μ. This formalism has been vigorously developed since 1970s in the Hamilton–De Donder, polysymplectic, multisymplectic, k-symplectic, k-cosymplectic and other variants [26, 27, 36, 45, 64, 69, 72, 81, 92, 95, 96, 103, 112, 114, 117]. We follow polysymplectic Hamiltonian formalism on a fibre bundle $Y \to X$ where the Legendre bundle Π (3.6) plays the role of a phase space (Sect. 3.5). A key point is that polysymplectic Hamiltonian formalism on a phase space Π is equivalent to particular first-order Lagrangian theory on a configuration space $\Pi \to X$. This fact enables us to describe symmetries of Hamiltonian field theory similarly to those in Lagrangian formalism (Sect. 3.7).

One can formulate non-autonomous nonrelativistic mechanics as particular field theory on fibre bundles $Q \to \mathbb{R}$ over the time axis $\mathbb{R}$ [62, 98, 121, 135]. Its velocity space is the first-order jet manifold J^1Q of sections of a configuration bundle $Q \to \mathbb{R}$, and its phase space Π (3.8) is the vertical cotangent bundle V^*Q of $Q \to \mathbb{R}$ (Chap. 4). A difference between mechanics and field theory however lies in the fact that fibre bundles over $\mathbb{R}$ always are trivial, and that all connections on these fibre bundles are flat. Consequently, they are not dynamic variables, but characterize nonrelativistic reference frames (Definition 4.1). By virtue of Noether's first theorem, any symmetry defines a symmetry current which also is an integral of motion in Lagrangian and Hamiltonian mechanics (Theorem 4.4). The converse is not true in Lagrangian mechanics where integrals of motion need not come from symmetries. We show that, in Hamiltonian mechanics, any integral of motion is a symmetry current (Theorem 4.15). One can think of this fact as being Noether's inverse first theorem.

The book presents a number of physically relevant models: commutative and noncommutative integrable Hamiltonian systems (Sect. 4.7), the global Kepler problem (Chap. 5), Yang–Mills gauge theory on principal bundles (Chap. 8), SUSY gauge theory (Chap. 9) on principal graded bundles, gauge gravitation theory on natural bundles (Chap. 10), topological Chern–Simons field theory (Chap. 11), and topological BF theory, exemplifying a reducible degenerate Lagrangian system (Chap. 12).

For the sake of convenience of the reader, a number of relevant mathematical topics are compiled in appendixes, thus making the exposition self-contained.

Chapter 1
Calculus of Variations on Fibre Bundles

There are different approaches to the calculus of variations on a fibre bundle $Y \to X$. It can be adequately formulated in algebraic terms of the variational bicomplex (1.21) of differential forms on the infinite order jet manifold $J^\infty Y$ of sections of $Y \to X$ [3, 15, 56, 61, 108, 133, 143]. In this framework, finite order Lagrangians (Definition 1.3) and Euler–Lagrange operators (Definition 1.4) are introduced as the elements (1.32) and (1.33) of this bicomplex, respectively. The cohomology of the variational bicomplex (Theorems 1.12 and 1.13) provides a solution of the global inverse problem of the calculus of variations (Theorems 1.15 and 1.16), and states the global variational decomposition (1.36) for Lagrangians and Euler–Lagrange operators (Theorem 1.17).

It should be emphasized that we deal with a variational bicomplex of the differential graded algebra (henceforth, DGA) $\mathscr{O}^*_\infty Y$ (1.7) of differential forms of finite jet order on an infinite order jet manifold $J^\infty Y$. They are exterior forms on finite order jet manifolds $J^r Y$ modulo the pull-back identification. One also considers both the variational bicomplex of the DGA $\mathscr{Q}^*_\infty$ (1.10) [3, 143] of differential forms of locally finite jet order on $J^\infty Y$ and different variants of variational sequences of finite jet order [2, 86, 89, 90, 150].

1.1 Infinite Order Jet Formalism

Let $Y \to X$ be a fibre bundle over an n-dimensional smooth manifold X (Definition B.2). Finite order jet manifolds $J^k Y$ of a fibre bundle $Y \to X$ (Definition B.11) form the inverse sequence

$$Y \xleftarrow{\pi} J^1 Y \longleftarrow \cdots J^{r-1} Y \xleftarrow{\pi^r_{r-1}} J^r Y \longleftarrow \cdots , \tag{1.1}$$

G. Sardanashvily, *Noether's Theorems*, Atlantis Studies
in Variational Geometry 3, DOI 10.2991/978-94-6239-171-0_1

where π^r_{r-1} are the affine bundles (B.77) modelled over the vector bundles (B.78). Its inverse limit $J^\infty Y$ is defined as a minimal set such that there exist surjections

$$\pi^\infty : J^\infty Y \to X, \quad \pi^\infty_0 : J^\infty Y \to Y, \qquad \pi^\infty_k : J^\infty Y \to J^k Y, \tag{1.2}$$

obeying the relations $\pi^\infty_r = \pi^k_r \circ \pi^\infty_k$ for all admissible k and $r < k$. An inverse limit of the inverse system (1.1) always exists. It consists of those elements

$$(\ldots, z_r, \ldots, z_k, \ldots), \qquad z_r \in J^r Y, \qquad z_k \in J^k Y,$$

of the Cartesian product $\prod\limits_k J^k Y$ which satisfy the relations $z_r = \pi^k_r(z_k)$ for all $k > r$. One can think of elements of $J^\infty Y$ as being *infinite order jets* of sections of $Y \to X$ identified by their Taylor series at points of X. A set $J^\infty Y$ is provided with the *inverse limit topology*. This is the coarsest topology such that the surjections π^∞_r (1.2) are continuous. Its base consists of inverse images of open subsets of $J^r Y$, $r = 0, \ldots$, under the maps π^∞_r. With the inverse limit topology, $J^\infty Y$ is a paracompact *Fréchet* (complete metrizable, but not Banach) manifold modelled over a locally convex vector space of formal number series $\{a^\lambda, a^i, a^i_\lambda, \cdots\}$ [143]. One can show that surjections π^∞_r are open maps admitting local sections, i.e., $J^\infty Y \to J^r Y$ are continuous bundles (Remark B.2). A bundle coordinate atlas $\{U_Y, (x^\lambda, y^i)\}$ of $Y \to X$ provides $J^\infty Y$ with a manifold coordinate atlas

$$\{(\pi^\infty_0)^{-1}(U_Y), (x^\lambda, y^i_\Lambda)\}_{0\le|\Lambda|}, \qquad y'^i_{\lambda+\Lambda} = \frac{\partial x^\mu}{\partial x'^\lambda} d_\mu y'^i_\Lambda. \tag{1.3}$$

Definition 1.1 One calls $J^\infty Y$, possessing the above mentioned properties, the *infinite order jet manifold.*

Theorem 1.1 *A fibre bundle Y is a strong deformation retract of the infinite order jet manifold $J^\infty Y$ [3, 56].*

Proof To show that Y is a strong deformation retract of $J^\infty Y$, let us construct a homotopy from $J^\infty Y$ to Y in an explicit form. Let $\gamma_{(k)}$, $k \le 1$, be global sections of an affine jet bundles $J^k Y \to J^{k-1} Y$. Then we have a global section

$$\gamma : Y \ni (x^\lambda, y^i) \to (x^\lambda, y^i, y^i_\Lambda = \gamma_{(|\Lambda|)}{}^i_\Lambda \circ \gamma_{(|\Lambda|-1)} \circ \cdots \circ \gamma_{(1)}) \in J^\infty Y \tag{1.4}$$

of an open surjection $\pi^\infty_0 : J^\infty Y \to Y$. Let us consider a map

$$\begin{aligned} &[0, 1] \times J^\infty Y \ni (t, x^\lambda, y^i, y^i_\Lambda) \to (x^\lambda, y^i, y'^i_\Lambda) \in J^\infty Y, \\ &y'^i_\Lambda = f_k(t) y^i_\Lambda + (1 - f_k(t))\gamma_{(k)}{}^i_\Lambda(x^\lambda, y^i, y^i_\Sigma), \quad |\Sigma| < k = |\Lambda|, \end{aligned} \tag{1.5}$$

where $f_k(t)$ is a continuous monotone real function on $[0, 1]$ such that

$$f_k(t) = \begin{cases} 0, & t \le 1 - 2^{-k}, \\ 1, & t \ge 1 - 2^{-(k+1)}. \end{cases}$$

A glance at the transition functions (1.3) shows that, although written in a coordinate form, this map is globally defined. It is continuous because, given an open subset $U_k \subset J^k Y$, the inverse image of an open set $(\pi^\infty_k)^{-1}(U_k) \subset J^\infty Y$ is an open subset

$$(t_k, 1] \times (\pi^\infty_k)^{-1}(U_k) \cup (t_{k-1}, 1] \times (\pi^\infty_{k-1})^{-1}(\pi^k_{k-1}[U_k \cap \gamma_{(k)}(J^{k-1}Y)]) \cup \cdots \cup [0, 1] \times (\pi^\infty_0)^{-1}(\pi^k_0[U_k \cap \gamma_{(k)} \circ \cdots \circ \gamma_{(1)}(Y)])$$

of $[0, 1] \times J^\infty Y$, where $[t_r, 1] = \operatorname{supp} f_r$. Then the map (1.5) is a desired homotopy from $J^\infty Y$ to Y which is identified with its image under the global section (1.4). □

Theorem 1.2 *By virtue of the Vietoris–Begle theorem [23], there is an isomorphism*

$$H^*(J^\infty Y; \mathbb{R}) = H^*(Y; \mathbb{R})$$

between the cohomology of $J^\infty Y$ with coefficients in the constant sheaf $\mathbb{R}$ and that of Y.

The inverse sequence (1.1) of jet manifolds yields the direct sequence of differential graded algebras (DGAs, Definition A.5) $\mathcal{O}^*_r = \mathcal{O}^*(J^r Y)$ of exterior forms on finite order jet manifolds

$$\mathcal{O}^*(X) \xrightarrow{\pi^*} \mathcal{O}^*(Y) \xrightarrow{\pi^{1}_0{}^*} \mathcal{O}^*_1 \longrightarrow \cdots \mathcal{O}^*_{r-1} \xrightarrow{\pi^{r}_{r-1}{}^*} \mathcal{O}^*_r \longrightarrow \cdots, \tag{1.6}$$

where $\pi^r_{r-1}{}^*$ are the pull-back monomorphisms. Its direct limit

$$\mathcal{O}^*_\infty Y = \overrightarrow{\lim}\, \mathcal{O}^*_r \tag{1.7}$$

(Definition A.1) exists and consists of all exterior forms on finite order jet manifolds modulo the pull-back identification. In accordance with Theorem A.4, $\mathcal{O}^*_\infty Y$ is a DGA which inherits operations of the exterior differential d and the exterior product $\wedge$ of exterior algebras $\mathcal{O}^*_r$.

If there is no danger of confusion, we further denote $\mathcal{O}^*_\infty = \mathcal{O}^*_\infty Y$.

Theorem 1.3 *The cohomology $H^*(\mathcal{O}^*_\infty)$ of the de Rham complex*

$$0 \longrightarrow \mathbb{R} \longrightarrow \mathcal{O}^0_\infty \xrightarrow{d} \mathcal{O}^1_\infty \xrightarrow{d} \cdots \tag{1.8}$$

of a DGA $\mathcal{O}^*_\infty$ equals the de Rham cohomology $H^*_{\rm DR}(Y)$ of a fibre bundle Y [2, 15].

Proof By virtue of Theorem A.8, the operation of taking homology groups of cochain complexes commutes with the passage to a direct limit. Since a DGA $\mathcal{O}^*_\infty$ is a direct limit of DGAs $\mathcal{O}^*_r$, its cohomology $H^*(\mathcal{O}^*_\infty)$ is isomorphic to the direct limit of a direct sequence

$$H^*_{\rm DR}(Y) \longrightarrow H^*_{\rm DR}(J^1Y) \longrightarrow \cdots \longrightarrow H^*_{\rm DR}(J^rY) \longrightarrow \cdots \tag{1.9}$$

of the de Rham cohomology groups $H^*_{\rm DR}(J^rY)$ of finite order jet manifolds J^rY. In accordance with Remark B.15, all these groups equal the de Rham cohomology $H^*_{\rm DR}(Y)$ of Y, and so is its direct limit $H^*(\mathcal{O}^*_\infty)$. □

Theorem 1.4 *Any closed form $\phi \in \mathcal{O}^*_\infty$ is decomposed into a sum $\phi = \sigma + d\xi$, where σ is a closed form on Y.*

One can think of elements of $\mathcal{O}^*_\infty$ as being *differential forms* on an infinite order jet manifold $J^\infty Y$ as follows. Let $\mathfrak{G}^*_r$ be a sheaf of germs of exterior forms on J^rY and $\overline{\mathfrak{G}}^*_r$ a canonical presheaf of local sections of $\mathfrak{G}^*_r$. Since π^r_{r-1} are open maps, there is the direct sequence of presheaves

$$\overline{\mathfrak{G}}^*_0 \xrightarrow{\pi^{1*}_0} \overline{\mathfrak{G}}^*_1 \cdots \xrightarrow{\pi^r_{r-1}{}^*} \overline{\mathfrak{G}}^*_r \longrightarrow \cdots .$$

Its direct limit $\overline{\mathfrak{G}}^*_\infty$ is a presheaf of DGAs on $J^\infty Y$. Let $\mathfrak{Q}^*_\infty$ be a sheaf of DGAs of germs of $\overline{\mathfrak{G}}^*_\infty$ on $J^\infty Y$. A structure module

$$\mathcal{Q}^*_\infty = \Gamma(\mathfrak{Q}^*_\infty) \tag{1.10}$$

of global sections of $\mathfrak{Q}^*_\infty$ is a DGA such that, given an element $\phi \in \mathcal{Q}^*_\infty$ and a point $z \in J^\infty Y$, there exist an open neighborhood U of z and an exterior form $\phi^{(k)}$ on some finite order jet manifold J^kY so that

$$\phi|_U = \pi^{\infty *}_k \phi^{(k)}|_U.$$

Therefore, one can think of $\mathcal{Q}^*_\infty$ as being an algebra of locally exterior forms on finite order jet manifolds. In particular, there is a monomorphism $\mathcal{O}^*_\infty \to \mathcal{Q}^*_\infty$.

Theorem 1.5 *A paracompact topological space $J^\infty Y$ admits the partition of unity by elements of a ring $\mathcal{Q}^0_\infty$ [143].*

Since elements of a DGA $\mathcal{Q}^*_\infty$ are locally exterior forms on finite order jet manifolds, the following Poincaré lemma holds.

Theorem 1.6 *Given a closed element $\phi \in \mathcal{Q}^*_\infty$, there exists a neighborhood U of each point $z \in J^\infty Y$ such that $\phi|_U$ is exact.*

Theorem 1.7 *The cohomology $H^*(\mathcal{Q}^*_\infty)$ of the de Rham complex*

$$0 \longrightarrow \mathbb{R} \longrightarrow \mathcal{Q}^0_\infty \xrightarrow{d} \mathcal{Q}^1_\infty \xrightarrow{d} \cdots \tag{1.11}$$

of a DGA $\mathcal{Q}^*_\infty$ equals the de Rham cohomology of a fibre bundle Y [3, 143].

Proof Let us consider the de Rham complex of sheaves

$$0 \longrightarrow \mathbb{R} \longrightarrow \mathfrak{O}^0_\infty \xrightarrow{d} \mathfrak{O}^1_\infty \xrightarrow{d} \cdots \tag{1.12}$$

on $J^\infty Y$. By virtue of Theorem 1.6, it is exact at all terms, except $\mathbb{R}$. Being sheaves of $\mathcal{Q}^0_\infty$-modules, the sheaves $\mathfrak{O}^r_\infty$ are fine and, consequently, acyclic because a paracompact space $J^\infty Y$ admits the partition of unity by elements of a ring $\mathcal{Q}^0_\infty$ (Theorems 1.5, C.7 and C.8). Thus, the complex (1.12) is a resolution of the constant sheaf $\mathbb{R}$ on $J^\infty Y$. In accordance with abstract de Rham Theorem C.6, cohomology $H^*(\mathcal{Q}^*_\infty)$ of the complex (1.11) equals the cohomology $H^*(J^\infty Y; \mathbb{R})$ of $J^\infty Y$ with coefficients in the constant sheaf $\mathbb{R}$. Since Y is a strong deformation retract of $J^\infty Y$, there is the isomorphism (B.79) and, consequently, a desired isomorphism

$$H^*(\mathcal{Q}^*_\infty) = H^*_{\mathrm{DR}}(Y).$$

□

Due to a monomorphism $\mathcal{O}^*_\infty \to \mathcal{Q}^*_\infty$, one can restrict $\mathcal{O}^*_\infty$ to the coordinate chart (1.3) where horizontal forms dx^λ and contact one-forms

$$\theta^i_\Lambda = dy^i_\Lambda - y^i_{\lambda+\Lambda} dx^\lambda$$

make up a local basis for a $\mathcal{O}^0_\infty$-algebra $\mathcal{O}^*_\infty$. Though $J^\infty Y$ is not a smooth manifold, elements of $\mathcal{O}^*_\infty$ are exterior forms on finite order jet manifolds and, therefore, their coordinate transformations are smooth. Moreover, there is a decomposition $\mathcal{O}^*_\infty = \oplus \mathcal{O}^{k,m}_\infty$ of $\mathcal{O}^*_\infty$ into $\mathcal{O}^0_\infty$-modules $\mathcal{O}^{k,m}_\infty$ of k-contact and m-horizontal forms together with the corresponding projections

$$h_k : \mathcal{O}^*_\infty \to \mathcal{O}^{k,*}_\infty, \qquad h^m : \mathcal{O}^*_\infty \to \mathcal{O}^{*,m}_\infty.$$

In particular, the projection h_0 takes a form

$$h_0 : \mathcal{O}^*_\infty \to \mathcal{O}^{0,*}_\infty, \qquad h_0(dx^\lambda) = dx^\lambda, \qquad h_0(\theta^i_\Lambda) = 0. \tag{1.13}$$

It is called the *horizontal projection*.

Accordingly, the exterior differential on $\mathcal{O}^*_\infty$ is decomposed into a sum $d = d_V + d_H$ of the *vertical differential*

$$d_V \circ h^m = h^m \circ d \circ h^m, \qquad d_V(\phi) = \theta^i_\Lambda \wedge \partial^\Lambda_i \phi, \qquad \phi \in \mathcal{O}^*_\infty,$$

and the *total differential*

$$d_H \circ h_k = h_k \circ d \circ h_k, \qquad d_H \circ h_0 = h_0 \circ d, \qquad d_H(\phi) = dx^\lambda \wedge d_\lambda(\phi),$$

where

$$d_\lambda = \partial_\lambda + y^i_\lambda \partial_i + \sum_{0<|\Lambda|} y^i_{\lambda+\Lambda} \partial^\Lambda_i \tag{1.14}$$

are the *infinite order total derivatives*. These differentials obey nilpotent conditions

$$d_H \circ d_H = 0, \qquad d_V \circ d_V = 0, \qquad d_H \circ d_V + d_V \circ d_H = 0, \tag{1.15}$$

and bring $\mathcal{O}^{*,*}_\infty$ into a bicomplex.

Let us consider the $\mathcal{O}^0_\infty$-module $\mathfrak{d}\mathcal{O}^0_\infty$ of derivations of a real ring $\mathcal{O}^0_\infty$ (Definition A.4).

Theorem 1.8 *The derivation module $\mathfrak{d}\mathcal{O}^0_\infty$ is isomorphic to the $\mathcal{O}^0_\infty$-dual $(\mathcal{O}^1_\infty)^*$ of a module of one-forms $\mathcal{O}^1_\infty$ [59].*

Proof At first, let us show that $\mathcal{O}^*_\infty$ is generated by elements df, $f \in \mathcal{O}^0_\infty$. It suffices to justify that any element of $\mathcal{O}^1_\infty$ is a finite $\mathcal{O}^0_\infty$-linear combination of elements df, $f \in \mathcal{O}^0_\infty$. Indeed, every $\phi \in \mathcal{O}^1_\infty$ is an exterior form on some finite order jet manifold J^rY. By virtue of Serre–Swan Theorem A.10, a $C^\infty(J^rY)$-module $\mathcal{O}^1_r$ of one-forms on J^rY is a projective module of finite rank, i.e., ϕ is represented by a finite $C^\infty(J^rY)$-linear combination of elements df, $f \in C^\infty(J^rY) \subset \mathcal{O}^0_\infty$. Any element $\Phi \in (\mathcal{O}^1_\infty)^*$ yields a derivation $\vartheta_\Phi(f) = \Phi(df)$ of a real ring $\mathcal{O}^0_\infty$. Since a module $\mathcal{O}^1_\infty$ is generated by elements df, $f \in \mathcal{O}^0_\infty$, different elements of $(\mathcal{O}^1_\infty)^*$ provide different derivations of $\mathcal{O}^0_\infty$, i.e., there is a monomorphism $(\mathcal{O}^1_\infty)^* \to \mathfrak{d}\mathcal{O}^0_\infty$. By the same formula, any derivation $\vartheta \in \mathfrak{d}\mathcal{O}^0_\infty$ sends $df \to \vartheta(f)$ and, since $\mathcal{O}^0_\infty$ is generated by elements df, it defines a morphism $\Phi_\vartheta : \mathcal{O}^1_\infty \to \mathcal{O}^0_\infty$. Moreover, different derivations ϑ provide different morphisms Φ_ϑ. Thus, we have a monomorphism $\mathfrak{d}\mathcal{O}^0_\infty \to (\mathcal{O}^1_\infty)^*$ and, consequently, an isomorphism $\mathfrak{d}\mathcal{O}^0_\infty = (\mathcal{O}^1_\infty)^*$. □

The proof of Theorem 1.8 gives something more. A DGA $\mathcal{O}^*_\infty$ is the minimal Chevalley–Eilenberg differential calculus $\mathcal{O}^*\mathcal{A}$ over a real ring $\mathcal{A} = \mathcal{O}^0_\infty$ of smooth real functions on finite order jet manifolds of $Y \to X$ (Definition A.7). Let $\vartheta \rfloor \phi$, $\vartheta \in \mathfrak{d}\mathcal{O}^0_\infty$, $\phi \in \mathcal{O}^1_\infty$, denote the interior product. Extended to a DGA $\mathcal{O}^*_\infty$, the interior product $\rfloor$ obeys a rule

$$\vartheta \rfloor (\phi \wedge \sigma) = (\vartheta \rfloor \phi) \wedge \sigma + (-1)^{|\phi|} \phi \wedge (\vartheta \rfloor \sigma).$$

Restricted to the coordinate chart (1.3), $\mathcal{O}^1_\infty$ is a free $\mathcal{O}^0_\infty$-module generated by one-forms dx^λ, θ^i_Λ. Since $\mathfrak{d}\mathcal{O}^0_\infty = (\mathcal{O}^1_\infty)^*$, any derivation of a real ring $\mathcal{O}^0_\infty$ takes a coordinate form

$$\vartheta = \vartheta^\lambda \partial_\lambda + \vartheta^i \partial_i + \sum_{0<|\Lambda|} \vartheta^i_\Lambda \partial^\Lambda_i, \tag{1.16}$$

where

$$\partial^\Lambda_i(y^j_\Sigma) = \partial^\Lambda_i \rfloor dy^j_\Sigma = \delta^j_i \delta^\Lambda_\Sigma$$

up to permutations of multi-indices Λ and Σ [59]. Its coefficients ϑ^λ, ϑ^i, ϑ^i_Λ are local smooth functions of finite jet order possessing the transformation law

$$\vartheta'^\lambda = \frac{\partial x'^\lambda}{\partial x^\mu}\vartheta^\mu, \qquad \vartheta'^i = \frac{\partial y'^i}{\partial y^j}\vartheta^j + \frac{\partial y'^i}{\partial x^\mu}\vartheta^\mu,$$
$$\vartheta'^i_\Lambda = \sum_{|\Sigma|\leq|\Lambda|} \frac{\partial y'^i_\Lambda}{\partial y^j_\Sigma}\vartheta^j_\Sigma + \frac{\partial y'^i_\Lambda}{\partial x^\mu}\vartheta^\mu. \tag{1.17}$$

Any derivation ϑ (1.16) of a ring $\mathcal{O}^0_\infty$ yields a derivation (called the Lie derivative) $\mathbf{L}_\vartheta$ of a DGA $\mathcal{O}^*_\infty$ given by relations

$$\mathbf{L}_\vartheta\phi = \vartheta\rfloor d\phi + d(\vartheta\rfloor\phi), \tag{1.18}$$
$$\mathbf{L}_\vartheta(\phi\wedge\phi') = \mathbf{L}_\vartheta(\phi)\wedge\phi' + \phi\wedge\mathbf{L}_\vartheta(\phi').$$

Remark 1.1 In particular, the total derivatives (1.14) are defined as the local derivations of $\mathcal{O}^0_\infty$ and the corresponding local Lie derivatives $d_\lambda\phi = \mathbf{L}_{d_\lambda}\phi$ of a DGA $\mathcal{O}^*_\infty$.

1.2 Variational Bicomplex

In order to transform a bicomplex $\mathcal{O}^{*,*}_\infty$ into the variational one, let us consider the following two operators acting on $\mathcal{O}^{*,n}_\infty$ [53, 147].

(i) There exists an $\mathbb{R}$-module endomorphism

$$\rho = \sum_{k>0}\frac{1}{k}\overline{\rho}\circ h_k\circ h^n : \mathcal{O}^{*>0,n}_\infty \to \mathcal{O}^{*>0,n}_\infty, \tag{1.19}$$
$$\overline{\rho}(\phi) = \sum_{0\leq|\Lambda|}(-1)^{|\Lambda|}\theta^i\wedge[d_\Lambda(\partial^\Lambda_i\rfloor\phi)], \qquad \phi\in\mathcal{O}^{>0,n}_\infty,$$

possessing the following properties.

Theorem 1.9 *For any $\phi\in\mathcal{O}^{>0,n}_\infty$, a form $\phi-\rho(\phi)$ is locally d_H-exact on each coordinate chart (1.3).*

Theorem 1.10 *The operator ρ (1.19) obeys a relation*

$$(\rho\circ d_H)(\psi) = 0, \qquad \psi\in\mathcal{O}^{>0,n-1}_\infty.$$

It follows from Theorems 1.9 and 1.10 that ρ (1.19) is a projector, i.e., $\rho\circ\rho=\rho$.

(ii) One defines a *variational operator*

$$\delta = \rho\circ d : \mathcal{O}^{*,n}_\infty \to \mathcal{O}^{*+1,n}_\infty. \tag{1.20}$$

Theorem 1.11 *The variational operator δ (1.20) is nilpotent, i.e., $\delta \circ \delta = 0$, and it obeys a relation $\delta \circ \rho = \delta$.*

Let us denote $\mathbf{E}_k = \rho(\mathcal{O}_\infty^{k,n})$.

Definition 1.2 Provided with the operators d_H, d_V, ρ and δ, a DGA $\mathcal{O}_\infty^*$ is decomposed into a *variational bicomplex*

$$\begin{array}{ccccccccc}
 & & \vdots & & \vdots & & \vdots & & \vdots \\
 & & {\scriptstyle d_V}\uparrow & & {\scriptstyle d_V}\uparrow & & {\scriptstyle d_V}\uparrow & & {\scriptstyle -\delta}\uparrow \\
 & 0 \to & \mathcal{O}_\infty^{1,0} & \overset{d_H}{\to} & \mathcal{O}_\infty^{1,1} & \overset{d_H}{\to}\cdots & \mathcal{O}_\infty^{1,n} & \overset{\rho}{\to} & \mathbf{E}_1 \to 0 \\
 & & {\scriptstyle d_V}\uparrow & & {\scriptstyle d_V}\uparrow & & {\scriptstyle d_V}\uparrow & & {\scriptstyle -\delta}\uparrow \\
0 \to \mathbb{R} \to & & \mathcal{O}_\infty^{0} & \overset{d_H}{\to} & \mathcal{O}_\infty^{0,1} & \overset{d_H}{\to}\cdots & \mathcal{O}_\infty^{0,n} & \equiv & \mathcal{O}_\infty^{0,n} \\
 & & \uparrow & & \uparrow & & \uparrow & & \\
0 \to \mathbb{R} \to & & \mathcal{O}^0(X) & \overset{d}{\to} & \mathcal{O}^1(X) & \overset{d}{\to}\cdots & \mathcal{O}^n(X) & \overset{d}{\to} & 0 \\
 & & \uparrow & & \uparrow & & \uparrow & & \\
 & & 0 & & 0 & & 0 & &
\end{array} \tag{1.21}$$

This bicomplex possesses the following cohomology [55, 56, 122].

Theorem 1.12 *The second row from the bottom and the last column of the variational bicomplex (1.21) make up the variational complex*

$$0 \to \mathbb{R} \to \mathcal{O}_\infty^0 \xrightarrow{d_H} \mathcal{O}_\infty^{0,1} \cdots \xrightarrow{d_H} \mathcal{O}_\infty^{0,n} \xrightarrow{\delta} \mathbf{E}_1 \xrightarrow{\delta} \mathbf{E}_2 \longrightarrow \cdots. \tag{1.22}$$

Its cohomology is isomorphic to the de Rham cohomology of a fibre bundle Y:

$$H^{k<n}(d_H; \mathcal{O}_\infty^*) = H_{\mathrm{DR}}^{k<n}(Y), \qquad H^{k\geq n}(\delta; \mathcal{O}_\infty^*) = H_{\mathrm{DR}}^{k\geq n}(Y). \tag{1.23}$$

Proof See the proof below. □

Theorem 1.13 *The rows of contact forms of the variational bicomplex (1.21) are exact sequences.*

Proof See the proof below. □

Let us note that the cohomology isomorphism (1.23) gives something more. Due to the relations $d_H \circ h_0 = h_0 \circ d$ and $\delta \circ \rho = \delta$, we have a cochain morphism

$$\begin{array}{ccccccccccc}
\cdots \to & \mathcal{O}_\infty^{n-1} & \overset{d}{\to} & \mathcal{O}_\infty^{n} & \overset{d}{\to} & \mathcal{O}_\infty^{n+1} & \overset{d}{\to} & \mathcal{O}_\infty^{n+2} & \to & \cdots \\
 & {\scriptstyle h_0}\downarrow & & {\scriptstyle h_0}\downarrow & & {\scriptstyle \rho}\downarrow & & {\scriptstyle \rho}\downarrow & & \\
\cdots \to & \mathcal{O}_\infty^{0,n-1} & \overset{d_H}{\to} & \mathcal{O}_\infty^{0,n} & \overset{\delta}{\to} & \mathbf{E}_1 & \overset{\delta}{\to} & \mathbf{E}_2 & \longrightarrow & \cdots
\end{array}$$

of the de Rham complex (1.8) of a DGA $\mathcal{O}^*_\infty$ to its variational complex (1.22). By virtue of Theorems 1.3 and 1.12, the corresponding homomorphism of their cohomology groups is an isomorphism. Then the splitting of a closed form $\phi \in \mathcal{O}^*_\infty$ in Theorem 1.4 leads to the following decompositions.

Theorem 1.14 *Any d_H-closed form $\phi \in \mathcal{O}^{0,m}$, $m < n$, is represented by a sum*

$$\phi = h_0\sigma + d_H\xi, \qquad \xi \in \mathcal{O}^{m-1}_\infty, \tag{1.24}$$

where σ is a closed m-form on Y. Any δ-closed form $\phi \in \mathcal{O}^{k,n}$ is split into

$$\phi = h_0\sigma + d_H\xi, \qquad k = 0, \qquad \xi \in \mathcal{O}^{0,n-1}_\infty, \tag{1.25}$$

$$\phi = \rho(\sigma) + \delta(\xi), \qquad k = 1, \qquad \xi \in \mathcal{O}^{0,n}_\infty, \tag{1.26}$$

$$\phi = \rho(\sigma) + \delta(\xi), \qquad k > 1, \qquad \xi \in \mathbf{E}_{k-1}, \tag{1.27}$$

where σ is a closed $(n + k)$-form on Y.

1.2.1 Cohomology of the Variational Bicomplex

This section is devoted to the proof of Theorems 1.12 and 1.13. At first, we obtain the corresponding cohomology of the DGA $\mathcal{Q}^*_\infty$ (1.10). For this purpose, one can use abstract de Rham Theorem C.5 because, as was mentioned above, a paracompact infinite order jet manifold $J^\infty Y$ admits the partition of unity by elements of $\mathcal{Q}^0_\infty$, but not $\mathcal{O}^0_\infty$ [3, 143]. After that, we show that cohomology of $\mathcal{O}^*_\infty \subset \mathcal{Q}^*_\infty$ equals that of $\mathcal{Q}^*_\infty$ [55, 56, 122].

Let us start with the so called *algebraic Poincaré lemma* [108, 147].

Lemma 1.1 *If Y is a contractible bundle $\mathbb{R}^{n+p} \to \mathbb{R}^n$, the variational bicomplex (1.21) is exact at all terms, except $\mathbb{R}$.*

Proof The homotopy operators for d_V, d_H, δ and ρ are given by the formulas (5.72), (5.109), (5.84) in [108] and (4.5) in [147], respectively. □

Let $\mathfrak{Q}^*_\infty$ be a sheaf of germs of differential forms $\phi \in \mathcal{O}^*_\infty$ on $J^\infty Y$. It is decomposed into the variational bicomplex $\mathfrak{Q}^{*,*}_\infty$. A DGA $\mathcal{Q}^*_\infty$ of global sections of $\mathfrak{Q}^*_\infty$ also is decomposed into the variational bicomplex $\mathcal{Q}^{*,*}_\infty$ similar to the bicomplex (1.21). Let us consider a variational subcomplex

$$0 \to \mathbb{R} \to \mathfrak{Q}^0_\infty \xrightarrow{d_H} \mathfrak{Q}^{0,1}_\infty \cdots \xrightarrow{d_H} \mathfrak{Q}^{0,n}_\infty \xrightarrow{\delta} \mathbf{E}_1 \xrightarrow{\delta} \mathbf{E}_2 \longrightarrow \cdots \tag{1.28}$$

of $\mathfrak{Q}^{*,*}_\infty$ and a subcomplexes of sheaves of contact forms

$$0 \to \mathfrak{Q}^{k,0}_\infty \xrightarrow{d_H} \mathfrak{Q}^{k,1}_\infty \cdots \xrightarrow{d_H} \mathfrak{Q}^{k,n}_\infty \xrightarrow{\rho} \mathbf{E}_k \to 0, \qquad k = 1, \ldots, \tag{1.29}$$

where $\mathbf{E}_k = \rho(\mathfrak{O}^{k,n}_\infty)$. By virtue of Lemma 1.1, these complexes are exact at all terms, except $\mathbb{R}$.

Since a paracompact space $J^\infty Y$ admits the partition of unity by elements of the ring $\mathcal{Q}^0_\infty$, the sheaves $\mathfrak{O}^{m,k}_\infty$ of $\mathcal{Q}^0_\infty$-modules are fine (Theorem C.8) and, consequently, acyclic (Theorem C.7). Let us show that the sheaves $\mathbf{E}_k$ also are fine [56, 61]. Though the $\mathbb{R}$-modules $\Gamma(\mathbf{E}_{k>1})$ fail to be $\mathcal{Q}^0_\infty$-modules [147], one can use the fact that the sheaves $\mathbf{E}_{k>0}$ are projections $\rho(\mathfrak{O}^{k,n}_\infty)$ of sheaves of $\mathcal{Q}^0_\infty$-modules. Let $\{U_i\}_{i\in I}$ be a locally finite open cover of $J^\infty Y$ and $\{f_i \in \mathcal{Q}^0_\infty\}$ the associated partition of unity. For any open subset $U \subset J^\infty Y$ and any section φ of the sheaf $\mathfrak{O}^{k,n}_\infty$ over U, let us put $g_i(\varphi) = f_i\varphi$. The endomorphisms g_i of $\mathfrak{O}^{k,n}_\infty$ yield the $\mathbb{R}$-module endomorphisms

$$\overline{g}_i = \rho \circ g_i : \mathbf{E}_k \xrightarrow{\text{in}} \mathfrak{O}^{k,n}_\infty \xrightarrow{g_i} \mathfrak{O}^{k,n}_\infty \xrightarrow{\rho} \mathbf{E}_k$$

of sheaves $\mathbf{E}_k$. They possess the properties required for $\mathbf{E}_k$ to be a fine sheaf. Indeed, for each $i \in I$, supp $f_i \subset U_i$ provides a closed set such that $\overline{g}_i$ is zero outside this set, while a sum $\sum\limits_{i\in I} \overline{g}_i$ is the identity morphism.

Consequently, all sheaves, except $\mathbb{R}$, in the complexes (1.28) and (1.29) are acyclic. Therefore, these complexes are resolutions of the constant sheaf $\mathbb{R}$ and the zero sheaf on $J^\infty Y$, respectively. Let us consider the corresponding subcomplexes

$$0 \to \mathbb{R} \to \mathcal{Q}^0_\infty \xrightarrow{d_H} \mathcal{Q}^{0,1}_\infty \cdots \xrightarrow{d_H} \mathcal{Q}^{0,n}_\infty \xrightarrow{\delta} \Gamma(\mathbf{E}_1) \xrightarrow{\delta} \Gamma(\mathbf{E}_2) \to \cdots, \tag{1.30}$$

$$0 \to \mathcal{Q}^{k,0}_\infty \xrightarrow{d_H} \mathcal{Q}^{k,1}_\infty \cdots \xrightarrow{d_H} \mathcal{Q}^{k,n}_\infty \xrightarrow{\rho} \Gamma(\mathbf{E}_k) \to 0, \quad k = 1, \ldots, \tag{1.31}$$

of DGA $\mathcal{Q}^*_\infty$. In accordance with abstract de Rham Theorem C.6, cohomology of the complex (1.30) equals the cohomology of $J^\infty Y$ with coefficients in the constant sheaf $\mathbb{R}$, while the complex (1.31) is exact. Since Y is a strong deformation retract of $J^\infty Y$, cohomology of the complex (1.30) equals the de Rham cohomology of Y (Remark B.15).

Thus, the following has been proved.

Lemma 1.2 *The cohomology of the variational complex (1.30) equals the de Rham cohomology of a fibre bundle Y. All the complexes (1.31) are exact.*

Now, let us show the following.

Lemma 1.3 *A subalgebra $\mathcal{O}^*_\infty \subset \mathcal{Q}^*_\infty$ has the same d_H- and δ-cohomology as $\mathcal{Q}^*_\infty$.*

Let the common symbol D stand for d_H and δ. Bearing in mind the decompositions (1.24)–(1.27), it suffices to show that, if an element $\phi \in \mathcal{O}^*_\infty$ is D-exact in an algebra $\mathcal{Q}^*_\infty$, then it is so in an algebra $\mathcal{O}^*_\infty$.

Lemma 1.1 states that, if Y is a contractible bundle and a D-exact form ϕ on $J^\infty Y$ is of finite jet order $[\phi]$ (i.e., $\phi \in \mathcal{O}^*_\infty$), there exists a differential form $\varphi \in \mathcal{O}^*_\infty$ on $J^\infty Y$ such that $\phi = D\varphi$. Moreover, a glance at the homotopy operators for d_H and δ shows that the jet order $[\varphi]$ of φ is bounded by an integer $N([\phi])$, depending only on the jet order of ϕ. Let us call this fact the *finite exactness* of an operator D. Lemma 1.1 shows that the finite exactness takes place on $J^\infty Y|_U$ over any domain $U \subset Y$. Let us prove the following.

Lemma 1.4 *Given a family $\{U_\alpha\}$ of disjoint open subsets of Y, let us suppose that the finite exactness takes place on $J^\infty Y|_{U_\alpha}$ over every subset U_α from this family. Then it is true on $J^\infty Y$ over the union $\bigcup_\alpha U_\alpha$ of these subsets.*

Proof Let $\phi \in \mathcal{O}^*_\infty$ be a D-exact form on $J^\infty Y$. The finite exactness on $(\pi^\infty_0)^{-1}(\cup U_\alpha)$ holds since $\phi = D\varphi_\alpha$ on every $(\pi^\infty_0)^{-1}(U_\alpha)$ and $[\varphi_\alpha] < N([\phi])$. □

Lemma 1.5 *Suppose that the finite exactness of an operator D takes place on $J^\infty Y$ over open subsets U, V of Y and their non-empty overlap $U \cap V$. Then it also is true on $J^\infty Y|_{U\cup V}$.*

Proof Let $\phi = D\varphi \in \mathcal{O}^*_\infty$ be a D-exact form on $J^\infty Y$. By assumption, it can be brought into the forms $D\varphi_U$ on $(\pi^\infty_0)^{-1}(U)$ and $D\varphi_V$ on $(\pi^\infty_0)^{-1}(V)$, where φ_U and φ_V are differential forms of bounded jet order. Let us consider their difference $\varphi_U - \varphi_V$ on $(\pi^\infty_0)^{-1}(U \cap V)$. It is a D-exact form of bounded jet order

$$[\varphi_U - \varphi_V] < N([\phi])$$

which, by assumption, can be written as $\varphi_U - \varphi_V = D\sigma$ where σ also is of bounded jet order $[\sigma] < N(N([\phi]))$. Lemma 1.6 below shows that $\sigma = \sigma_U + \sigma_V$ where σ_U and σ_V are differential forms of bounded jet order on $(\pi^\infty_0)^{-1}(U)$ and $(\pi^\infty_0)^{-1}(V)$, respectively. Then, putting

$$\varphi'|_U = \varphi_U - D\sigma_U, \qquad \varphi'|_V = \varphi_V + D\sigma_V,$$

we have the form ϕ, equal to $D\varphi'_U$ on $(\pi^\infty_0)^{-1}(U)$ and $D\varphi'_V$ on $(\pi^\infty_0)^{-1}(V)$, respectively. Since the difference $\varphi'_U - \varphi'_V$ on $(\pi^\infty_0)^{-1}(U \cap V)$ vanishes, we obtain $\phi = D\varphi'$ on $(\pi^\infty_0)^{-1}(U \cup V)$ where

$$\varphi' = \begin{cases} \varphi'|_U = \varphi'_U \\ \varphi'|_V = \varphi'_V \end{cases}$$

is of bounded jet order $[\varphi'] < N(N([\phi]))$. □

Lemma 1.6 *Let U and V be open subsets of a bundle Y and $\sigma \in \mathfrak{G}^*_\infty$ a differential form of bounded jet order on*

$$(\pi^\infty_0)^{-1}(U \cap V) \subset J^\infty Y.$$

Then σ is decomposed into a sum $\sigma_U + \sigma_V$ of differential forms σ_U and σ_V of bounded jet order on $(\pi^\infty_0)^{-1}(U)$ and $(\pi^\infty_0)^{-1}(V)$, respectively.

Proof By taking the smooth partition of unity on $U \cup V$ subordinate to the cover $\{U, V\}$ and passing to the function with support in V, one gets a smooth real function f on $U \cup V$ which equals 0 on a neighborhood of $U \setminus V$ and 1 on a neighborhood of $V \setminus U$ in $U \cup V$. Let $(\pi_0^\infty)^* f$ be the pull-back of f onto $(\pi_0^\infty)^{-1}(U \cup V)$. A differential form $((\pi_0^\infty)^* f)\sigma$ equals 0 on a neighborhood of $(\pi_0^\infty)^{-1}(U)$ and, therefore, can be extended by 0 to $(\pi_0^\infty)^{-1}(U)$. Let us denote it σ_U. Accordingly, a differential form $(1 - (\pi_0^\infty)^* f)\sigma$ has an extension σ_V by 0 to $(\pi_0^\infty)^{-1}(V)$. Then $\sigma = \sigma_U + \sigma_V$ is a desired decomposition because σ_U and σ_V are of the jet order which does not exceed that of σ. □

To prove the finite exactness of D on $J^\infty Y$, it remains to choose an appropriate cover of Y. A smooth manifold Y admits a countable cover $\{U_\xi\}$ by domains U_ξ, $\xi \in \mathbf{N}$, and its refinement $\{U_{ij}\}$, where $j \in \mathbf{N}$ and i runs through a finite set, such that $U_{ij} \cap U_{ik} = \emptyset$, $j \neq k$ [67]. Then Y has a finite cover $\{U_i = \cup_j U_{ij}\}$. Since the finite exactness of an operator D takes place over any domain U_ξ, it also holds over any member U_{ij} of the refinement $\{U_{ij}\}$ of $\{U_\xi\}$ and, in accordance with Lemma 1.4, over any member of the finite cover $\{U_i\}$ of Y. Then by virtue of Lemma 1.5, the finite exactness of D takes place on $J^\infty Y$ over Y.

Similarly, one can show that:

Lemma 1.7 *Restricted to $\mathscr{O}_\infty^{k,n}$, an operator ρ remains exact.*

Lemmas 1.2 and 1.7 result in Theorems 1.12 and 1.13.

1.3 Lagrangian Formalism

Lagrangian theory on fibre bundles is formulated in terms of a variational bicomplex (Definition 1.2) as follows.

Definition 1.3 A finite order *Lagrangian* is defined as horizontal densities

$$L = \mathscr{L}\omega \in \mathscr{O}_\infty^{0,n} \tag{1.32}$$

of the variational complex (1.22) (see the notation (B.32)).

Definition 1.4 The variational operator

$$\delta : \mathscr{O}_\infty^{0,n} \to \mathbf{E}_1, \quad \delta L = \mathscr{E}_L = \mathscr{E}_i \theta^i \wedge \omega = \sum_{0 \leq |\Lambda|} (-1)^{|\Lambda|} d_\Lambda(\partial_i^\Lambda \mathscr{L})\theta^i \wedge \omega, \tag{1.33}$$

is called the *Euler–Lagrange operator*, and its coefficients $\mathscr{E}_i$ (1.33) are said to be the *variational derivatives*.

The Lagrangian L (1.32) is termed *variationally trivial* if it is δ-closed, i.e., $\delta(L) = 0$. The following is a corollary of Theorem 1.14.

Theorem 1.15 *The finite order Lagrangian L (1.32) is variationally trivial if and only if*

$$L = h_0\sigma + d_H\xi, \qquad \xi \in \mathcal{O}_\infty^{0,n-1},$$

where σ is a closed n-form on Y.

In particular, a variationally trivial Lagrangian necessarily is d_H-exact if the de Rham cohomology group $H^n_{\mathrm{DR}}(Y)$ of Y is trivial.

A variational operator

$$\delta : \mathbf{E}_1 \to \mathbf{E}_2, \quad \delta(\mathcal{E}) = \sum_{0\leq|\Lambda|} [\partial_j^\Lambda \mathcal{E}_i \theta^j_\Lambda \wedge \theta^i + (-1)^{|\Lambda|}\theta^j \wedge d_\Lambda(\partial_j^\Lambda \mathcal{E}_i \theta^i)] \wedge \omega,$$

is called the *Helmholtz–Sonin map*. An element $\mathcal{E} \in \mathbf{E}_1$ is said to be the *Euler–Lagrange-type operator* if it satisfies the *Helmholtz condition* $\delta(\mathcal{E}) = 0$, i.e., if it is δ-closed.

The following is a corollary of Theorem 1.14, too.

Theorem 1.16 *An Euler–Lagrange-type operator $\mathcal{E} \in \mathbf{E}_1$ reads*

$$\mathcal{E} = \delta L + \rho(\sigma), \qquad L \in \mathcal{O}_\infty^{0,n},$$

where σ is a closed $(n+1)$-form on Y.

In particular, any Euler–Lagrange-type operator $\mathcal{E} \in \mathbf{E}_1$ is the Euler–Lagrange one $\mathcal{E} = \delta L$ (1.33) if the de Rham cohomology group $H^{n+1}_{\mathrm{DR}}(Y)$ of Y is trivial. For instance, this is the case of an affine bundle $Y \to X$ (Remark B.14). Certainly, every Euler–Lagrange-type operator locally is an Euler–Lagrange operator over some open subset of Y.

Remark 1.2 Theorems 1.15 and 1.16 provide a solution of the so called global inverse problem of the calculus of variations. This solution agrees with that of [2] obtained by computing cohomology of a variational sequence of bounded jet order, but without minimizing an order of a Lagrangian (see also particular results of [87, 149]). A solution of the global inverse problem of the calculus of variations in the case of a graded differential algebra $\mathcal{Q}^*_\infty$ (1.10) has been found in [3, 143] (Theorem 1.2).

A glance at the expression (1.33) shows that, if a Lagrangian L (1.32) is of r-order, its Euler–Lagrange operator $\mathcal{E}_L$ is of $2r$-order. Its kernel

$$\mathfrak{E}_L = \operatorname{Ker} \mathcal{E}_L \subset J^{2r}Y \tag{1.34}$$

is called the *Euler–Lagrange equation*. It is locally given by the equalities

$$\mathscr{E}_i = \sum_{0\leq|\Lambda|} (-1)^{|\Lambda|} d_\Lambda(\partial_i^\Lambda \mathscr{L}) = 0. \tag{1.35}$$

However, it may happen that the Euler–Lagrange equation (1.34) is not a differential equation in the strict sense of Definition B.12 because Ker $\mathscr{E}_L$ need not be a closed subbundle of $J^{2r}Y \to X$.

The Euler–Lagrange equation (1.35) traditionally is derived from the *variational formula*

$$dL = \delta L - d_H \Xi_L \tag{1.36}$$

of the calculus of variations. In formalism of a variational bicomplex, this formula is a corollary of Theorem 1.13.

Theorem 1.17 *The exactness of the row of one-contact forms of the variational bicomplex (1.21) at a term $\mathscr{O}_\infty^{1,n}$ relative to the projector ρ provides a global $\mathbb{R}$-module splitting*

$$\mathscr{O}_\infty^{1,n} = \mathbf{E}_1 \oplus d_H(\mathscr{O}_\infty^{1,n-1}).$$

In particular, any Lagrangian L admits the decomposition (1.36).

Defined with accuracy to a d_H-closed summand, a form $\Xi_L \in \mathscr{O}_\infty^n$ in the variational formula (1.36) reads

$$\begin{aligned}
\Xi_L &= L + [(\partial_i^\lambda \mathscr{L} - d_\mu F_i^{\mu\lambda})\theta^i + \sum_{s=1} F_i^{\lambda\nu_s\ldots\nu_1}\theta^i_{\nu_s\ldots\nu_1}] \wedge \omega_\lambda, \qquad (1.37)\\
F_i^{\nu_k\ldots\nu_1} &= \partial_i^{\nu_k\ldots\nu_1}\mathscr{L} - d_\mu F_i^{\mu\nu_k\ldots\nu_1} + \psi_i^{\nu_k\ldots\nu_1}, \qquad k = 2, 3, \ldots,
\end{aligned}$$

where $\psi_i^{\nu_k\ldots\nu_1}$ are local functions such that $\psi_i^{(\nu_k\nu_{k-1})\ldots\nu_1} = 0$. It is readily observed that the form Ξ_L (1.37) possesses the following properties:

- $h_0(\Xi_L) = L$,
- $h_0(\vartheta\rfloor d\Xi_L) = \vartheta^i\mathscr{E}_i\omega$ for any derivation ϑ (1.16).

Consequently, Ξ_L is a Lepage equivalent of a Lagrangian L.

Remark 1.3 Following the terminology of finite order jet formalism [53, 64, 88, 89], we call an exterior n-form $\rho \in \mathscr{O}_\infty^n$ the *Lepage form* if, for any derivation ϑ (1.16), the density $h_0(\vartheta\rfloor d\rho)$ depends only on the restriction of ϑ to a derivation $\vartheta^\lambda\partial_\lambda + \vartheta^i\partial_i$ of the subring $C^\infty(Y) \subset \mathscr{O}_\infty^0$. The Lepage forms constitute a real vector space. In particular, closed n-forms and $(2 \leq k)$-contact n-forms are Lepage forms. Given a Lagrangian L, a Lepage form ρ is called the *Lepage equivalent* of L if $h_0(\rho) = L$. Any Lepage form ρ is a Lepage equivalent of the Lagrangian $h_0(\rho)$. Conversely, any r-order Lagrangian possesses a Lepage equivalent of $(2r - 1)$-order [64]. The

Lepage equivalents of a Lagrangian L constitute an affine space modelled over a vector space of contact Lepage forms. In particular, one can locally put $\psi_i^{\nu_k \ldots \nu_1} = 0$ in the formula (1.37).

Definition 1.5 We agree to call a pair $(\mathcal{O}^*_\infty Y, L)$ the *Lagrangian system* on a fibre bundle Y.

Hereafter, base manifolds X of fibre bundles are assumed to be oriented and connected.

Chapter 2
Noether's First Theorem

We here are concerned with Lagrangian theory on fibre bundles (Chap. 1). In this case, Noether's first theorem (Theorem 2.7) and Noether's direct second theorem (Theorem 2.6) are corollaries of the global variational formula (1.36).

2.1 Lagrangian Symmetries

Noether's theorems deal with infinitesimal transformations of Lagrangian systems.

Definition 2.1 Given a Lagrangian system $(\mathcal{O}^*_\infty Y, L)$ (Definition 1.5), its *infinitesimal transformations* are defined to be contact derivations of a real ring $\mathcal{O}^0_\infty Y$ [59, 61].

The derivation $\vartheta \in \mathfrak{d}\mathcal{O}^0_\infty Y$ (1.16) is termed *contact* if the Lie derivative $\mathbf{L}_\vartheta$ (1.18) along ϑ preserves an ideal of contact forms of a DGA $\mathcal{O}^*_\infty Y$, i.e., the Lie derivative $\mathbf{L}_\vartheta$ of a contact form is a contact form.

Theorem 2.1 *The derivation ϑ (1.16) is contact if and only if it takes a form*

$$\vartheta = \upsilon^\lambda \partial_\lambda + \upsilon^i \partial_i + \sum_{0<|\Lambda|} [d_\Lambda(\upsilon^i - y^i_\mu \upsilon^\mu) + y^i_{\mu+\Lambda}\upsilon^\mu]\partial_i^\Lambda. \tag{2.1}$$

Proof The expression (2.1) results from a direct computation similar to that of the first part of Bäcklund's theorem [76]. One can then justify that local functions (2.1) satisfy the transformation law (1.17). □

Comparing the expressions (2.1) and (B.81) enables one to regard the contact derivation ϑ (2.1) as the infinite order jet prolongation

$$\vartheta = J^\infty \upsilon \tag{2.2}$$

G. Sardanashvily, *Noether's Theorems*, Atlantis Studies in Variational Geometry 3, DOI 10.2991/978-94-6239-171-0_2

of its restriction

$$\upsilon = \upsilon^\lambda \partial_\lambda + \upsilon^i \partial_i \tag{2.3}$$

to a ring $C^\infty(Y)$. Since coefficients υ^λ and υ^i of υ (2.3) generally depend on jet coordinates y^i_Λ of bounded jet order $0 < |\Lambda| \leq N$, one calls υ (2.3) the *generalized vector field*. It can be represented as a section of the pull-back bundle

$$J^N Y \underset{Y}{\times} TY \to J^N Y.$$

Let ϑ and ϑ' be contact derivations (2.1) whose restrictions to a real ring $C^\infty(Y)$ are generalized vector fields υ and υ', respectively. Certainly, their Lie bracket $[\vartheta, \vartheta']$ is a contact derivation. Its restriction to $C^\infty(Y)$ reads

$$[\upsilon, \upsilon']_J = [\vartheta(\upsilon'^\lambda) - \vartheta'(\upsilon^\lambda)]\partial_\lambda + [\vartheta(\upsilon'^i) - \vartheta'(\upsilon^i)]\partial_i. \tag{2.4}$$

We call it the *bracket of generalized vector fields* $[\upsilon, \upsilon']_J$. It obeys a relation

$$[J^\infty \upsilon, J^\infty \upsilon'] = J^\infty([\upsilon, \upsilon']_J). \tag{2.5}$$

If υ and υ' are vector fields on Y, their bracket (2.4) is the familiar Lie one.

The contact derivation ϑ (2.1) is said to be *projectable*, if the generalized vector field υ (2.3) projects onto a vector field $\upsilon^\lambda \partial_\lambda$ on X, i.e., its components ϑ^λ depend only on coordinates on X. In particular, it is readily observed that the bracket (2.4) of projectable generalized vector fields also is projectable.

Every contact derivation ϑ (2.1) admits the canonical splitting

$$\vartheta = \vartheta_H + \vartheta_V = \upsilon^\lambda d_\lambda + J^\infty \upsilon_V = \upsilon^\lambda d_\lambda + [\upsilon^i_V \partial_i + \sum_{0<|\Lambda|} d_\Lambda \upsilon^i_V \partial^\Lambda_i], \tag{2.6}$$

$$\upsilon = \upsilon_H + \upsilon_V = \upsilon^\lambda d_\lambda + (\upsilon^i - y^i_\mu \upsilon^\mu)\partial_i = \upsilon^\lambda d_\lambda + \upsilon^i_V \partial_i, \tag{2.7}$$

into the horizontal and vertical parts ϑ_H and ϑ_V, respectively [59].

Theorem 2.2 *Any vertical contact derivation*

$$\vartheta = J^\infty \upsilon = \upsilon^i \partial_i + \sum_{0<|\Lambda|} d_\Lambda \upsilon^i \partial^\Lambda_i, \qquad \upsilon = \upsilon^i \partial_i,$$

obeys the relations

$$\vartheta \rfloor d_H \phi = -d_H(\vartheta \rfloor \phi), \tag{2.8}$$

$$\mathbf{L}_\vartheta (d_H \phi) = d_H(\mathbf{L}_\vartheta \phi), \qquad \phi \in \mathcal{O}^*_\infty Y. \tag{2.9}$$

Proof It is easily justified that, if ϕ and ϕ' satisfy the relation (2.8), then $\phi \wedge \phi'$ does well. Then it suffices to prove the relation (2.8) when ϕ is a function and $\phi = \theta^i_\Lambda$.

The result follows from the equalities

$$\vartheta\rfloor\theta^i_\Lambda = \upsilon^i_\Lambda, \quad d_H(\upsilon^i_\Lambda) = \upsilon^i_{\lambda+\Lambda}dx^\lambda, \quad d_H\theta^i_\lambda = dx^\lambda \wedge \theta^i_{\lambda+\Lambda},$$
$$d_\lambda \circ \upsilon^i_\Lambda \partial^\Lambda_i = \upsilon^i_\Lambda \partial^\Lambda_i \circ d_\lambda.$$

The relation (2.9) is a corollary of the equality (2.8). □

Given a Lagrangian system $(\mathcal{O}^*_\infty Y, L)$, let us consider the Lie derivative $\mathbf{L}_\vartheta L$ (1.18) of its Lagrangian L along the contact derivation ϑ (1.9). The global decomposition (1.36) results in the following corresponding splitting of $\mathbf{L}_\vartheta L$.

Theorem 2.3 *Given a Lagrangian $L \in \mathcal{O}^{0,n}_\infty Y$, its Lie derivative $\mathbf{L}_\vartheta L$ along the contact derivation ϑ (2.6) fulfils the first variational formula*

$$\mathbf{L}_\vartheta L = \upsilon_V\rfloor\delta L + d_H(h_0(\vartheta\rfloor\Xi_L)) + \mathcal{L}d_V(\upsilon_H\rfloor\omega), \tag{2.10}$$

where Ξ_L is the Lepage equivalent (1.37) of L and h_0 is the horizontal projection (1.13).

Proof The formula (2.10) comes from the variational formula (1.36) and the relations (2.8)–(2.9) as follows:

$$\mathbf{L}_\vartheta L = \vartheta\rfloor dL + d(\vartheta\rfloor L) = [\vartheta_V\rfloor dL - d_V\mathcal{L}\wedge\upsilon_H\rfloor\omega] + [d_H(\upsilon_H\rfloor L) +$$
$$d_V(\mathcal{L}\upsilon_H\rfloor\omega)] = \vartheta_V\rfloor dL + d_H(\upsilon_H\rfloor L) + \mathcal{L}d_V(\upsilon_H\rfloor\omega) =$$
$$\upsilon_V\rfloor\delta L - \vartheta_V\rfloor d_H\Xi_L + d_H(\upsilon_H\rfloor L) + \mathcal{L}d_V(\upsilon_H\rfloor\omega) =$$
$$\upsilon_V\rfloor\delta L + d_H(\vartheta_V\rfloor\Xi_L + \upsilon_H\rfloor L) + \mathcal{L}d_V(\upsilon_H\rfloor\omega),$$

where

$$\vartheta_V\rfloor\Xi_L = h_0(\vartheta_V\rfloor\Xi_L), \qquad \upsilon_H\rfloor L = h_0(\upsilon_H\rfloor\Xi_L)$$

since $\Xi_L - L$ is a one-contact form and $\upsilon_H = \vartheta_H$. □

Definition 2.2 The generalized vector field υ (2.7) on Y is called the *Lagrangian symmetry* (or, shortly, the *symmetry*) of a Lagrangian L if the Lie derivative $\mathbf{L}_{J^\infty\upsilon}L$ of L along the contact derivation $J^\infty\upsilon$ (2.2) is d_H-exact, i.e.,

$$\mathbf{L}_{J^\infty\upsilon}L = d_H\sigma, \tag{2.11}$$

where σ is a horizontal $(n-1)$-form.

Remark 2.1 Certainly, the jet prolongation $J^\infty\upsilon$ of a generalized vector field υ in the expression (2.11) always is of finite jet order because a Lagrangian L is of finite order (Sect. 3.3). Therefore, we further use the notation $J^*\upsilon$ for the finite order jet prolongation of υ whose order however is nor specified.

Theorem 2.4 *A glance at the expression (2.10) shows the following.*

(i) A generalized vector field υ is a Lagrangian symmetry only if it is projectable.

(ii) Any projectable generalized vector field is a symmetry of a variationally trivial Lagrangian.

(iii) A projectable generalized vector field υ is a Lagrangian symmetry if and only if its vertical part υ_V (2.7) is well.

(iv) A projectable generalized vector field υ is a symmetry if and only if the density $\upsilon_V \rfloor \delta L$ is d_H-exact.

Remark 2.2 In accordance with the standard terminology, symmetries represented by generalized vector fields (2.3) are called *generalized symmetries* because they depend on derivatives of variables. Generalized symmetries of differential equations and Lagrangian systems have been intensively investigated [25, 40, 59, 76, 85, 108]. Accordingly, by symmetries one means only those represented by vector fields $\upsilon = u$ on Y. We agree to call them *classical symmetries*.

Remark 2.3 Owing to the relation (2.5), the bracket (2.4) of Lagrangian symmetries is a Lagrangian symmetry. It follows that symmetries constitute a real Lie algebra $\mathcal{G}_L$ with respect to this bracket, and their jet prolongation $\upsilon \to J^*\upsilon$ (2.2) provides a monomorphism of this Lie algebra to the Lie algebra $\mathfrak{d}\mathcal{O}^0_\infty Y$ (1.16).

Remark 2.4 Let υ be a classical symmetry of a Lagrangian L, i.e., it is a vector field on Y. Then the relation

$$\mathbf{L}_{J^*\upsilon}\mathcal{E}_L = \delta(\mathbf{L}_{J^*\upsilon}L) \tag{2.12}$$

holds [53, 108]. It follows that υ also is a symmetry of the Euler–Lagrange operator $\mathcal{E}_L$ of L (Definition 4.5), i.e., $\mathbf{L}_{J^*\upsilon}\mathcal{E}_L = 0$, and as a consequence it is an infinitesimal symmetry of the Euler–Lagrange equation $\mathfrak{E}_L$ (1.34) (Definition 4.4). However, the equality (2.12) fails to be true in the case of generalized symmetries.

2.2 Gauge Symmetries: Noether's Direct Second Theorem

As was mentioned above, the notion of gauge symmetries comes from Yang–Mills gauge theory on principal bundles (Sect. 8.2). It is generalized to Lagrangian theory on an arbitrary fibre bundle $Y \to X$ as follows [9, 10].

Definition 2.3 Let $E \to X$ be a vector bundle and $E(X)$ a $C^\infty(X)$-module of sections of $E \to X$. Let ζ be a linear differential operator on $E(X)$ (Definition A.3) with values into a vector space $\mathcal{G}_L$ of symmetries of a Lagrangian L (Remark 2.3). Elements

$$u_\xi = \zeta(\xi) \tag{2.13}$$

of Im ζ are termed the *gauge symmetries* of a Lagrangian L parameterized by sections ξ of $E \to X$. The latter are regarded as the *gauge parameters*.

Remark 2.5 A differential operator ζ in Definition 2.3 takes its values into a vector space $\mathcal{G}_L$ as a subspace of a $C^\infty(X)$-module $\mathfrak{d}\mathcal{O}^0_\infty Y$ seen as a real vector space. The differential operator ζ is assumed to be at least of first order (Remark 2.7).

Equivalently, the gauge symmetry (2.13) is given by a section $\widetilde{\zeta}$ of a fibre bundle

$$(J^rY \underset{Y}{\times} J^mE) \underset{Y}{\times} TY \to J^rY \underset{Y}{\times} J^mE$$

(Definition B.14) such that $u_\xi = \zeta(\xi) = \widetilde{\zeta} \circ \xi$ for any section ξ of $E \to X$. Hence, it is a generalized vector field u_ζ on a bundle product $Y \times_X E$ represented by a section of the pull-back bundle

$$J^k(Y \underset{X}{\times} E) \underset{Y}{\times} T(Y \underset{X}{\times} E) \to J^k(Y \underset{X}{\times} E), \qquad k = \max(r, m),$$

which lives in $TY \subset T(Y \times_X E)$. This generalized vector field yields the contact derivation $J^\infty u_\zeta$ (2.2) of a real ring $\mathcal{O}^0_\infty[Y \times_X E]$ which obeys the following condition.

- Given a Lagrangian $L \in \mathcal{O}^{0,n}_\infty E \subset \mathcal{O}^{0,n}_\infty[Y \times_X E]$, let us consider its Lie derivative

$$\mathbf{L}_{J^* u_\zeta} L = J^\infty u_\zeta \rfloor dL + d(J^* u_\zeta \rfloor L) \tag{2.14}$$

where d is the exterior differential on $\mathcal{O}^*_\infty[Y \times_X E]$. Then, for any section ξ of $E \to X$, the pull-back $\xi^* \mathbf{L}_{J^* u_\zeta} L$ is d_H-exact.

It follows from the first variational formula (2.10) for the Lie derivative (2.14) that the above mentioned condition holds only if u_ζ is projected onto a generalized vector field on Y and, in this case, if and only if the density $(u_\zeta)_V \rfloor \mathcal{E}_L$ is d_H-exact. Thus, we come to the following equivalent definition of gauge symmetries.

Definition 2.4 Let $E \to X$ be a vector bundle. A gauge symmetry of a Lagrangian L parameterized by sections ξ of $E \to X$ is defined as a generalized vector field u on $Y \times_X E$ such that:

(i) a contact derivation $\vartheta = J^\infty u$ of a ring $\mathcal{O}^0_\infty[Y \times_X E]$ vanishes on a subring $\mathcal{O}^0_\infty E$,

(ii) a generalized vector field u is linear in coordinates χ^a_Λ on $J^\infty E$, and it is projected onto a generalized vector field on E, i.e., it takes a form

$$u = \left(\sum_{0 \le |\Lambda| \le m} u^{\lambda\Lambda}_a(x^\mu) \chi^a_\Lambda \right) \partial_\lambda + \left(\sum_{0 \le |\Lambda| \le m} u^{i\Lambda}_a(x^\mu, y^j_\Sigma) \chi^a_\Lambda \right) \partial_i, \tag{2.15}$$

(iii) the vertical part of u (2.15) obeys the equality

$$u_V \rfloor \delta L = d_H \sigma. \tag{2.16}$$

Theorem 2.5 *By virtue of item (iii) of Definition 2.4, u (2.15) is a gauge symmetry if and only if its vertical part is so.*

Gauge symmetries possess the following particular properties.

(i) Let $E' \to X$ be another vector bundle and ζ' a linear $E(X)$-valued differential operator on a $C^\infty(X)$-module $E'(X)$ of sections of $E' \to X$. Then $u_{\zeta'(\xi')} = (\zeta \circ \zeta')(\xi')$ also is a gauge symmetry of L parameterized by sections ξ' of $E' \to X$. It factorizes through the gauge symmetries u_χ (2.13).

(ii) The conserved symmetry current $\mathcal{J}_u$ (2.21) associated to a gauge symmetry in accordance with Noether's first theorem (Theorem 2.7) is reduced to a superpotential (Theorem 2.8).

(iii) *Noether's direct second theorem* (Theorem 2.6) associates to a gauge symmetry of a Lagrangian L the Noether identities (NI) of its Euler–Lagrange operator.

Theorem 2.6 *Let u (2.15) be a gauge symmetry of a Lagrangian L, then its Euler–Lagrange operator δL obeys the NI (2.17).*

Proof The density (2.16) is variationally trivial and, therefore, its variational derivatives with respect to variables χ^a vanish, i.e.,

$$\mathcal{E}_a = \sum_{0\leq|\Lambda|} (-1)^{|\Lambda|} d_\Lambda[(u_a^{i\Lambda} - y_\lambda^i u_a^{\lambda\Lambda})\mathcal{E}_i] = \sum_{0\leq|\Lambda|} \eta(u_a^i - y_\lambda^i u_a^\lambda)^\Lambda d_\Lambda \mathcal{E}_i = 0 \quad (2.17)$$

(see Remark 7.6 for the notation). In accordance with Definition D.1, the equalities (2.17) are the NI for the Euler–Lagrange operator δL. □

Remark 2.6 If the gauge symmetry u (2.15) is of second jet order in gauge parameters, i.e.,

$$u_V = (u_a^i \chi^a + u_a^{i\mu} \chi_\mu^a + u_a^{i\nu\mu} \chi_{\nu\mu}^a)\partial_i, \quad (2.18)$$

the corresponding NI (2.17) take a form

$$u_a^i \mathcal{E}_i - d_\mu(u_a^{i\mu} \mathcal{E}_i) + d_{\nu\mu}(u_a^{i\nu\mu} \mathcal{E}_i) = 0. \quad (2.19)$$

Let us note that the NI (2.17) need not be independent (Sect. 6.2).

Remark 2.7 A glance at the expression (2.19) shows that, if a gauge symmetry is independent of derivatives of gauge parameters (i.e., a differential operator ζ in Definition 2.3 is of zero order), then all variational derivatives of a Lagrangian equals zero, i.e., this Lagrangian is variationally trivial. Therefore, such gauge symmetries usually are not considered.

Remark 2.8 The notion of gauge symmetries can be generalized as follows. Let a differential operator ζ in Definition 2.3 need not be linear. Then elements of $\mathrm{Im}\,\zeta$ are called the *generalized gauge symmetry*. However, Noether's direct second Theorem 2.6 is not relevant to generalized gauge symmetries because, in this case, an Euler–Lagrange operator satisfies the identities depending on gauge parameters.

It follows from Noether's direct second Theorem 2.6 that gauge symmetries of Lagrangian field theory characterize its degeneracy. A problem is that any Lagrangian possesses gauge symmetries and, therefore, one must separate them into the trivial and nontrivial ones. Moreover, gauge symmetries can be reducible, i.e., $\operatorname{Ker}\zeta \neq 0$. Another problem is that gauge symmetries need not form an algebra [48, 60, 63]. The Lie bracket $[u_\phi, u_{\phi'}]$ of gauge symmetries $u_\phi, u_{\phi'} \in \operatorname{Im}\zeta$ is a symmetry, but it need not belong to $\operatorname{Im}\zeta$. To solve these problems, we follow a different definition of gauge symmetries (Definitions 7.4–7.5) as those associated to nontrivial NI by means of Noether's inverse second Theorem 7.9. They are parameterized by Grassmann-graded ghosts, but not gauge parameters (Remark 7.7).

2.3 Noether's First Theorem: Conservation Laws

Let $(\mathcal{O}^*_\infty Y, L)$ be a Lagrangian system (Definition 1.5). The following is *Noether's first theorem.*

Theorem 2.7 *Let the generalized vector field υ (2.7) be a symmetry of a Lagrangian L (Definition 2.2), i.e., let it obey the equality (2.11). Then the first variational formula (2.10) restricted to the Euler–Lagrange equation $\mathfrak{E}_L$ (1.34) takes a form of the weak conservation law on-shell*

$$0 \approx -d_H(-h_0(\vartheta \rfloor \Xi_L) + \sigma) \approx -d_H \mathcal{J}_\upsilon \tag{2.20}$$

of a symmetry current

$$\mathcal{J}_\upsilon = \mathcal{J}^\mu_\upsilon \omega_\mu = -h_0(\vartheta \rfloor \Xi_L) + \sigma \tag{2.21}$$

along a generalized vector field υ. It is called the Lagrangian conservation law.

The weak conservation law (2.20) leads to a *differential conservation law*

$$\partial_\lambda(\mathcal{J}^\lambda_\upsilon \circ s) = 0$$

on classical solutions s of the Euler–Lagrange equation (1.35) (Definition B.13). This differential conservation law, in turn, yields an *integral conservation law*

$$\int_{\partial M} s^* \mathcal{J}_\upsilon = 0, \tag{2.22}$$

where M is an n-dimensional compact submanifold of X with a boundary ∂M.

Remark 2.9 Of course, the symmetry current $\mathcal{J}_\upsilon$ (2.21) is defined with the accuracy to a d_H-closed term. For instance, if we choose a different Lepage equivalent Ξ_L

(1.37) in the variational formula (1.36), the corresponding symmetry current differs from $\mathcal{J}_\upsilon$ (2.21) in a d_H-exact term. This term is independent of a Lagrangian, and it does not contribute to the integral conservation law (2.22).

Obviously, the symmetry current $\mathcal{J}_u$ (2.21) is linear in a generalized vector field υ. Therefore, one can consider a superposition of symmetry currents

$$\mathcal{J}_\upsilon + \mathcal{J}_{\upsilon'} = \mathcal{J}_{\upsilon+\upsilon'}, \qquad \mathcal{J}_{c\upsilon} = c\mathcal{J}_\upsilon, \qquad c \in \mathbb{R},$$

associated to different symmetries υ and that of weak conservation laws (2.20).

A symmetry υ of a Lagrangian L is called *exact* if the Lie derivative $\mathbf{L}_{J^*\upsilon}L$ of L along $J^*\upsilon$ vanishes, i.e., $\mathbf{L}_{J^*\upsilon}L = 0$.

In this case, the first variational formula (2.10) takes a form

$$0 = \upsilon_V \rfloor \delta L + d_H(h_0(J^*\upsilon \rfloor \Xi_L)),$$

and results in the weak conservation law (2.20):

$$0 \approx d_H(h_0(J^*\upsilon \rfloor \Xi_L)) \approx -d_H \mathcal{J}_\upsilon, \tag{2.23}$$

of the symmetry current $\mathcal{J}_\upsilon = -h_0(J^*\upsilon \rfloor \Xi_L)$.

For instance, let $\upsilon = \upsilon^i \partial_i$ be a vertical generalized vector field on $Y \to X$. If it is an exact symmetry of L, the weak conservation law (2.23) takes a form

$$0 \approx -d_H(J^*\upsilon \rfloor \Xi_L). \tag{2.24}$$

Definition 2.5 The equality (2.24) is called the *Noether conservation law* of a *Noether current*

$$\mathcal{J} = -J^*\upsilon \rfloor \Xi_L. \tag{2.25}$$

If a Lagrangian L admits the gauge symmetry u (2.15), the weak conservation law (2.20) of the corresponding symmetry current $\mathcal{J}_u$ (2.21) holds. We call it the *gauge conservation law*. Because gauge symmetries depend on parameter variables and their jets, all gauge conservation laws possess the following peculiarity.

Theorem 2.8 *If u (2.15) is a gauge symmetry of a Lagrangian L, the corresponding conserved symmetry current $\mathcal{J}_u$ (2.21) along u takes a form*

$$\mathcal{J}_u = W + d_H U = (W^\mu + d_\nu U^{\nu\mu})\omega_\mu, \tag{2.26}$$

where the term W vanishes on-shell, and $U = U^{\nu\mu}\omega_{\nu\mu}$ is a horizontal $(n-2)$-form.

Proof Theorem 2.8 is a particular variant of Theorem 7.11. □

Definition 2.6 A term U in the expression (2.26) is called the *superpotential*.

If a symmetry current admits the decomposition (2.26), one says that it is reduced to a superpotential [39, 61, 66, 128]. If a symmetry current $\mathcal{J}$ reduces to a superpotential, the integral conservation law (2.22) becomes tautological.

Remark 2.10 Theorem 2.8 generalizes the result in [66] for gauge symmetries u whose gauge parameters $\chi^\lambda = u^\lambda$ are components of a projection $u^\lambda \partial_\lambda$ of u onto X.

Remark 2.11 In mechanics on a configuration bundle over $\mathbb{R}$ (Chap. 4), the whole conserved current (2.26) along a gauge symmetry vanishes on-shell (Theorem 4.7).

Sometimes, Theorem 2.8 is called *Noether's third theorem* on a superpotential.

Chapter 3
Lagrangian and Hamiltonian Field Theories

We focus our attention on first order Lagrangian and polysymplectic Hamiltonian theories on fibre bundles since the most of relevant field models is of this type [53, 61, 133]. These theories are equivalent in the case of hyperregular Lagrangians (Definition 3.1), and one can state comprehensive relations between them for semi-regular and almost regular Lagrangians (Sect. 3.6). Moreover, polysymplectic Hamiltonian formalism on the Legendre bundle Π (3.6) is equivalent to particular first order Lagrangian theory on a configuration space Π that enables us to describe symmetries of covariant Hamiltonian theory similarly to those in Lagrangian formalism (Sect. 3.7).

3.1 First Order Lagrangian Formalism

Let $\pi : Y \to X$ be a fibre bundle over a $(1 < n)$-dimensional base X (see Chap. 4 for the case of $n = 1$). Let Y be provided with an atlas of bundle coordinates (x^λ, y^i).

In Lagrangian formalism on a *configuration space*. $Y \to X$ (Sect. 1.3), a *first order Lagrangian* is defined as a horizontal density

$$L = \mathcal{L}\omega : J^1 Y \to \overset{n}{\wedge} T^* X \tag{3.1}$$

on the first order jet manifold $J^1 Y$ of Y. Its pull-back onto higher order jet manifolds $J^* Y$ can be considered (Definition 1.3).

The corresponding *second-order Euler–Lagrange operator* (1.33) (Definition 1.4) reads

$$\begin{aligned}
&\mathcal{E}_L : J^2 Y \to T^* Y \wedge (\overset{n}{\wedge} T^* X), \\
&\mathcal{E}_L = (\partial_i \mathcal{L} - d_\lambda \pi_i^\lambda)\theta^i \wedge \omega, \qquad \pi_i^\lambda = \partial_i^\lambda \mathcal{L}.
\end{aligned} \tag{3.2}$$

G. Sardanashvily, *Noether's Theorems*, Atlantis Studies
in Variational Geometry 3, DOI 10.2991/978-94-6239-171-0_3

Its kernel defines the *second order Euler–Lagrange equation* (1.35):

$$(\partial_i - d_\lambda \partial_i^\lambda)\mathcal{L} = 0, \tag{3.3}$$

on a configuration space $Y \to X$.

Given a first order Lagrangian L, let us consider the vertical tangent morphism (B.17):

$$VL : V_Y J^1 Y \underset{Y}{\longrightarrow} V_Y J^1 Y \underset{Y}{\times} \overset{n}{\wedge} T^* X \tag{3.4}$$

over Y to L (3.1), where $V_Y J^1 Y$ denotes the vertical tangent bundle (Definition B.3) of $J^1 Y \to Y$. Since $J^1 Y \to Y$ is an affine bundle modelled over the vector bundle (B.47), we have the canonical vertical splitting

$$V_Y J^1 Y = J^1 Y \underset{Y}{\times} (T^* X \underset{Y}{\otimes} VY).$$

Accordingly, the vertical tangent map VL (3.4) yields a linear morphism

$$J^1 Y \underset{Y}{\times} (T^* X \underset{Y}{\otimes} VY) \longrightarrow J^1 Y \underset{Y}{\times} (\overset{n}{\wedge} T^* X)$$

over $J^1 Y$ and the corresponding morphism

$$\widehat{L} : J^1 Y \to V^* Y \underset{Y}{\otimes} (\overset{n}{\wedge} T^* X) \underset{Y}{\otimes} TX \tag{3.5}$$

over Y, where $V^* Y$ is the vertical cotangent bundle of $Y \to X$ (Definition B.4). The latter is termed the *Legendre map* associated to a Lagrangian L.

A fibre bundle

$$\Pi = V^* Y \underset{Y}{\otimes} (\overset{n}{\wedge} T^* X) \underset{Y}{\otimes} TX \tag{3.6}$$

over Y is named the *Legendre bundle*. It is a composite bundle $\Pi \to Y \to X$, called the *composite Legendre bundle* and provided with the adapted *holonomic coordinates* $(x^\lambda, y^i, p_i^\lambda)$ possessing transition functions

$$p'^\lambda_i = \det\left(\frac{\partial x^\varepsilon}{\partial x'^\nu}\right) \frac{\partial y^j}{\partial y'^i} \frac{\partial x'^\lambda}{\partial x^\mu} p_j^\mu. \tag{3.7}$$

With respect to these coordinates, the Legendre map (3.5) reads $p_i^\lambda \circ \widehat{L} = \pi_i^\lambda$.

Remark 3.1 There is the canonical isomorphism

$$\Pi = V^* Y \underset{Y}{\wedge} (\overset{n-1}{\wedge} T^* X), \qquad (p_i^\lambda) \to p_i^\lambda \overline{d} y^i \omega_\lambda, \tag{3.8}$$

where $\{\overline{d} y^i\}$ are fibre bases for the vertical cotangent bundle $V^* Y$ of $Y \to X$.

Certainly, the Legendre map (3.5) need not be a bundle isomorphism. Its range

$$N_L = \widehat{L}(J^1Y) \subset \Pi \tag{3.9}$$

is termed the *Lagrangian constraint space*.

Definition 3.1 A Lagrangian L is said to be:

- *hyperregular* if the Legendre map $\widehat{L}$ is a diffeomorphism;
- *regular* if $\widehat{L}$ is a local diffeomorphism, i.e., $\det(\partial_i^\mu \partial_j^\nu \mathcal{L}) \neq 0$;
- *semiregular* if the inverse image $\widehat{L}^{-1}(q)$ of any point $q \in N_L$ is a connected submanifold of J^1Y;
- *almost regular* if the Lagrangian constraint space N_L is a closed imbedded subbundle $i_N : N_L \to \Pi$ of a Legendre bundle $\Pi \to Y$ and a Legendre map

$$\widehat{L} : J^1Y \to N_L \tag{3.10}$$

is a fibred manifold (Definition B.1) with connected fibres (i.e., a Lagrangian L is semiregular).

Given a first order Lagrangian L, its Lepage equivalents Ξ_L (1.37) in the variational formula (1.36) read

$$\Xi_L = L + (\pi_i^\lambda - d_\mu \psi_i^{\mu\lambda})\theta^i \wedge \omega_\lambda + \psi_i^{\lambda\mu}\theta_\mu^i \wedge \omega_\lambda, \tag{3.11}$$

where $\psi_i^{\mu\lambda} = -\sigma_i^{\lambda\mu}$ are skew-symmetric local functions on Y. These Lepage equivalents constitute an affine space modelled over a vector space of d_H-exact one-contact Lepage forms

$$\rho = -d_\mu \psi_i^{\mu\lambda}\theta^i \wedge \omega_\lambda + \psi_i^{\lambda\mu}\theta_\mu^i \wedge \omega_\lambda.$$

Let us choose the *Poincaré–Cartan form*

$$H_L = \mathcal{L}\omega + \pi_i^\lambda \theta^i \wedge \omega_\lambda \tag{3.12}$$

as the origin of this affine space because it is defined on J^1Y. In a general setting, one also considers other Lepage equivalents of L [91, 92].

The Poincaré–Cartan form (3.12) takes its values into a subbundle

$$J^1Y \underset{Y}{\times} (T^*Y \underset{Y}{\wedge} (\overset{n-1}{\wedge} T^*X))$$

of $\overset{n}{\wedge} T^*J^1Y$. Hence, it defines a bundle morphism

$$\widehat{H}_L : J^1Y \to Z_Y = T^*Y \wedge (\overset{n-1}{\wedge} T^*X) \tag{3.13}$$

over Y whose range

$$Z_L = \widehat{H}_L(J^1Y) \tag{3.14}$$

is an imbedded subbundle $i_L : Z_L \to Z_Y$ of the fibre bundle $Z_Y \to Y$. This morphism is called the *homogeneous Legendre map*. Accordingly, the fibre bundle $Z_Y \to Y$ (3.13) is said to be the *homogeneous Legendre bundle*. It is equipped with the *holonomic coordinates* $(x^\lambda, y^i, p_i^\lambda, p)$ possessing transition functions

$$p'^\lambda_i = \det\left(\frac{\partial x^\varepsilon}{\partial x'^\nu}\right)\frac{\partial y^j}{\partial y'^i}\frac{\partial x'^\lambda}{\partial x^\mu}p_j^\mu, \tag{3.15}$$

$$p' = \det\left(\frac{\partial x^\varepsilon}{\partial x'^\nu}\right)\left(p - \frac{\partial y^j}{\partial y'^i}\frac{\partial y'^i}{\partial x^\mu}p_j^\mu\right).$$

With respect to these coordinates, the morphism $\widehat{H}_L$ (3.13) reads

$$(p_i^\mu, p)\circ \widehat{H}_L = (\pi_i^\mu, \mathcal{L} - y_\mu^i\pi_i^\mu).$$

A glance at the transition functions (3.15) shows that Z_Y (3.13) is a one-dimensional affine bundle

$$\pi_{Z\Pi} : Z_Y \to \Pi \tag{3.16}$$

over the Legendre bundle Π (3.8) modelled over the pull-back vector bundle

$$\Pi \underset{X}{\times} \overset{n}{\wedge} T^*X \to \Pi.$$

Moreover, the Legendre map $\widehat{L}$ (3.5) is exactly the composition of morphisms

$$\widehat{L} = \pi_{Z\Pi}\circ H_L : J^1Y \underset{Y}{\to} \Pi.$$

3.2 Cartan and Hamilton–De Donder Equations

Let us note that the Euler–Lagrange equation (3.3) does not exhaust all equations considered in first order Lagrangian theory.

Being a Lepage equivalent of L, the Poincaré–Cartan form H_L (3.12) also is a Lepage equivalent of the first order Lagrangian

$$\overline{L} = \widehat{h}_0(H_L) = (\mathcal{L} + (\widehat{y}_\lambda^i - y_\lambda^i)\pi_i^\lambda)\omega, \qquad \widehat{h}_0(dy^i) = \widehat{y}_\lambda^i dx^\lambda, \tag{3.17}$$

on the repeated jet manifold J^1J^1Y, coordinated by $(x^\lambda, y^i, y_\lambda^i, \widehat{y}_\mu^i, y_{\mu\lambda}^i)$. The Euler–Lagrange operator for $\overline{L}$ (called the *Euler–Lagrange–Cartan operator*) reads

$$\mathscr{E}_{\overline{L}} : J^1J^1Y \to T^*J^1Y \wedge (\overset{n}{\wedge} T^*X),$$
$$\mathscr{E}_{\overline{L}} = [(\partial_i \mathscr{L} - \widehat{d}_\lambda \pi_i^\lambda + \partial_i \pi_j^\lambda (\widehat{y}_\lambda^j - y_\lambda^j))dy^i + \partial_i^\lambda \pi_j^\mu (\widehat{y}_\mu^j - y_\mu^j) dy_\lambda^i] \wedge \omega,$$
$$\widehat{d}_\lambda = \partial_\lambda + \widehat{y}_\lambda^i \partial_i + y_{\lambda\mu}^i \partial_i^\mu.$$

Its kernel $\operatorname{Ker} \mathscr{E}_{\overline{L}} \subset J^1J^1Y$ is the *Cartan equation* on a configuration space J^1Y, which is locally given by equalities

$$\partial_i^\lambda \pi_j^\mu (\widehat{y}_\mu^j - y_\mu^j) = 0, \tag{3.18}$$
$$\partial_i \mathscr{L} - \widehat{d}_\lambda \pi_i^\lambda + (\widehat{y}_\lambda^j - y_\lambda^j) \partial_i \pi_j^\lambda = 0. \tag{3.19}$$

Since $\mathscr{E}_{\overline{L}}|_{J^2Y} = \mathscr{E}_L$, the Cartan equations (3.18) and (3.19) are equivalent to the Euler–Lagrange equation (3.3) on integrable sections of $J^1Y \to X$. These equations are equivalent if a Lagrangian is regular. The Cartan equations (3.18) and (3.19) on sections $\overline{s} : X \to J^1Y$ are equivalent to the relation

$$\overline{s}^*(u \rfloor dH_L) = 0, \tag{3.20}$$

which is assumed to hold for all vertical vector fields u on $J^1Y \to X$.

Definition 3.2 The homogeneous Legendre bundle Z_Y (3.13) admits the canonical exterior form

$$\Xi_Y = p\omega + p_i^\lambda dy^i \wedge \omega_\lambda. \tag{3.21}$$

It is termed the *multisymplectic Liouville form*

Accordingly, the imbedded subbundle Z_L (3.14) of Z_Y is provided with the pull-back *De Donder form* $\Xi_L = i_L^* \Xi_Y$. There is the equality

$$H_L = \widehat{H}_L^* \Xi_L = \widehat{H}_L^*(i_L^* \Xi_Y). \tag{3.22}$$

By analogy with the Cartan equation (3.20), the *Hamilton–De Donder equation* for sections $\overline{r}$ of $Z_L \to X$ is written as

$$\overline{r}^*(u \rfloor d\Xi_L) = 0, \tag{3.23}$$

where u is an arbitrary vertical vector field on $Z_L \to X$. Then the following holds [64].

Theorem 3.1 *Let the homogeneous Legendre map $\widehat{H}_L$ be a submersion. Then a section $\overline{s}$ of $J^1Y \to X$ is a solution of the Cartan equation (3.20) if and only if $\widehat{H}_L \circ \overline{s}$ is a solution of the Hamilton–De Donder equation (3.23), i.e., the Cartan and Hamilton–De Donder equations are quasi-equivalent.*

Remark 3.2 The Legendre bundle Π (3.6) and the homogeneous Legendre bundle Z_Y (3.13) play the role of a phase space and homogeneous phase space in covariant Hamiltonian field theory, respectively (Sect. 3.5).

3.3 Noether's First Theorem: Energy-Momentum Currents

We restrict our study of symmetries of first order Lagrangian field theory to classical symmetries, represented by projectable vector fields on a fibre bundle $Y \to X$. This is the case of all basic classical field models.

Let

$$u = u^\lambda(x^\mu)\partial_\lambda + u^i(x^\mu, y^j)\partial_i \tag{3.24}$$

be a projectable vector field on a fibre bundle $Y \to X$. Its canonical decomposition (2.7) into the horizontal and vertical parts over J^1Y reads

$$u = u_H + u_V = u^\lambda(\partial_\lambda + y^i_\lambda\partial_i) + (u^i\partial_i - y^i_\lambda u^\lambda\partial_i).$$

Given the first order Lagrangian L (3.1), it is sufficient to consider the first order jet prolongation (B.52):

$$J^1u = u^\lambda\partial_\lambda + u^i\partial_i + (d_\lambda u^i - y^i_\mu\partial_\lambda u^\mu)\partial^\lambda_i,$$

of u (3.24) onto J^1Y. Then the first variational formula (2.10) takes a form

$$\mathbf{L}_{J^1u}L = u_V\rfloor\mathcal{E}_L + d_H(h_0(u\rfloor H_L)), \tag{3.25}$$

where $\Xi_L = H_L$ is the Poincaré–Cartan form (3.12). Its coordinate expression reads

$$\begin{aligned}\partial_\lambda u^\lambda\mathcal{L} + [u^\lambda\partial_\lambda + u^i\partial_i + (d_\lambda u^i - y^i_\mu\partial_\lambda u^\mu)\partial^\lambda_i]\mathcal{L} = \\ (u^i - y^i_\lambda u^\lambda)\mathcal{E}_i - d_\lambda[\pi^\lambda_i(u^\mu y^i_\mu - u^i) - u^\lambda\mathcal{L}].\end{aligned} \tag{3.26}$$

If u is a symmetry of L (Definition 2.2), we obtain the weak Lagrangian conservation law (2.20):

$$0 \approx -d_\lambda[\pi^\lambda_i(u^\mu y^i_\mu - u^i) - u^\lambda\mathcal{L} + \sigma^\lambda] \approx -d_\lambda\mathcal{J}^\lambda_u, \tag{3.27}$$

of the symmetry current (2.21):

$$\mathcal{J}_u = \mathcal{J}^\lambda_u\omega_\lambda = [\pi^\lambda_i(u^\mu y^i_\mu - u^i) - u^\lambda\mathcal{L} + \sigma^\lambda]\omega_\lambda, \tag{3.28}$$

on J^1Y along the vector field u (3.24). If u is an exact symmetry, the symmetry current (3.28) reads

$$\mathcal{J}_u = \mathcal{J}^\lambda_u\omega_\lambda = [\pi^\lambda_i(u^\mu y^i_\mu - u^i) - u^\lambda\mathcal{L}]\omega_\lambda. \tag{3.29}$$

Remark 3.3 If we choose a different Lepage equivalent Ξ_L (3.11) in the first variational formula (Remark 2.9), the corresponding symmetry current differs from $\mathcal{J}_u$ (3.28) in a d_H-exact term

$$\psi = d_\mu(\psi^{\mu\lambda}_i(u^i - y^i_\nu u^\nu))\omega_\lambda.$$

As was mentioned above, the symmetry current $\mathcal{J}_u$ (3.28) is linear in a vector field u. Therefore, one can consider a superposition of symmetry currents

$$\mathcal{J}_u + \mathcal{J}_{u'} = \mathcal{J}_{u+u'}, \qquad \mathcal{J}_{cu} = c\mathcal{J}_u, \qquad c \in \mathbb{R}, \tag{3.30}$$

and a superposition of weak conservation laws (3.27) associated to different symmetries u.

For instance, let $\upsilon = \upsilon^i \partial_i$ be a vertical vector field on $Y \to X$. If it is an exact symmetry of L, the weak conservation law (3.27) takes a form of the Noether conservation law (2.24):

$$0 \approx d_\lambda(\pi_i^\lambda \upsilon^i) \tag{3.31}$$

of the Noether current (2.25):

$$\mathcal{J}_\upsilon^\lambda = -\pi_i^\lambda \upsilon^i. \tag{3.32}$$

This is the case of so called *internal symmetries* in field theory.

In contrast to a case of internal symmetries, let $\tau = \tau^\lambda \partial_\lambda$ be a vector field on X and $\gamma\tau$ its lift (B.27) onto Y (Definition B.2).

Definition 3.3 A symmetry current

$$\mathcal{J}_{\gamma\tau}^\lambda = \pi_i^\lambda(\tau^\mu y_\mu^i - (\gamma\tau)^i) - \tau^\lambda \mathcal{L} \tag{3.33}$$

along $\gamma\tau$ (B.27) is called the *energy-momentum current*. If $\gamma\tau$ (B.27) is an exact symmetry of a Lagrangian L, the *energy-momentum conservation law*

$$0 \approx -d_\lambda[\pi_i^\lambda(\tau^\mu y_\mu^i - (\gamma\tau)^i) - \tau^\lambda \mathcal{L}] \tag{3.34}$$

holds.

In particular, given the connection Γ (B.55) on a fibre bundle $Y \to X$, let $\Gamma\tau$ (B.56) be the horizontal lift of a vector field τ on X. The corresponding energy-momentum current (3.33) along $\Gamma\tau$ reads

$$\mathcal{J}_{\Gamma\tau} = \tau^\mu \mathcal{J}_\Gamma{}^\lambda{}_\mu \omega_\lambda = \tau^\mu(\pi_i^\lambda(y_\mu^i - \Gamma_\mu^i) - \delta_\mu^\lambda \mathcal{L})\omega_\lambda.$$

Its coefficients $\mathcal{J}_\Gamma{}^\lambda{}_\mu$ are components of the tensor field

$$\mathcal{J}_\Gamma = \mathcal{J}_\Gamma{}^\lambda{}_\mu dx^\mu \otimes \omega_\lambda, \qquad \mathcal{J}_\Gamma{}^\lambda{}_\mu = \pi_i^\lambda(y_\mu^i - \Gamma_\mu^i) - \delta_\mu^\lambda \mathcal{L}, \tag{3.35}$$

termed the *energy-momentum tensor* relative to a connection Γ [42, 100, 119]. If $\Gamma\tau$ (B.56) is an exact symmetry of a Lagrangian L, we obtain the energy-momentum conservation law (3.34):

$$0 \approx -d_\lambda[\pi_i^\lambda \tau^\mu(y_\mu^i - \Gamma_\mu^i) - \delta_\mu^\lambda \tau^\mu \mathcal{L}].$$

For instance, let a fibre bundle $Y \to X$ admit a flat connection Γ. By virtue of Theorem B.15, there exist bundle coordinates such that $\Gamma^i_\lambda = 0$, and the corresponding energy-momentum tensor (3.35) takes a form

$$\mathcal{J}_0{}^\lambda{}_\mu = \pi^\lambda_i y^i_\mu - \delta^\lambda_\mu \mathcal{L}$$

of the familiar *canonical energy–momentum tensor.*

In particular, this is just the case of nonrelativistic mechanics on fibre bundles over $\mathbb{R}$ where $\tau = \partial_t$ is the canonical vector field on $\mathbb{R}$ and $\mathcal{J}_\Gamma$ (3.35) is the energy function E_Γ (4.41) with respect to a reference frame Γ (Sect. 4.3).

The lift $\gamma\tau$, considered for an arbitrary vector field τ on X, exemplifies a gauge transformation in accordance with Definition 2.4. It is parameterized by vector fields τ as sections of the vector bundle $TX \to X$.

Then it should be emphasized that, since the horizontal lift $\Gamma\tau$ (B.56) is independent of derivatives of components of a vector field τ, it can not be an exact symmetry of a Lagrangian L for any vector field τ, unless a Lagrangian L is variationally trivial (Remark 2.7).

At the same time, a canonical functorial lift $\widetilde{\tau}$, e.g., the $\widetilde{\tau}$ (B.29) (Definition B.6) depends on derivatives of gauge parameters τ, and can become a gauge symmetry. In this case, the energy-momentum current (3.33) is reduced to a superpotential (Theorem 2.8). This is just the case of gravitation theory on natural bundles (Sect. 10.4).

Any projectable vector field u (3.24) can be represented as a sum

$$u = \gamma(u^\lambda \partial_\lambda) + \upsilon \tag{3.36}$$

of some lift $\gamma(u^\lambda)\partial_\lambda)$ (B.27) of its projection $u^\lambda\partial_\lambda$ onto X and a vertical vector field υ. Let u (3.36) be an exact symmetry and $\mathcal{J}_u$ the corresponding symmetry current (3.29). In view of superposition (3.30), it falls into a sum

$$\mathcal{J}_u = \mathcal{J}_{\gamma\tau} + \mathcal{J}_\upsilon$$

of the energy-momentum current $\mathcal{J}_{\gamma\tau}$ (3.33) and the Noether one $\mathcal{J}_\upsilon$ (3.32).

3.4 Conservation Laws in the Presence of a Background Field

In Lagrangian field theory on a fibre bundle $Y \to X$, by *background fields* are meant classical fields which do not obey Euler–Lagrange equations. Let these fields be represented by global sections h of a fibre bundle $\Sigma \to X$ endowed with bundle coordinates (x^λ, σ^m).

For instance, Lagrangians of gauge and matter fields (Chap. 8), except the topological ones (Chap. 11), depend on a pseudo-Riemannian world metric (Remark B.12) as a background field.

In order to formulate Lagrangian field theory in the presence of background fields, let us consider a bundle product

$$Y_{\text{tot}} = \Sigma \underset{X}{\times} Y \to X \tag{3.37}$$

coordinated by $(x^\lambda, \sigma^m, y^i)$ and its jet manifold

$$J^1 Y_{\text{tot}} = J^1 \Sigma \underset{X}{\times} J^1 Y. \tag{3.38}$$

Let L be a first order Lagrangian on the jet manifold (3.38). It can be considered as a total Lagrangian of field theory on the configuration space (3.37) where background fields are treated as dynamic variables.

Given a global section h of $\Sigma \to X$, we obtain the pull-back Lagrangian

$$L_h = (J^1 h)^* L \tag{3.39}$$

on $J^1 Y$. It can be regarded as a Lagrangian of first order field theory on a configuration space Y in the presence of a background field h.

Let us consider the variational formula (1.36) for a total Lagrangian L:

$$dL - \delta L - d_H \Xi = 0, \qquad \delta L = (\mathcal{E}_i \theta^i + \mathcal{E}_m \theta^m) \wedge \omega.$$

Its pull-back

$$(J^2 h)^*(dL - \delta L - d_H \Xi) = dL_h - \delta L_h - d_H \Xi_h \tag{3.40}$$

is exactly the variational formula for the Lagrangian L_h (3.39) in the presence of a background field h. The corresponding Euler–Lagrange operator in the presence of a background field h reads

$$\delta L_h = (J^2 h)^*(\delta L) = (J^2 h)^*(\mathcal{E}_i \theta^i + \mathcal{E}_m \theta^m) \wedge \omega = (J^1 h)^*(\mathcal{E}_i) \theta^i \omega.$$

The variational formula (3.40) enables us to obtain conservation laws in the presence of a background field. Let

$$u = u^\lambda (x^\mu) \partial_\lambda + u^m (x^\mu, \sigma^n) \partial_m + u^i (x^\mu, \sigma^n, y^j) \partial_i \tag{3.41}$$

be a vector field on Y_{tot} (3.37) projected onto Σ and X. Its restriction

$$\begin{aligned} &u_h : Y \underset{X}{\times} h(X) \xrightarrow{u} TY \underset{X}{\times} T\Sigma \longrightarrow TY, \\ &u_h = u = u^\lambda \partial_\lambda + u_h^i \partial_i, \qquad u_h^i (x^\mu, y^j) = u^i (x^\mu, h^n(x), y^j), \end{aligned} \tag{3.42}$$

is a vector field on Y. Let us suppose that u (3.41) is an exact symmetry of a total Lagrangian L. The corresponding first variational formula (3.25) leads to an equality

$$0 = (u^m - y^m_\lambda u^\lambda)\partial_m\mathcal{L} + \pi^\lambda_m d_\lambda(u^m - y^m_\mu u^\mu) + \\ (u^i - y^i_\lambda u^\lambda)\delta_i\mathcal{L} - d_\lambda[\partial^\lambda_i\mathcal{L}(u^\mu y^i_\mu - u^i) - u^\lambda\mathcal{L}].$$

Putting $\sigma^m = h^m(x)$, we obtain an equality

$$0 = (J^1h)^*[(u^m - y^m_\lambda u^\lambda)\partial_m\mathcal{L} + \pi^\lambda_m d_\lambda(u^m - y^m_\mu u^\mu)] + \\ (u^i_h - y^i_\lambda u^\lambda)\delta_i\mathcal{L}_h - d_\lambda[\partial^\lambda_i\mathcal{L}_h(u^\mu y^i_\mu - u^i_h) - u^\lambda\mathcal{L}_h].$$

On the shell $\delta_i L_h = 0$, this equality is brought into the *weak transformation law*

$$0 \approx (J^1h)^*[(u^m - y^m_\lambda u^\lambda)\partial_m\mathcal{L} + \pi^\lambda_m d_\lambda(u^m - y^m_\mu u^\mu)] - \\ d_\lambda[\partial^\lambda_i\mathcal{L}_h(u^\mu y^i_\mu - u^i_h) - u^\lambda\mathcal{L}_h] \tag{3.43}$$

of a symmetry current

$$\mathcal{J}^\lambda_{u_h} = \partial^\lambda_i\mathcal{L}_h(u^\mu y^i_\mu - u^i_h) - u^\lambda\mathcal{L}_h$$

of dynamic fields y^i along the vector field u_h (3.42) in the presence of a background field h.

3.5 Covariant Hamiltonian Formalism

As was mentioned above, there are different variants of covariant Hamiltonian formalism on fibre bundles. We follow polysymplectic Hamiltonian formalism where the Legendre bundle Π (3.6) plays the role of a phase space [54, 61, 118]. If $X = \mathbb{R}$, this is the case of Hamiltonian nonrelativistic mechanics (Chap. 4).

A key point is that polysymplectic Hamiltonian formalism on a phase space Π is equivalent to particular first order Lagrangian theory on a configuration space $\Pi \to X$. Given the Hamiltonian form H (3.52) on Π, the corresponding covariant Hamilton equation (3.61) is the Euler–Lagrange equation of the affine first order Lagrangian L_H (3.59) (Theorem 3.4). Moreover, it follows from the equality (3.83) that a Hamiltonian form H possesses the same classical symmetries as a Lagrangian L_H. This fact enables us to describe symmetries of Hamiltonian field theory similarly to those in Lagrangian formalism.

Treated as a *phase space*, the Legendre bundle Π (3.6) is endowed with the following *polysymplectic structure* (Definitions 3.4–3.10).

Definition 3.4 There is the canonical bundle monomorphism

$$\Theta_Y : \Pi \underset{Y}{\longrightarrow} \overset{n+1}{\wedge} T^*Y \underset{Y}{\otimes} TX, \qquad \Theta_Y = p_i^\lambda dy^i \wedge \omega \otimes \partial_\lambda, \tag{3.44}$$

called the *polysymplectic Liouville form* on Π.

It should be emphasized that Θ_Y (3.44) is a TX-valued, but not tangent-valued (i.e., $T\Pi$-valued) form on Π. Therefore, standard technique of tangent-valued forms, as like as that of exterior forms (e.g., the exterior differential d) is not applied to Θ_Y (3.44).

At the same time, there is a unique TX-valued $(n+2)$-form

$$\Upsilon_Y = dp_i^\lambda \wedge dy^i \wedge \omega \otimes \partial_\lambda \tag{3.45}$$

on Π so that a relation $\Upsilon_Y \rfloor \phi = d(\Theta_Y \rfloor \phi)$ holds for an arbitrary exterior one-form ϕ on X.

Definition 3.5 The form Υ_Y (3.45) is termed the *polysymplectic form.*

Definition 3.6 A Legendre bundle Π endowed with the polysymplectic form (3.45) is said to be the *polysymplectic manifold.*

Remark 3.4 Following [69], one sometimes provides Π (3.6) with an exterior form $dp_i^\lambda \wedge dy^i \wedge \omega_\lambda$ which however globally is ill defined because it is not maintained under transition functions (3.7).

By analogy with the notion of Hamiltonian vector fields on a symplectic manifold (Remark 4.17), we define Hamiltonian connections on a polysymplectic manifold.

Let $J^1\Pi$ be the first order jet manifold of a composite Legendre bundle $\Pi \to X$. It is equipped with the adapted coordinates $(x^\lambda, y^i, p_i^\lambda, y_\mu^i, p_{\mu i}^\lambda)$.

Definition 3.7 A connection

$$\gamma = dx^\lambda \otimes (\partial_\lambda + \gamma_\lambda^i \partial_i + \gamma_{\lambda i}^\mu \partial_\mu^i) \tag{3.46}$$

on $\Pi \to X$ is called the *Hamiltonian connection* if an exterior form

$$\gamma \rfloor \Upsilon_Y = (\partial_\lambda + \gamma_\lambda^i \partial_i + \gamma_{\lambda i}^\mu \partial_\mu^i) \rfloor (dp_i^\lambda \wedge dy^i \wedge \omega)$$

on $J^1\Pi$ is exact, i.e.,

$$\gamma \rfloor \Upsilon_Y = dp_i^\lambda \wedge dy^i \wedge \omega_\lambda - (\gamma_\lambda^i dp_i^\lambda - \gamma_{\lambda i}^\lambda dy^i) \wedge \omega = dH_\gamma. \tag{3.47}$$

Components of a Hamiltonian connection satisfy the conditions

$$\partial_\lambda^i \gamma_\mu^j - \partial_\mu^j \gamma_\lambda^i = 0, \qquad \partial_i \gamma_{\mu j}^\mu - \partial_j \gamma_{\mu i}^\mu = 0, \qquad \partial_j \gamma_\lambda^i + \partial_\lambda^i \gamma_{\mu j}^\mu = 0. \tag{3.48}$$

It is readily observed that, by virtue of the conditions (3.48), the second term in the right-hand side of the equality (3.47) also is an exact form. In accordance with the relative Poincaré lemma [53], this term can be brought into the form $d\mathcal{H}\wedge\omega$ where $\mathcal{H}$ is a function on Π. Then a form H_γ in the expression (3.47) reads

$$H_\gamma = p^\lambda_i dy^i \wedge \omega_\lambda - \mathcal{H}\omega. \tag{3.49}$$

On another side, let us consider the homogeneous Legendre bundle Z_Y (3.13) and the affine bundle $Z_Y \to \Pi$ (3.16). This affine bundle is modelled over the pull-back vector bundle $\Pi \times_X \overset{n}{\wedge} T^*X \to \Pi$ in accordance with the exact sequence

$$0 \longrightarrow \Pi \underset{X}{\times} \overset{n}{\wedge} T^*X \longrightarrow Z_Y \longrightarrow \Pi \longrightarrow 0. \tag{3.50}$$

A homogeneous Legendre bundle Z_Y is provided with the canonical multisymplectic Liouville form Ξ_Y (3.21). Its exterior differential is the *multisymplectic form*

$$\Omega_Y = d\Xi_Y = dp\wedge\omega + dp^\lambda_i \wedge dy^i \wedge \omega_\lambda. \tag{3.51}$$

Let $h = -\mathcal{H}\omega$ be a section of the affine bundle $Z_Y \to \Pi$ (3.16). A glance at the transformation law (3.15) shows that it is not a density.

Definition 3.8 By analogy with Hamiltonian mechanics (Sect. 4.5), $-h$ is said to be the *covariant Hamiltonian* of covariant Hamiltonian formalism. It defines the pull-back

$$H = h^*\Xi_Y = p^\lambda_i dy^i \wedge \omega_\lambda - \mathcal{H}\omega \tag{3.52}$$

(cf. (3.49)) of a multisymplectic Liouville form Ξ_Y onto a Legendre bundle Π which is called the *Hamiltonian form* on Π.

The following is a straightforward corollary of this definition.

Theorem 3.2 *(i) Hamiltonian forms constitute a non-empty affine space modelled over a linear space of horizontal densities* $\widetilde{H} = \widetilde{\mathcal{H}}\omega$ *on* $\Pi \to X$.
(ii) Every connection Γ *on a fibre bundle* $Y \to X$ *yields the splitting (B.57) of the exact sequence (3.50) and defines a Hamiltonian form*

$$H_\Gamma = \Gamma^*\Xi_Y = p^\lambda_i dy^i \wedge \omega_\lambda - p^\lambda_i \Gamma^i_\lambda \omega. \tag{3.53}$$

(iii) Given a connection Γ *on* $Y \to X$*, every Hamiltonian form* H *admits the decomposition*

$$H = H_\Gamma - \widetilde{H}_\Gamma = p^\lambda_i dy^i \wedge \omega_\lambda - p^\lambda_i \Gamma^i_\lambda \omega - \widetilde{\mathcal{H}}_\Gamma \omega. \tag{3.54}$$

Remark 3.5 In polysymplectic Hamiltonian formalism, the homogeneous Legendre bundle Z_Y (3.13) plays the role of a *homogeneous phase space*. If $X = \mathbb{R}$, a phase space Π and a homogeneous phase space Z_Y are the vertical cotangent bundle V^*Y and the cotangent bundle T^*Y, respectively (Sect. 4.5). In this case, the multisymplectic form Ω_Y (3.51) is exactly the canonical symplectic form (4.58) on the cotangent bundle T^*Y of Y.

One can generalize item (ii) of Theorem 3.2 as follows. We agree to call any bundle morphism

$$\Phi = dx^\lambda \otimes (\partial_\lambda + \Phi^i_\lambda \partial_i) : \Pi \underset{Y}{\to} J^1Y \tag{3.55}$$

over Y the *Hamiltonian map*. In particular, let Γ be a connection on $Y \to X$. Then the composition

$$\widehat{\Gamma} = \Gamma \circ \pi_{\Pi Y} = dx^\lambda \otimes (\partial_\lambda + \Gamma^i_\lambda \partial_i) : \Pi \to Y \to J^1Y \tag{3.56}$$

is a Hamiltonian map.

Theorem 3.3 *Every Hamiltonian map (3.55) defines a Hamiltonian form*

$$H_\Phi = -\Phi \rfloor \Theta_Y = p^\lambda_i dy^i \wedge \omega_\lambda - p^\lambda_i \Phi^i_\lambda \omega. \tag{3.57}$$

Proof Given an arbitrary connection Γ on a fibre bundle $Y \to X$, the corresponding Hamiltonian map (3.56) defines a form $-\widehat{\Gamma} \rfloor \Theta_Y$ which is exactly the Hamiltonian form H_Γ (3.53). Since $\Phi - \widehat{\Gamma}$ is a VY-valued basic one-form on $\Pi \to X$, $H_\Phi - H_\Gamma$ is a horizontal density on Π. Then the result follows from item (i) of Theorem 3.2.

Theorem 3.4 *Every Hamiltonian form H (3.52) admits the Hamiltonian connection γ_H (3.46) which obeys the condition*

$$\begin{aligned} &\gamma_H \rfloor \Upsilon_Y = dH, \\ &\gamma^i_\lambda = \partial^i_\lambda \mathcal{H}, \qquad \gamma^\lambda_{\lambda i} = -\partial_i \mathcal{H}. \end{aligned} \tag{3.58}$$

Proof It is readily observed that the Hamiltonian form H (3.52) is the Poincaré–Cartan form (3.12) of a first order Lagrangian

$$L_H = h_0(H) = (p^\lambda_i y^i_\lambda - \mathcal{H})\omega \tag{3.59}$$

on the jet manifold $J^1\Pi$ [53, 81]. The Euler–Lagrange operator (3.2) associated to this Lagrangian reads

$$\begin{aligned} &\mathcal{E}_H : J^1\Pi \to T^*\Pi \wedge (\overset{n}{\wedge} T^*X), \\ &\mathcal{E}_H = [(y^i_\lambda - \partial^i_\lambda \mathcal{H})dp^\lambda_i - (p^\lambda_{\lambda i} + \partial_i \mathcal{H})dy^i] \wedge \omega. \end{aligned} \tag{3.60}$$

It is termed the *Hamilton operator* for H. A glance at the expression (3.60) shows that this operator is an affine morphism over Π of constant rank. It follows that its kernel

$$y^i_\lambda = \partial^i_\lambda \mathcal{H}, \qquad p^\lambda_{\lambda i} = -\partial_i \mathcal{H} \tag{3.61}$$

is an affine closed imbedded subbundle of the jet bundle $J^1\Pi \to \Pi$. Therefore, it admits a global section γ_H which is a desired Hamiltonian connection obeying the relation (3.58).

It follows from the expression (3.49) that, conversely, a Hamiltonian form is associated to any Hamiltonian connection.

Definition 3.9 The Lagrangian L_H (3.59) is called the *characteristic Lagrangian* of a Hamiltonian system.

Remark 3.6 In fact, the Lagrangian (3.59) is the pull-back onto $J^1\Pi$ of an exterior form L_H on a product $\Pi \times_Y J^1Y$.

It should be emphasized that, if $\dim X > 1$, there is a set of Hamiltonian connections associated to the same Hamiltonian form H. They differ from each other in soldering forms σ on $\Pi \to X$ which fulfill the equation $\sigma \rfloor \Upsilon_Y = 0$. Every Hamiltonian form H yields a Hamiltonian map

$$\widehat{H} = J^1\pi_{\Pi Y} \circ \gamma_H : \Pi \to J^1\Pi \to J^1Y, \qquad y^i_\lambda \circ \widehat{H} = \partial^i_\lambda \mathcal{H}, \tag{3.62}$$

which is the same for all Hamiltonian connections γ_H associated to H.

Being a closed imbedded subbundle of the jet bundle $J^1\Pi \to X$, the kernel (3.61) of the Euler–Lagrange operator $\mathcal{E}_H$ (3.60) defines a first order Euler–Lagrange equation on Π in accordance with Definition B.12.

Definition 3.10 It is called the *covariant Hamilton equation*.

Every integral section $J^1r = \gamma_H \circ r$ of a Hamiltonian connection γ_H associated to a Hamiltonian form H is obviously a solution of the covariant Hamilton equation (3.61). By virtue of Theorem B.4, if $r : X \to \Pi$ is a global classical solution, there exists an extension of a local section

$$J^1r : r(X) \to \operatorname{Ker} \mathcal{E}_H$$

of the jet bundle $J^1\Pi \to \Pi$ over $r(X) \subset \Pi$ to a Hamiltonian connection γ_H which has r as an integral section. Substituting J^1r in (3.62), we obtain an equality

$$J^1(\pi_{\Pi Y} \circ r) = \widehat{H} \circ r, \tag{3.63}$$

which is equivalent to the covariant Hamilton equation (3.61).

Remark 3.7 Similarly to the Cartan equation (3.20), the covariant Hamilton equation (3.61) is equivalent to the condition

$$r^*(u \rfloor dH) = 0 \tag{3.64}$$

for any vertical vector field u on $\Pi \to X$.

3.6 Associated Lagrangian and Hamiltonian Systems

Let us study the relations between first order Lagrangian and covariant Hamiltonian formalisms on fibre bundles [54, 61, 118].

We are based on the fact that any Lagrangian L (3.1) on a jet manifold J^1Y defines the Legendre map $\widehat{L}$ (3.5) and its jet prolongation

$$J^1\widehat{L} : J^1J^1Y \underset{Y}{\longrightarrow} J^1\Pi, \qquad (p_i^\lambda, y_\mu^i, p_{\mu i}^\lambda) \circ J^1\widehat{L} = (\pi_i^\lambda, \widehat{y}_\mu^i, \widehat{d}_\mu \pi_i^\lambda),$$
$$\widehat{d}_\lambda = \partial_\lambda + \widehat{y}_\lambda^j \partial_j + y_{\lambda\mu}^j \partial_j^\mu,$$

and that any Hamiltonian form H (3.52) on a phase space Π defines the Hamiltonian map $\widehat{H}$ (3.62) and its jet prolongation

$$J^1\widehat{H} : J^1\Pi \underset{Y}{\longrightarrow} J^1J^1Y, \qquad (y_\mu^i, \widehat{y}_\lambda^i, y_{\lambda\mu}^i) \circ J^1\widehat{H} = (\partial_\mu^i \mathcal{H}, y_\lambda^i, d_\lambda \partial_\mu^i \mathcal{H}),$$
$$d_\lambda = \partial_\lambda + y_\lambda^j \partial_j + p_{\lambda j}^\nu \partial_\nu^j.$$

Let us start with the case of a hyperregular Lagrangian L, i.e., when the Legendre map $\widehat{L}$ is a diffeomorphism (Definition 3.1). Then $\widehat{L}^{-1}$ is a Hamiltonian map. Let us consider a Hamiltonian form

$$H = H_{\widehat{L}^{-1}} + \widehat{L}^{-1*}L, \qquad \mathcal{H} = p_i^\mu \widehat{L}^{-1i}{}_\mu - \mathcal{L}(x^\mu, y^j, \widehat{L}^{-1j}{}_\mu), \tag{3.65}$$

where $H_{\widehat{L}^{-1}}$ is the Hamiltonian form (3.57) associated to a Hamiltonian map $\widehat{L}^{-1}$. Let s be a classical solution of the Euler–Lagrange equation (3.3) for a Lagrangian L. A direct computation shows that $\widehat{L} \circ J^1 s$ is a solution of the covariant Hamilton equation (3.61) for the Hamiltonian form H (3.65). Conversely, if r is a classical solution of the covariant Hamilton equation (3.61) for the Hamiltonian form H (3.65), then $s = \pi_{\Pi Y} \circ r$ is a classical solution of the Euler–Lagrange equation (3.3) for L (see the equality (3.63)). It follows that, in the case of hyperregular Lagrangians, covariant Hamiltonian formalism is equivalent to the Lagrangian one.

Let now L be an arbitrary Lagrangian on a jet manifold J^1Y.

Definition 3.11 A Hamiltonian form H is said to be *associated* to a Lagrangian L if H satisfies the relations

$$\widehat{L} \circ \widehat{H} \circ \widehat{L} = \widehat{L}, \tag{3.66}$$
$$H = H_{\widehat{H}} + \widehat{H}^* L. \tag{3.67}$$

A glance at the relation (3.66) shows that $\widehat{L} \circ \widehat{H}$ is the projector

$$p_i^\mu(p) = \partial_i^\mu \mathcal{L}(x^\mu, y^i, \partial_\lambda^j \mathcal{H}(p)), \qquad p \in N_L, \tag{3.68}$$

from Π onto the Lagrangian constraint space $N_L = \widehat{L}(J^1Y)$ (3.9). Accordingly, $\widehat{H} \circ \widehat{L}$ is the projector from J^1Y onto $\widehat{H}(N_L)$.

Definition 3.12 A Hamiltonian form is termed *weakly associated* to a Lagrangian L if the condition (3.67) holds on a Lagrangian constraint space N_L.

Theorem 3.5 *If a Hamiltonian map Φ (3.55) obeys a relation*

$$\widehat{L} \circ \Phi \circ \widehat{L} = \widehat{L}, \tag{3.69}$$

then a Hamiltonian form $H = H_\Phi + \Phi^* L$ is weakly associated to a Lagrangian L. If $\Phi = \widehat{H}$, then H is associated to L.

Theorem 3.6 *Any Hamiltonian form H weakly associated to a Lagrangian L fulfills a relation*

$$H|_{N_L} = \widehat{H}^* H_L|_{N_L}, \tag{3.70}$$

where H_L is the Poincaré–Cartan form (3.12).

Proof The relation (3.67) takes a coordinate form

$$\mathcal{H}(p) = p_i^\mu \partial_\mu^i \mathcal{H} - \mathcal{L}(x^\mu, y^i, \partial_\lambda^j \mathcal{H}(p)), \qquad p \in N_L. \tag{3.71}$$

Substituting (3.68) and (3.71) in (3.52), we obtain the relation (3.70).

The difference between associated and weakly associated Hamiltonian forms lies in the following. Let H be an associated Hamiltonian form, i.e., the equality (3.71) holds everywhere on Π. Acting on this equality by the exterior differential, we obtain the relations

$$\begin{aligned}
&\partial_\mu \mathcal{H}(p) = -(\partial_\mu \mathcal{L}) \circ \widehat{H}(p), \qquad p \in N_L,\\
&\partial_i \mathcal{H}(p) = -(\partial_i \mathcal{L}) \circ \widehat{H}(p), \qquad p \in N_L,\\
&(p_i^\mu - (\partial_i^\mu \mathcal{L})(x^\mu, y^i, \partial_\lambda^j \mathcal{H}))\partial_\mu^i \partial_\alpha^a \mathcal{H} = 0.
\end{aligned} \tag{3.72}$$

The relation (3.72) shows that an associated Hamiltonian form (i.e., a Hamiltonian map $\widehat{H}$) is not regular outside a Lagrangian constraint space N_L.

Remark 3.8 Any Hamiltonian form is weakly associated to a Lagrangian $L = 0$, while associated Hamiltonian forms are only H_Γ (3.53).

Remark 3.9 A hyperregular Lagrangian has a unique weakly associated Hamiltonian form (3.65). In the case of a regular Lagrangian L, a Lagrangian constraint space N_L is an open subbundle of a Legendre bundle $\Pi \to Y$. If $N_L \neq \Pi$, a weakly associated Hamiltonian form fails to be defined everywhere on Π in general. At the same time, N_L itself can be provided with the pull-back polysymplectic structure with respect to the imbedding $N_L \to \Pi$, so that one may consider Hamiltonian forms on N_L.

One can say something more in a case of semiregular Lagrangians (Definition 3.1).

Theorem 3.7 *The Poincaré–Cartan form H_L for a semiregular Lagrangian L is constant on the connected inverse image $\widehat{L}^{-1}(p)$ of any point $p \in N_L$.*

Proof Let u be a vertical vector field on the affine jet bundle $J^1Y \to Y$ which takes its values into the kernel of the tangent map $T\widehat{L}$ to $\widehat{L}$. Then $\mathbf{L}_u H_L = 0$.

A corollary of Theorem 3.7 is the following.

Theorem 3.8 *Hamiltonian forms weakly associated to a semiregular Lagrangian L coincide with each other on the Lagrangian constraint space N_L, and the Poincaré–Cartan form H_L (3.12) for L is the pull-back*

$$H_L = \widehat{L}^* H, \qquad (\pi_i^\lambda y_\lambda^i - \mathcal{L})\omega = \mathcal{H}(x^\mu, y^j, \pi_j^\mu)\omega, \tag{3.73}$$

of any such a Hamiltonian form H.

Proof Given a vector $v \in T_p\Pi$, the value $T\widehat{H}(v)\rfloor H_L(\widehat{H}(p))$ is the same for all Hamiltonian maps $\widehat{H}$ satisfying the relation (3.66). Then the result follows from the relation (3.70).

Theorem 3.8 enables us to relate the Euler–Lagrange equation for an almost regular Lagrangian L with the covariant Hamilton equation for Hamiltonian forms weakly associated to L [54, 117, 118].

Theorem 3.9 *Let a section r of $\Pi \to X$ be a classical solution of the covariant Hamilton equation (3.61) for a Hamiltonian form H weakly associated to a semiregular Lagrangian L. If r lives in a Lagrangian constraint space N_L, a section $s = \pi_{\Pi Y} \circ r$ of $Y \to X$ satisfies the Euler–Lagrange equation (3.3), while its first order jet prolongation $\overline{s} = \widehat{H} \circ r = J^1 s$ obeys the Cartan equation (3.18) and (3.19).*

Proof Put $\overline{s} = \widehat{H} \circ r$. Since $r(X) \subset N_L$, then

$$r = \widehat{L} \circ \overline{s}, \qquad J^1 r = J^1\widehat{L} \circ J^1\overline{s}.$$

If r is a classical solution of the covariant Hamilton equation, an exterior form $\mathcal{E}_H$ vanishes at points of $J^1 r(X)$. Hence, the pull-back form $\mathcal{E}_{\overline{L}} = (J^1\widehat{L})^*\mathcal{E}_H$ vanishes at points $J^1\overline{s}(X)$. It follows that the section $\overline{s}$ of a jet bundle $J^1Y \to X$ obeys the Cartan equation (3.18) and (3.19). By virtue of the relation (3.63), we have $\overline{s} = J^1 s$. Hence, s is a classical solution of the Euler–Lagrange equation.

The converse assertion is more intricate [53].

Theorem 3.10 *Given a semiregular Lagrangian L, let a section $\overline{s}$ of the jet bundle $J^1Y \to X$ be a solution of the Cartan equation (3.18) and (3.19). Let H be a*

Hamiltonian form weakly associated to L so that the associated Hamiltonian map satisfies a condition

$$\widehat{H} \circ \widehat{L} \circ \overline{s} = J^1(\pi_0^1 \circ \overline{s}).$$

Then a section

$$r = \widehat{L} \circ \overline{s}, \qquad r_i^\lambda = \pi_i^\lambda(x^\lambda, \overline{s}^j, \overline{s}_\lambda^j), \qquad r^i = \overline{s}^i,$$

of a composite Legendre bundle $\Pi \to X$ is a classical solution of the Hamilton equation (3.61) for H.

Being restricted to classical solutions of Euler–Lagrange equations, Theorem 3.10 comes to the following.

Theorem 3.11 *Given a semiregular Lagrangian L, let a section s of a fibre bundle $Y \to X$ be a classical solution of the Euler–Lagrange equation (3.3) (i.e., $J^1 s$ is a solution of the Cartan equation (3.18) and (3.19), and $s = \pi_0^1 \circ J^1 s$). Let H be a Hamiltonian form weakly associated to L, and let H satisfy a relation*

$$\widehat{H} \circ \widehat{L} \circ J^1 s = J^1 s. \tag{3.74}$$

Then a section $r = \widehat{L} \circ J^1 s$ of a composite Legendre bundle $\Pi \to X$ is a classical solution of the covariant Hamilton equation (3.61) for H.

Remark 3.10 Let $L = 0$. This Lagrangian is semiregular. Its Euler–Lagrange equation comes to the identity $0 = 0$. Every section s of a fibre bundle $Y \to X$ is a solution of this equation. Given a section s, let Γ be a connection on Y such that s is its integral section. The Hamiltonian form H_Γ (3.53) is associated to $L = 0$, and a Hamiltonian map $\widehat{H}_\Gamma$ satisfies the relation (3.74). The corresponding Hamilton equation has a classical solution

$$r = \widehat{L} \circ J^1 s, \qquad r^i = s^i, \qquad r_i^\lambda = 0.$$

In view of Theorem 3.11, one may try to consider a set of Hamiltonian forms associated to a semiregular Lagrangian L in order to exhaust all classical solutions of the Euler–Lagrange equation for L.

Definition 3.13 Let us say that a set of Hamiltonian forms H weakly associated to a semiregular Lagrangian L is complete if, for each classical solution s of the Euler–Lagrange equation for L, there exists a classical solution r of the covariant Hamilton equation for a Hamiltonian form H from this set such that $s = \pi_{\Pi Y} \circ r$.

A complete family of Hamiltonian forms associated to a given Lagrangian need not exist, or it fails to be defined uniquely. For instance, Remark 3.10 shows that the Hamiltonian forms (3.53) constitute a complete family associated to the zero Lagrangian, but this family is not minimal.

By virtue of Theorem 3.11, a set of weakly associated Hamiltonian forms is complete if, for every classical solution s of the Euler–Lagrange equation for L, there is a Hamiltonian form H from this set which fulfills the relation (3.74).

In the case of almost regular Lagrangians (Definition 3.1), one can formulate the following necessary and sufficient conditions of the existence of weakly associated Hamiltonian forms. An immediate consequence of Theorem 3.5 is the following.

Theorem 3.12 *A Hamiltonian form H weakly associated to an almost regular Lagrangian L exists if and only if the fibred manifold $J^1Y \to N_L$ (3.10) admits a global section.*

Proof A global section of $J^1Y \to N_L$ can be extended to a Hamiltonian map $\Phi : \Pi \to J^1Y$ which obeys the relation (3.69).

In particular, on an open neighborhood $U \subset \Pi$ of each point $p \in N_L \subset \Pi$, there exists a complete set of local Hamiltonian forms weakly associated to an almost regular Lagrangian L. Moreover, one can always construct a complete set of associated Hamiltonian forms [118]. At the same time, a complete set of associated Hamiltonian forms may exist when a Lagrangian is not necessarily semiregular [53].

Given a global section Ψ of the fibred manifold

$$\widehat{L} : J^1Y \to N_L, \tag{3.75}$$

let us consider the pull-back form

$$H_N = \Psi^* H_L = i_N^* H \tag{3.76}$$

on N_L called the *constrained Hamiltonian form*. By virtue of Theorem 3.7, it does not depend on the choice of a section of the fibred manifold (3.75) and, consequently, $H_L = \widehat{L}^* H_N$. For sections r of the fibre bundle $N_L \to X$, one can write the *constrained Hamilton equation*

$$r^*(u_N \rfloor dH_N) = 0, \tag{3.77}$$

where u_N is an arbitrary vertical vector field on $N_L \to X$. These equations possess the following important properties.

Theorem 3.13 *For any Hamiltonian form H weakly associated to an almost regular Lagrangian L, every classical solution r of the covariant Hamilton equation which lives in the Lagrangian constraint space N_L is a solution of the constrained Hamilton equation (3.77).*

Proof Such a Hamiltonian form H defines a global section $\Psi = \widehat{H} \circ i_N$ of the fibred manifold (3.75). Since $H_N = i_N^* H$ due to the relation (3.73), the constrained Hamilton equation can be written as

$$r^*(u_N \rfloor d i_N^* H) = r^*(u_N \rfloor dH|_{N_L}) = 0. \tag{3.78}$$

Let us note that these equations differ from the Hamilton equation (3.64) restricted to N_L. These read

$$r^*(u\rfloor dH|_{N_L}) = 0, \tag{3.79}$$

where r is a section of $N_L \to X$ and u is an arbitrary vertical vector field on $\Pi \to X$. A solution r of the equation (3.79) obviously satisfies the weaker condition (3.78).

Theorem 3.14 *The constrained Hamilton equation (3.77) is proved to be equivalent to the Hamilton–De Donder equation (3.23).*

Proof It is readily observed that $\widehat{L} = \pi_{Z\Pi} \circ \widehat{H}_L$. Hence, the projection $\pi_{Z\Pi}$ (3.16) yields a surjection of Z_L onto N_L. Given a section Ψ of the fibred manifold (3.75), we have a morphism $\widehat{H}_L \circ \Psi : N_L \to Z_L$. By virtue of Theorem (3.7), this is a surjection such that

$$\pi_{Z\Pi} \circ \widehat{H}_L \circ \Psi = \mathrm{Id}\, N_L.$$

Hence, $\widehat{H}_L \circ \Psi$ is a bundle isomorphism over Y which is independent of the choice of a global section Ψ. Combination of (3.22) and (3.76) results in $H_N = (\widehat{H}_L \circ \Psi)^* \Xi_L$ that leads to a desired equivalence.

This proof gives something more. Namely, since Z_L and N_L are isomorphic, the homogeneous Legendre map $\widehat{H}_L$ fulfils the conditions of Theorem 3.1. Then combining Theorems 3.1 and 3.14, we obtain the following.

Theorem 3.15 *Let L be an almost regular Lagrangian such that the fibred manifold (3.75) has a global section. A section $\overline{s}$ of the jet bundle $J^1Y \to X$ is a solution of the Cartan equation (3.20) if and only if $\widehat{L} \circ \overline{s}$ is a solution of the constrained Hamilton equation (3.77).*

Theorem 3.15 also is a corollary of Theorem 3.16 below. The constrained Hamiltonian form H_N (3.76) defines the *constrained Lagrangian*

$$L_N = h_0(H_N) = (J^1 i_N)^* L_H \tag{3.80}$$

on the jet manifold J^1N_L of the fibre bundle $N_L \to X$.

Theorem 3.16 *There are the relations*

$$\overline{L} = (J^1\widehat{L})^* L_N, \qquad L_N = (J^1\Psi)^*\overline{L},$$

where $\overline{L}$ is the Lagrangian (3.17).

The Euler–Lagrange equation for the constrained Lagrangian L_N (3.80) is equivalent to the constrained Hamilton equation (3.77) and, by virtue of Theorem 3.16, is quasi-equivalent to the Cartan equation.

3.7 Noether's First Theorem: Hamiltonian Conservation Laws

In order to describe symmetries of covariant Hamiltonian theory, let us use the fact that this theory can be reformulated as particular Lagrangian theory with the characteristic Lagrangian L_H (3.59) (Sect. 3.5). Moreover, it follows from the equality (3.83) that a Hamiltonian form H possesses the same classical symmetries as a Lagrangian L_H. This fact enables us to analyze symmetries of polysymplectic Hamiltonian theory similarly to those in Lagrangian formalism (Sect. 3.3) in accordance with Noether's theorems.

We restrict our consideration to classical symmetries defined by projectable vector fields u (3.24) on a fibre bundle $Y \to X$.

In accordance with the canonical lift (B.29), every projectable vector field u on $Y \to X$ gives rise to a vector field

$$\widetilde{u} = u^\mu \partial_\mu + u^i \partial_i + (-\partial_i u^j p^\lambda_j - \partial_\mu u^\mu p^\lambda_i + \partial_\mu u^\lambda p^\mu_i)\partial^i_\lambda \tag{3.81}$$

on the Legendre bundle Π and then to the vector field

$$J\widetilde{u} = \widetilde{u} + J^1 u \tag{3.82}$$

on $\Pi \times_Y J^1Y$. Then we have

$$\mathbf{L}_{\widetilde{u}} H = \mathbf{L}_{J\widetilde{u}} L_H = (-u^i \partial_i \mathcal{H} - \partial_\mu (u^\mu \mathcal{H}) - u^\lambda_i \partial^i_\lambda \mathcal{H} + p^\lambda_i \partial_\lambda u^i)\omega. \tag{3.83}$$

It follows that a Hamiltonian form H and the characteristic Lagrangian L_H have the same classical symmetries.

Remark 3.11 Given the splitting (3.54) of a Hamiltonian form H, the Lie derivative (3.83) takes a form

$$\begin{aligned}\mathbf{L}_{\widetilde{u}} H = {} & p^\lambda_j([\partial_\lambda + \Gamma^i_\lambda \partial_i, u]^j - [\partial_\lambda + \Gamma^i_\lambda \partial_i, u]^\nu \Gamma^j_\nu)\omega - \\ & (\partial_\mu u^\mu \widetilde{\mathcal{H}_\Gamma} + u \rfloor d\widetilde{\mathcal{H}_\Gamma})\omega,\end{aligned} \tag{3.84}$$

where $[.,.]$ is the Lie bracket of vector fields.

Let us apply the first variational formula (3.25) to the Lie derivative $\mathbf{L}_{J\widetilde{u}} L_H$ (3.59) [53]. It reads

$$\begin{aligned}-u^i \partial_i \mathcal{H} - \partial_\mu (u^\mu \mathcal{H}) - u^\lambda_i \partial^i_\lambda \mathcal{H} + p^\lambda_i \partial_\lambda u^i = {} & -(u^i - y^i_\mu u^\mu)(p^\lambda_{\lambda i} + \partial_i \mathcal{H}) + \\ & (-\partial_i u^j p^\lambda_j - \partial_\mu u^\mu p^\lambda_i + \partial_\mu u^\lambda p^\mu_i - p^\lambda_{\mu i} u^\mu)(y^i_\lambda - \partial^i_\lambda \mathcal{H}) - \\ & d_\lambda [p^\lambda_i (\partial^i_\mu \mathcal{H} u^\mu - u^i) - u^\lambda (p^\mu_i \partial^i_\mu \mathcal{H} - \mathcal{H})].\end{aligned}$$

On the shell (3.61), this identity takes a form

$$-u^i\partial_i\mathcal{H} - \partial_\mu(u^\mu\mathcal{H}) - u^\lambda_i\partial^i_\lambda\mathcal{H} + p^\lambda_i\partial_\lambda u^i \approx - \\ d_\lambda[p^\lambda_i(\partial^i_\mu\mathcal{H}u^\mu - u^i) - u^\lambda(p^\mu_i\partial^i_\mu\mathcal{H} - \mathcal{H})]. \tag{3.85}$$

If $\mathbf{L}_{J^1\widetilde{u}}L_H = 0$, we obtain the weak *Hamiltonian conservation law*

$$0 \approx -d_\lambda[p^\lambda_i(u^\mu\partial^i_\mu\mathcal{H} - u^i) - u^\lambda(p^\mu_i\partial^i_\mu\mathcal{H} - \mathcal{H})] \tag{3.86}$$

of a *Hamiltonian symmetry current*

$$\widetilde{\mathcal{J}}^\lambda_u = p^\lambda_i(u^\mu\partial^i_\mu\mathcal{H} - u^i) - u^\lambda(p^\mu_i\partial^i_\mu\mathcal{H} - \mathcal{H}) \tag{3.87}$$

along a vector field u. On classical solutions r of the covariant Hamilton equation (3.61), the weak equality (3.86) leads to the differential conservation law

$$\partial_\lambda(\widetilde{\mathcal{J}}^\lambda_u(r)) = 0.$$

There is the following relation between differential conservation laws in Lagrangian and Hamiltonian formalisms.

Theorem 3.17 *Let a Hamiltonian form H be associated to an almost regular Lagrangian L. Let r be a classical solution of the covariant Hamilton equation (3.61) for H which lives in the Lagrangian constraint space N_L. Let $s = \pi_{\Pi Y}\circ r$ be the corresponding classical solution of the Euler–Lagrange equation for L so that the relation (3.74) holds. Then, for any projectable vector field u on a fibre bundle $Y \to X$, we have*

$$\widetilde{\mathcal{J}}_u(r) = \mathcal{J}_u(\pi_{\Pi Y}\circ r), \qquad \widetilde{\mathcal{J}}_u(\widehat{L}\circ J^1s) = \mathcal{J}_u(s), \tag{3.88}$$

where $\mathcal{J}_u$ is the symmetry current (3.29) on J^1Y and $\widetilde{\mathcal{J}}_u$ is the symmetry current (3.87) on Π.

Proof The proof follows from the relations (3.68), (3.71) and (3.73).

By virtue of Theorems 3.9–3.11, it follows that:

- if $\mathcal{J}_u$ in Theorem 3.17 is a conserved symmetry, then the symmetry current $\widetilde{\mathcal{J}}_u$ (3.88) is conserved on classical solutions of the covariant Hamilton equation which live in a Lagrangian constraint space,
- if $\widetilde{\mathcal{J}}_u$ in Theorem 3.17 is a conserved symmetry current, then the symmetry current $\mathcal{J}_u$ (3.88) is conserved on classical solutions s of the Euler–Lagrange equation which obey the condition (3.74).

In particular, let $u = u^i\partial_i$ be a vertical vector field on $Y \to X$. Then the Lie derivative $\mathbf{L}_{\widetilde{u}}H$ (3.84) takes a form

$$\mathbf{L}_{\widetilde{u}}H = (p^\lambda_j[\partial_\lambda + \Gamma^i_\lambda\partial_i, u]^j - u\rfloor d\widetilde{\mathcal{H}}_\Gamma)\omega.$$

The corresponding Hamiltonian symmetry current (3.87) reads $\widetilde{\mathcal{J}}_u^\lambda = -u^i p_i^\lambda$ (cf. $\mathcal{J}_u$ (2.25)).

Let $\tau = \tau^\lambda \partial_\lambda$ be a vector field on X and $\Gamma\tau$ (B.56) its horizontal lift onto Y by means of a connection Γ on $Y \to X$. In this case, the weak identity (3.85) takes a form

$$-(\partial_\mu + \Gamma_\mu^j \partial_j - p_i^\lambda \partial_j \Gamma_\mu^i \partial_\lambda^j)\widetilde{\mathcal{H}_\Gamma} + p_i^\lambda R_{\lambda\mu}^i \approx -d_\lambda \widetilde{\mathcal{J}_\Gamma}^\lambda{}_\mu,$$

where the Hamiltonian symmetry current (3.87) reads

$$\widetilde{\mathcal{J}}_\Gamma^\lambda = \tau^\mu \widetilde{\mathcal{J}_\Gamma}^\lambda{}_\mu = \tau^\mu (p_i^\lambda \partial_\mu^i \widetilde{\mathcal{H}_\Gamma} - \delta_\mu^\lambda (p_i^\nu \partial_\nu^i \widetilde{\mathcal{H}_\Gamma} - \widetilde{\mathcal{H}_\Gamma})). \tag{3.89}$$

The relations (3.88) show that, on the Lagrangian constraint space N_L, the Hamiltonian symmetry current (3.89) can be treated as the *Hamiltonian energy-momentum current* relative to a connection Γ.

However, Theorem 3.17 fails to provide straightforward relations between symmetries of Lagrangians and associated Hamiltonian forms. In Sect. 3.8, we can obtain such relations between symmetries of almost regular quadratic Lagrangians L (3.90) and the corresponding constrained Lagrangians L_N (3.80) (Theorem 3.21).

3.8 Quadratic Lagrangian and Hamiltonian Systems

Field theories with almost regular quadratic Lagrangians L (3.90) admit a comprehensive polysymplectic Hamiltonian formulation [53, 54, 61]. There exists a complete set of Hamiltonian forms weakly associated to L (Theorem 3.20). Theorem 3.22 also states that Lagrangian symmetries under a certain condition yield the Hamiltonian ones restricted to a Lagrangian constraint space. At the same time, a Hamiltonian system can possess additional symmetries (Remark 3.15). In Sect. 8.5, an example of the almost regular quadratic Yang–Mills Lagrangian (8.62) is analyzed in detail.

Given a fibre bundle $Y \to X$, let us consider a quadratic Lagrangian L given by a coordinate expression

$$\mathcal{L} = \frac{1}{2} a_{ij}^{\lambda\mu}(x^\nu, y^k) y_\lambda^i y_\mu^j + b_i^\lambda(x^\nu, y^k) y_\lambda^i + c(x^\nu, y^k), \tag{3.90}$$

where a, b and c are local functions on Y. This property is coordinate-independent owing to the affine transformation law (B.44) of the jet coordinates y_λ^i. The associated Legendre map $\widehat{L}$ (3.5) is given by a coordinate expression

$$p_i^\lambda \circ \widehat{L} = a_{ij}^{\lambda\mu} y_\mu^j + b_i^\lambda, \tag{3.91}$$

and it is an affine morphism over Y. It yields the corresponding linear morphism

$$\widehat{a} : T^*X \underset{Y}{\otimes} VY \underset{Y}{\to} \Pi, \qquad p_i^\lambda \circ \widehat{a} = a_{ij}^{\lambda\mu} \overline{y}_\mu^j, \tag{3.92}$$

where $\overline{y}_\mu^j$ are fibred coordinates on the vector bundle (B.47).

Let the Lagrangian L (3.90) be almost regular, i.e., the morphism $\widehat{a}$ (3.92) is of constant rank. Then the Lagrangian constraint space N_L (3.91) is an affine subbundle of a Legendre bundle $\Pi \to Y$, modelled over the vector subbundle $\overline{N}_L$ (3.92) of $\Pi \to Y$. Hence, $N_L \to Y$ has a global section s. For the sake of simplicity, let us assume that $s = \widehat{0}$ is the canonical zero section of $\Pi \to Y$. Then $\overline{N}_L = N_L$. Accordingly, the kernel of the Legendre map (3.91) is an affine subbundle of the affine jet bundle $J^1Y \to Y$, modelled over the kernel of the linear morphism $\widehat{a}$ (3.92). Then there exists a connection

$$\Gamma : Y \to \operatorname{Ker} \widehat{L} \subset J^1Y, \qquad a_{ij}^{\lambda\mu} \Gamma_\mu^j + b_i^\lambda = 0, \tag{3.93}$$

on $Y \to X$. Connections (3.93) constitute an affine space modelled over a linear space of soldering forms $\phi = \phi_\lambda^i dx^\lambda \otimes \partial_i$ on $Y \to X$, satisfying the conditions

$$a_{ij}^{\lambda\mu} \phi_\mu^j = 0 \tag{3.94}$$

and, as a consequence, the conditions $\phi_\lambda^i b_i^\lambda = 0$. If the Lagrangian (3.90) is regular, the connection (3.93) is unique.

Remark 3.12 If $s \neq \widehat{0}$, one can consider connections Γ with values into $\operatorname{Ker}_s \widehat{L}$.

A matrix a in the Lagrangian L (3.90) can be seen as a global section of constant rank of a tensor bundle

$$\overset{n}{\wedge} T^*X \underset{Y}{\otimes} [\overset{2}{\vee}(TX \underset{Y}{\otimes} V^*Y)] \to Y.$$

Then it satisfies the following corollary of Theorem B.13.

Theorem 3.18 *Given a k-dimensional vector bundle $E \to Z$, let a be a fibre metric of rank r in E. There is a splitting $E = \operatorname{Ker} a \oplus_Z E'$ where $E' = E/\operatorname{Ker} a$, and a is a nondegenerate fibre metric in E'.*

Theorem 3.19 *There exists a linear bundle map*

$$\sigma : \Pi \underset{Y}{\to} T^*X \underset{Y}{\otimes} VY, \qquad \overline{y}_\lambda^i \circ \sigma = \sigma_{\lambda\mu}^{ij} p_j^\mu, \tag{3.95}$$

such that $\widehat{a} \circ \sigma \circ i_N = i_N$.

Proof The map (3.95) is a solution of the algebraic equations

$$a_{ij}^{\lambda\mu}\sigma_{\mu\alpha}^{jk}a_{kb}^{\alpha\nu} = a_{ib}^{\lambda\nu}. \tag{3.96}$$

By virtue of Theorem 3.18, there exists the bundle splitting

$$TX^* \underset{Y}{\otimes} VY = \operatorname{Ker} a \underset{Y}{\oplus} E' \tag{3.97}$$

and an atlas of this bundle such that transition functions of $\operatorname{Ker} a$ and E' are independent. Since a is a nondegenerate section of

$$\overset{n}{\wedge} T^*X \underset{Y}{\otimes} (\overset{2}{\vee} E'^*) \to Y,$$

there exist fibre coordinates $(\overline{y}^A)$ on E' such that a is brought into a diagonal matrix with nonvanishing components a_{AA}. Due to the splitting (3.97), we have the corresponding bundle splitting

$$TX \underset{Y}{\otimes} V^*Y = (\operatorname{Ker} a)^* \underset{Y}{\oplus} E'^*.$$

Then a desired map σ is represented by a direct sum $\sigma_1 \oplus \sigma_0$ of an arbitrary sections σ_1 of a fibre bundle $\overset{n}{\wedge} TX \otimes_Y (\overset{2}{\vee} \operatorname{Ker} a) \to Y$ and a section σ_0 of a fibre bundle $\overset{n}{\wedge} TX \otimes_Y (\overset{2}{\vee} E') \to Y$ which has nonvanishing components $\sigma^{AA} = (a_{AA})^{-1}$ with respect to the coordinates $(\overline{y}^A)$ on E'. We have relations

$$\sigma_0 = \sigma_0 \circ a \circ \sigma_0, \qquad a \circ \sigma_1 = 0, \qquad \sigma_1 \circ a = 0. \tag{3.98}$$

Remark 3.13 Using the relations (3.98), one can write the above assumption, that the Lagrangian constraint space $N_L \to Y$ admits a global zero section, in the form

$$b_i^\mu = a_{ij}^{\mu\lambda}\sigma_{\lambda\nu}^{jk}b_k^\nu. \tag{3.99}$$

With the relations (3.93), (3.96) and (3.98), we obtain the splitting

$$J^1Y = \mathscr{S}(J^1Y) \underset{Y}{\oplus} \mathscr{F}(J^1Y) = \operatorname{Ker} \widehat{L} \underset{Y}{\oplus} \operatorname{Im}(\sigma \circ \widehat{L}), \tag{3.100}$$
$$y_\lambda^i = \mathscr{S}_\lambda^i + \mathscr{F}_\lambda^i = [y_\lambda^i - \sigma_{\lambda\alpha}^{ik}(a_{kj}^{\alpha\mu}y_\mu^j + b_k^\alpha)] + [\sigma_{\lambda\alpha}^{ik}(a_{kj}^{\alpha\mu}y_\mu^j + b_k^\alpha)],$$

where, in fact, $\sigma = \sigma_0$ owing to the relations (3.98) and (3.99). Then with respect to the coordinates $\mathscr{S}_\lambda^i$ and $\mathscr{F}_\lambda^i$, the Lagrangian (3.90) reads

$$\mathscr{L} = \frac{1}{2}a_{ij}^{\lambda\mu}\mathscr{F}_\lambda^i\mathscr{F}_\mu^j + c', \tag{3.101}$$

where

$$\mathcal{F}^i_\lambda = \sigma_0{}^{ik}_{\lambda\alpha} a^{\alpha\mu}_{kj}(y^j_\mu - \Gamma^j_\mu) \tag{3.102}$$

for some $(\operatorname{Ker}\widehat{L})$-valued connection Γ (3.93) on $Y \to X$. Thus, the Lagrangian (3.90), written in the form (3.101), factorizes through the covariant differential relative to any such connection.

Remark 3.14 Let us note that, in gauge theory of principal connections (Chap. 7), we have the canonical (independent of a Lagrangian) variant (8.38) of the splitting (3.100) where $\mathcal{F}$ is the strength form (8.36). The Yang–Mills Lagrangian (8.62) of gauge field theory is exactly of the form (3.101) where $c' = 0$.

Field theories with almost regular quadratic Lagrangians admit comprehensive Hamiltonian formulation.

Let L (3.90) be an almost regular quadratic Lagrangian brought into the form (3.101), $\sigma = \sigma_0 + \sigma_1$ the linear map (3.95), and Γ the connection (3.93). Similarly to the splitting (3.100) of a jet bundle $J^1Y \to Y$, we have the following decomposition of a phase space:

$$\Pi = \mathcal{R}(\Pi) \underset{Y}{\oplus} \mathcal{P}(\Pi) = \operatorname{Ker}\sigma_0 \underset{Y}{\oplus} N_L, \tag{3.103}$$

$$p^\lambda_i = \mathcal{R}^\lambda_i + \mathcal{P}^\lambda_i = [p^\lambda_i - a^{\lambda\mu}_{ij}\sigma^{jk}_{\mu\alpha} p^\alpha_k] + [a^{\lambda\mu}_{ij}\sigma^{jk}_{\mu\alpha} p^\alpha_k]. \tag{3.104}$$

The relations (3.98) lead to the equalities

$$\sigma_0{}^{jk}_{\mu\alpha}\mathcal{R}^\alpha_k = 0, \qquad \sigma_1{}^{jk}_{\mu\alpha}\mathcal{P}^\alpha_k = 0, \qquad \mathcal{R}^\lambda_i \mathcal{F}^i_\lambda = 0. \tag{3.105}$$

Relative to the coordinates (3.104), the Lagrangian constraint space N_L (3.91) is given by the equations

$$\mathcal{R}^\lambda_i = [p^\lambda_i - a^{\lambda\mu}_{ij}\sigma^{jk}_{\mu\alpha} p^\alpha_k] = 0. \tag{3.106}$$

Let the splitting (3.97) be provided with the adapted fibre coordinates $(\overline{y}^a, \overline{y}^A)$ such that the matrix function a (3.92) is brought into a diagonal matrix with non-vanishing components a_{AA}. Then the Legendre bundle Π (3.103) is endowed with the dual (nonholonomic) fibre coordinates (p_a, p_A) where p_A are coordinates on the Lagrangian constraint manifold N_L, given by the equalities $p_a = 0$. Relative to these coordinates, σ_0 becomes a diagonal matrix $\sigma_0^{AA} = (a_{AA})^{-1}$, $\sigma_0^{aa} = 0$, while $\sigma_1^{Aa} = \sigma_1^{AB} = 0$. Let us write $p_a = M_a{}^i_\lambda p^\lambda_i$, $p_A = M_A{}^i_\lambda p^\lambda_i$, where M is a matrix function on Y which obeys the relations

$$\begin{aligned} &M_a{}^i_\lambda a^{\lambda\mu}_{ij} = 0, \qquad (M^{-1})^{a\lambda}{}_i \sigma_0{}^{ij}_{\lambda\mu} = 0, \\ &M_A{}^i_\lambda (a \circ \sigma_0)^{\lambda j}_{i\mu} = M_A{}^j_\mu, \qquad (M^{-1})^{A\mu}{}_j M_A{}^i_\lambda = a^{\mu\nu}_{jk}\sigma_0{}^{ki}_{\nu\lambda}. \end{aligned} \tag{3.107}$$

Let us consider an affine Hamiltonian map

$$\Phi = \widehat{\Gamma} + \sigma : \Pi \to J^1 Y, \qquad \Phi^i_\lambda = \Gamma^i_\lambda + \sigma^{ij}_{\lambda\mu} p^\mu_j, \tag{3.108}$$

and a Hamiltonian form

$$H(\Gamma, \sigma_1) = H_\Phi + \Phi^* L = p^\lambda_i dy^i \wedge \omega_\lambda - \tag{3.109}$$
$$[\Gamma^i_\lambda p^\lambda_i + \frac{1}{2}\sigma_0{}^{ij}_{\lambda\mu} p^\lambda_i p^\mu_j + \sigma_1{}^{ij}_{\lambda\mu} p^\lambda_i p^\mu_j - c']\omega = (\mathcal{R}^\lambda_i + \mathcal{P}^\lambda_i) dy^i \wedge \omega_\lambda -$$
$$[(\mathcal{R}^\lambda_i + \mathcal{P}^\lambda_i)\Gamma^i_\lambda + \frac{1}{2}\sigma_0{}^{ij}_{\lambda\mu} \mathcal{P}^\lambda_i \mathcal{P}^\mu_j + \sigma_1{}^{ij}_{\lambda\mu} \mathcal{R}^\lambda_i \mathcal{R}^\mu_j - c']\omega.$$

Theorem 3.20 *The Hamiltonian forms $H(\Gamma, \sigma_1)$ (3.109) parameterized by connections Γ (3.93) are weakly associated to the Lagrangian (3.90), and they constitute a complete set.*

Proof By the very definitions of Γ and σ, the Hamiltonian map (3.108) satisfies the condition (3.66). Then $H(\Gamma, \sigma_1)$ is weakly associated to L (3.90) in accordance with Theorem 3.5. Let us write the first Hamilton equation (3.61) for a section r of the composite Legendre bundle $\Pi \to X$. It reads

$$J^1 s = (\widehat{\Gamma} + \sigma) \circ r, \qquad s = \pi_{\Pi Y} \circ r. \tag{3.110}$$

Due to the surjections $\mathcal{S}$ and $\mathcal{F}$ (3.100), the Hamilton equation (3.110) is brought into the two parts

$$\mathcal{S} \circ J^1 s = \Gamma \circ s, \qquad \partial_\lambda r^i - \sigma_0{}^{ik}_{\lambda\alpha}(a^{\alpha\mu}_{kj} \partial_\mu r^j + b^\alpha_k) = \Gamma^i_\lambda \circ s, \tag{3.111}$$
$$\mathcal{F} \circ J^1 s = \sigma \circ r, \qquad \sigma_0{}^{ik}_{\lambda\alpha}(a^{\alpha\mu}_{kj} \partial_\mu r^j + b^\alpha_k) = \sigma^{ik}_{\lambda\alpha} r^\alpha_k. \tag{3.112}$$

Let s be an arbitrary section of $Y \to X$, e.g., a classical solution of the Euler–Lagrange equation. There exists the connection Γ (3.93) such that the relation (3.111) holds, namely, $\Gamma = \mathcal{S} \circ \Gamma'$ where Γ' is a connection on $Y \to X$ which has s as an integral section. It is easily seen that, in this case, the Hamiltonian map (3.108) satisfies the relation (3.74) for s. Hence, the Hamiltonian forms (3.109) constitute a complete set.

It is readily observed that, if $\sigma_1 = 0$, then $\Phi = \widehat{H}(\Gamma)$, and the Hamiltonian forms $H(\Gamma, \sigma_1 = 0)$ (3.109) are associated to the Lagrangian (3.90).

For different σ_1, we have different complete sets of Hamiltonian forms (3.109). Hamiltonian forms $H(\Gamma, \sigma_1)$ and $H(\Gamma', \sigma_1)$ (3.109) of such a complete set differ from each other in the term $\phi^i_\lambda \mathcal{R}^\lambda_i$, where ϕ are the soldering forms (3.94). This term vanishes on the Lagrangian constraint space (3.106). Accordingly, the covariant Hamilton equation for different Hamiltonian forms $H(\Gamma, \sigma_1)$ and $H(\Gamma', \sigma_1)$ (3.109) differs from each other in the equations (3.111). These equations are independent of momenta and play the role of gauge-type conditions.

Since the Lagrangian constraint space N_L (3.106) is an imbedded subbundle of $\Pi \to Y$, all Hamiltonian forms $H(\Gamma, \sigma_1)$ (3.109) define a unique constrained Hamiltonian form H_N (3.76) on N_L which reads

$$H_N = i_N^* H(\Gamma, \sigma_1) = \mathcal{P}_i^\lambda dy^i \wedge \omega_\lambda - [\mathcal{P}_i^\lambda \Gamma_\lambda^i + \frac{1}{2}\sigma_{0\lambda\mu}^{ij} \mathcal{P}_i^\lambda \mathcal{P}_j^\mu - c']\omega.$$

In view of the relations (3.105), the corresponding constrained Lagrangian L_N (3.80) on $J^1 N_L$ takes a form

$$L_N = h_0(H_N) = (\mathcal{P}_i^\lambda \mathcal{F}_\lambda^i - \frac{1}{2}\sigma_{0\lambda\mu}^{ij} \mathcal{P}_i^\lambda \mathcal{P}_j^\mu + c')\omega. \tag{3.113}$$

It is the pull-back onto $J^1 N_L$ of a Lagrangian

$$L_{H(\Gamma,\sigma_1)} = \mathcal{R}_i^\lambda (\mathcal{S}_\lambda^i - \Gamma_\lambda^i) + \mathcal{P}_i^\lambda \mathcal{F}_\lambda^i - \frac{1}{2}\sigma_{0\lambda\mu}^{ij} \mathcal{P}_i^\lambda \mathcal{P}_j^\mu - \frac{1}{2}\sigma_{1\lambda\mu}^{ij} \mathcal{R}_i^\lambda \mathcal{R}_j^\mu + c'$$

on $J^1 \Pi$ for any Hamiltonian form $H(\Gamma, \sigma_1)$ (3.109).

In fact, the constrained Lagrangian L_N (3.113) is defined on a product $N_L \times_Y J^1 Y$ (Remark 3.6). Since the phase space Π (3.103) is a trivial bundle $\mathrm{pr}_2 : \Pi \to N_L$ over the Lagrangian constraint space N_L, one can consider the pull-back

$$L_\Pi = (\mathcal{P}_i^\lambda \mathcal{F}_\lambda^i - \frac{1}{2}\sigma_{0\lambda\mu}^{ij} \mathcal{P}_i^\lambda \mathcal{P}_j^\mu + c')\omega \tag{3.114}$$

of the constrained Lagrangian L_N (3.113) onto $\Pi \times_Y J^1 Y$.

Let us study symmetries of the constrained Lagrangians L_N (3.113) and L_Π (3.114). We aim to show that, under certain conditions, they inherit symmetries of an original Lagrangian L (Theorems 3.21 and 3.22).

Let a vertical vector field $u = u^i \partial_i$ on $Y \to X$ be an exact symmetry of the Lagrangian L (3.101), i.e.,

$$\mathbf{L}_{J^1 u} L = (u^i \partial_i + d_\lambda u^i \partial_i^\lambda)\mathcal{L}\omega = 0. \tag{3.115}$$

Since

$$J^1 u(y_\lambda^i - \Gamma_\lambda^i) = \partial_k u^i (y_\lambda^k - \Gamma_\lambda^k), \tag{3.116}$$

one easily obtains from the equality (3.115) that

$$u^k \partial_k a_{ij}^{\lambda\mu} + \partial_i u^k a_{kj}^{\lambda\mu} + a_{ik}^{\lambda\mu} \partial_j u^k = 0. \tag{3.117}$$

It follows that the summands of the Lagrangian (3.101) are separately invariant, i.e.,

$$J^1 u(a_{ij}^{\lambda\mu} \mathcal{F}_\lambda^i \mathcal{F}_\mu^j) = 0, \qquad J^1 u(c') = u^k \partial_k c' = 0. \tag{3.118}$$

The equalities (3.102), (3.116) and (3.117) give the transformation law

$$J^1u(a^{\lambda\mu}_{ij}\mathcal{F}^j_\mu) = -\partial_i u^k a^{\lambda\mu}_{kj}\mathcal{F}^j_\mu. \tag{3.119}$$

The relations (3.98) and (3.117) lead to the equality

$$a^{\lambda\mu}_{ij}[u^k\partial_k\sigma_0{}^{jn}_{\mu\alpha} - \partial_k u^j\sigma_0{}^{kn}_{\mu\alpha} - \sigma_0{}^{jk}_{\mu\alpha}\partial_k u^n]a^{\alpha\nu}_{nb} = 0. \tag{3.120}$$

Let us compare symmetries of the Lagrangian L (3.101) and the constrained Lagrangian L_N (3.113). Given the Legendre map $\widehat{L}$ (3.91) and the tangent morphism

$$T\widehat{L} : TJ^1Y \to TN_L, \qquad \dot{p}_A = (\dot{y}^i\partial_i + \dot{y}^k_\nu\partial^\nu_k)(M_A{}^i_\lambda a^{\lambda\mu}_{ij}\mathcal{F}^j_\mu),$$

let us consider the map

$$\begin{aligned}
&T\widehat{L}\circ J^1u : J^1Y \ni (x^\lambda, y^i, y^i_\lambda) \to \\
&\quad u^i\partial_i + (u^k\partial_k + \partial_\nu u^k\partial^\nu_k)(M_A{}^i_\lambda a^{\lambda\mu}_{ij}\mathcal{F}^j_\mu)\partial^A = \\
&\quad u^i\partial_i + [u^k\partial_k(M_A{}^i_\lambda)a^{\lambda\mu}_{ij}\mathcal{F}^j_\mu + M_A{}^i_\lambda J^1u(a^{\lambda\mu}_{ij}\mathcal{F}^j_\mu)]\partial^A = \\
&\quad u^i\partial_i + [u^k\partial_k(M_A{}^i_\lambda)a^{\lambda\mu}_{ij}\mathcal{F}^j_\mu - M_A{}^i_\lambda\partial_i u^k a^{\lambda\mu}_{kj}\mathcal{F}^j_\mu]\partial^A = \\
&\quad u^i\partial_i + [u^k\partial_k(a\circ\sigma_0)^{\mu i}_{j\lambda}\mathcal{P}^\lambda_i - (a\circ\sigma_0)^{\mu i}_{j\lambda}\partial_i u^k\mathcal{P}^\lambda_k]\partial^j_\mu \in TN_L,
\end{aligned} \tag{3.121}$$

where the relations (3.107) and (3.119) have been used. Let us assign to a point $(x^\lambda, y^i, \mathcal{P}^\lambda_i) \in N_L$ some point

$$(x^\lambda, y^i, y^i_\lambda) \in \widehat{L}^{-1}(x^\lambda, y^i, \mathcal{P}^\lambda_i) \tag{3.122}$$

and then an image of the point (3.122) under the morphism (3.121). We get a map

$$\upsilon_N : (x^\lambda, y^i, \mathcal{P}^\lambda_i) \to u^i\partial_i + [u^k\partial_k(a\circ\sigma_0)^{\mu i}_{j\lambda}\mathcal{P}^\lambda_i - (a\circ\sigma_0)^{\mu i}_{j\lambda}\partial_i u^k\mathcal{P}^\lambda_k]\partial^j_\mu \tag{3.123}$$

which is independent of the choice of a point (3.122). Therefore, it is a vector field on the Lagrangian constraint space N_L. This vector field gives rise to a vector field

$$J\upsilon_N = u^i\partial_i + [u^k\partial_k(a\circ\sigma_0)^{\mu i}_{j\lambda}\mathcal{P}^\lambda_i - (a\circ\sigma_0)^{\mu i}_{j\lambda}\partial_i u^k\mathcal{P}^\lambda_k]\partial^j_\mu + d_\lambda u^i\partial^\lambda_i \tag{3.124}$$

on $N_L \times_Y J^1Y$.

Theorem 3.21 *The Lie derivative $\mathbf{L}_{J\upsilon_N}L_N$ of the constrained Lagrangian L_N (3.113) along the vector field $J\upsilon_N$ (3.124) vanishes.*

Proof One can show that

$$\upsilon_N(\mathcal{P}^\lambda_i) = -\partial_i u^k\mathcal{P}^\lambda_k \tag{3.125}$$

on the constraint manifold $\mathscr{R}_i^\lambda = 0$. Then the invariance condition $J\upsilon_N(\mathscr{L}_N) = 0$ falls into the three equalities

$$J\upsilon_N(\sigma_0{}^{ij}_{\lambda\mu}\mathscr{P}_i^\lambda\mathscr{P}_j^\mu) = 0, \qquad J\upsilon_N(\mathscr{P}_i^\lambda\mathscr{F}_\lambda^i) = 0, \qquad J\upsilon_N(c') = 0. \tag{3.126}$$

The latter is exactly the second equality (3.118). The first equality (3.126) is satisfied due to the relations (3.120) and (3.125). The second one takes a form

$$J\upsilon_N(\mathscr{P}_i^\lambda(y_\lambda^i - \Gamma_\lambda^i)) = 0.$$

It holds owing to the relations (3.116) and (3.125).

Thus, any exact vertical symmetry u of the Lagrangian L (3.101) yields the exact symmetry υ_N (3.123) of the constrained Lagrangian L_N (3.113).

Turn now to symmetries of the Lagrangian L_Π (3.114). Since L_Π is the pull-back of L_N onto $\Pi \times_Y J^1Y$, its symmetry must be an appropriate lift of the vector field υ_N (3.123) onto Π.

Given a vertical vector field u on $Y \to X$, let us consider its canonical lift (3.81):

$$\widetilde{u} = u^i\partial_i - \partial_i u^j p_j^\lambda \partial_\lambda^i, \tag{3.127}$$

onto a Legendre bundle Π. It readily observed that the vector field $\widetilde{u}$ is projected onto the vector field υ_N (3.123).

Let us additionally suppose that the one-parameter group of automorphisms of Y generated by u preserves the splitting (3.100), i.e., u obeys a condition

$$u^k\partial_k(\sigma_0{}^{im}_{\lambda\nu}a^{\nu\mu}_{mj}) + \sigma_0{}^{im}_{\lambda\nu}a^{\nu\mu}_{mk}\partial_j u^k - \partial_k u^i\sigma_0{}^{km}_{\lambda\nu}a^{\nu\mu}_{mj} = 0. \tag{3.128}$$

The relations (3.116) and (3.128) lead to a transformation law

$$J^1u(\mathscr{F}_\mu^i) = \partial_j u^i\mathscr{F}_\mu^j. \tag{3.129}$$

Theorem 3.22 *If the condition (3.128) holds, the vector field $\widetilde{u}$ (3.127) is an exact symmetry of the Lagrangian L_Π (3.114) if and only if u is an exact symmetry of the Lagrangian L (3.101).*

Proof Due to the condition (3.128), the vector field $\widetilde{u}$ (3.127) preserves the splitting (3.103), i.e.,

$$\widetilde{u}(\mathscr{P}_i^\lambda) = -\partial_i u^k\mathscr{P}_k^\lambda, \qquad \widetilde{u}(\mathscr{R}_i^\lambda) = -\partial_i u^k\mathscr{R}_k^\lambda.$$

The vector field $\widetilde{u}$ gives rise to the vector field (3.82):

$$J\widetilde{u} = u^i\partial_i - \partial_i u^j p_j^\lambda\partial_\lambda^i + d_\lambda u^i\partial_i^\lambda,$$

on $\Pi \times_Y J^1Y$, and we obtain the Lagrangian symmetry condition

$$(u^i\partial_i - \partial_j u^i p_i^\lambda \partial_\lambda^j + d_\lambda u^i \partial_i^\lambda)\mathcal{L}_\Pi = 0. \tag{3.130}$$

It is readily observed that the first and third terms of the Lagrangian L_Π are separately invariant due to the relations (3.118) and (3.129). Its second term is invariant owing to the equality (3.120). Conversely, let the invariance condition (3.130) hold. It falls into the independent equalities

$$J\widetilde{u}(\sigma_0{}^{ij}_{\lambda\mu} p_i^\lambda p_j^\mu) = 0, \qquad J\widetilde{u}(p_i^\lambda \mathcal{F}^i_\lambda) = 0, \qquad u^i\partial_i c' = 0, \tag{3.131}$$

i.e., the Lagrangian L_Π is invariant if and only if its three summands are separately invariant. One obtains at once from the second condition (3.131) that the quantity $\mathcal{F}$ is transformed as the dual of momenta p. Then the first condition (3.131) shows that the quantity $\sigma_0 p$ is transformed by the same law as $\mathcal{F}$. It follows that the term $a\mathcal{F}\mathcal{F}$ in the Lagrangian L (3.101) is transformed as $a(\sigma_0 p)(\sigma_0 p) = \sigma_0 pp$, i.e., it is invariant. Then this Lagrangian is invariant due to the third equality (3.131).

Remark 3.15 A Lagrangian L_Π may possess symmetries which do not come from symmetries of an original Lagrangian L. They are represented by vector fields on Π which are not the canonical lift (3.127) of vector fields on Y (Remark 4.16).

Chapter 4
Lagrangian and Hamiltonian Nonrelativistic Mechanics

By mechanics throughout the book is meant classical non-autonomous nonrelativistic mechanics subject to time-dependent coordinate and reference frame transformations. This mechanics is formulated adequately as Lagrangian and Hamiltonian theory on fibre bundles $Q \to \mathbb{R}$ over the time axis $\mathbb{R}$ [62, 98, 121, 135].

Since equations of motion of mechanics almost always are of first and second order, we restrict our consideration to first order Lagrangian and Hamiltonian theory. Its velocity space is the first order jet manifold J^1Q of sections of a configuration bundle $Q \to \mathbb{R}$, and its phase space Π (3.8) is the vertical cotangent bundle V^*Q of $Q \to \mathbb{R}$.

This formulation of mechanics is similar to that of classical field theory on fibre bundles over a smooth manifold X of dimension $n > 1$ (Chap. 3). A difference between mechanics and field theory however lies in the fact that fibre bundles over $\mathbb{R}$ always are trivial, and that all connections on these fibre bundles are flat. Consequently, they are not dynamic variables, but characterize nonrelativistic reference frames (Definition 4.1).

In Lagrangian mechanics, Noether's first theorem (Theorem 2.7) is formulated as a straightforward corollary of the first variational formula (4.30). It associates to any classical Lagrangian symmetry υ (4.27) the conserved current (4.32) whose total differential vanishes on-shell.

In particular, an energy function relative to a reference frame is the symmetry current (4.41) along a connection Γ on a configuration bundle $Q \to \mathbb{R}$ which characterizes this reference frame (Definition 4.1).

A key point is that, in Lagrangian mechanics, any conserved current is an integral of motion (Theorem 4.4), but the converse need not be true (e.g., the Rung–Lenz vector (5.6) in a Lagrangian Kepler model).

Hamiltonian formulation of non-autonomous nonrelativistic mechanics is similar to covariant Hamiltonian field theory on fibre bundles [61, 137] in the particular case of fibre bundles over $\mathbb{R}$ [62, 121, 136]. A key point is that a non-autonomous Hamiltonian system on a phase space V^*Q is equivalent to a particular first order Lagrangian system with the characteristic Lagrangian (4.76) on V^*Q as a configuration space. This fact enables one to apply Noether's first theorem to study symmetries

G. Sardanashvily, *Noether's Theorems*, Atlantis Studies
in Variational Geometry 3, DOI 10.2991/978-94-6239-171-0_4

in Hamiltonian mechanics (Sect. 4.6). In particular, we show that, since Hamiltonian symmetries are vector fields on a phase space V^*Q (Definition 4.7), any integral of motion in Hamiltonian mechanics (Definition 4.6) is some conserved symmetry current (Theorem 4.15).

Therefore, it may happen that symmetries and the corresponding integrals of motion define a Hamiltonian system in full. This is the case of commutative and noncommutative completely integrable systems (Sect. 4.7).

4.1 Geometry of Fibre Bundles over $\mathbb{R}$

This Section summarizes peculiarities of geometry of fibre bundles over $\mathbb{R}$ [62, 98].

Let $\pi : Q \to \mathbb{R}$ be a fibred manifold (Definition B.1) whose base is regarded as the time axis $\mathbb{R}$ parameterized by the *Cartesian coordinate* t with transition functions $t' = t+$const. Relative to the Cartesian coordinate t, the time axis $\mathbb{R}$ is provided with the global *standard vector field* ∂_t and the global *standard one-form* dt which also is a global volume form on $\mathbb{R}$. The symbol dt also stands for any pull-back of the standard one-form dt onto a fibre bundle over $\mathbb{R}$.

Remark 4.1 Point out one-to-one correspondence between the vector fields $f\partial_t$, the densities fdt and the real functions f on $\mathbb{R}$. Roughly speaking, we can neglect the contribution of $T\mathbb{R}$ and $T^*\mathbb{R}$ to some expressions. In particular, the canonical imbedding (B.48) of J^1Q takes the form (4.6).

In order that the dynamics of a mechanical system can be defined at any instant $t \in \mathbb{R}$, we further assume that a fibred manifold $Q \to \mathbb{R}$ is a fibre bundle (Definition B.2) with a typical fibre M.

Remark 4.2 In accordance with Remark B.11, a fibred manifold $Q \to \mathbb{R}$ is a fibre bundle if and only if it admits an Ehresmann connection Γ, i.e., the horizontal lift $\Gamma\partial_t$ onto Q of the standard vector field ∂_t on $\mathbb{R}$ is complete.

Given bundle coordinates (t, q^i) on a fibre bundle $Q \to \mathbb{R}$, the first order jet manifold J^1Q of $Q \to \mathbb{R}$ is provided with the adapted coordinates (t, q^i, q^i_t) possessing transition functions (B.44) which read

$$t' = t + \text{const.}, \qquad q'^i = q'^i(t, q^j), \qquad q'^i_t = (\partial_t + q^j_t\partial_j)q'^i. \tag{4.1}$$

In mechanics on a configuration space $Q \to \mathbb{R}$, the jet manifold J^1Q plays the role of a *velocity space*.

Remark 4.3 By virtue of Theorem B.7, any fibre bundle $Q \to \mathbb{R}$ is trivial. Its different trivializations

$$\psi : Q = \mathbb{R} \times M \tag{4.2}$$

differ from each other in fibrations $Q \to M$. Given the trivialization (4.2) coordinated by $(t, \widetilde{q}^i)$, there is a canonical isomorphism

$$J^1(\mathbb{R} \times M) = \mathbb{R} \times TM, \qquad \widetilde{q}^i_t = \dot{\widetilde{q}}^i, \tag{4.3}$$

that one can justify by inspection of transition functions of coordinates $\widetilde{q}^i_t$ and $\dot{\widetilde{q}}^i$ when transition functions of q^i are time-independent. Due to the isomorphism (4.3), every trivialization (4.2) yields the corresponding trivialization of the jet manifold

$$J^1 Q = \mathbb{R} \times TM. \tag{4.4}$$

As a palliative variant, one develops nonrelativistic mechanics on the configuration space (4.2) and the velocity space (4.4) [35, 94]. Its phase space $\mathbb{R} \times T^*M$ is provided with the presymplectic form

$$\mathrm{pr}_2^* \Omega_T = dp_i \wedge dq^i \tag{4.5}$$

which is the pull-back of the canonical symplectic form Ω_T (4.102) on T^*M. A problem is that the presymplectic form (4.5) is broken by time-dependent transformations.

With respect to the bundle coordinates (4.1), the canonical imbedding (B.48) of $J^1 Q$ takes a form

$$\begin{aligned} &\lambda_{(1)} : J^1 Q \ni (t, q^i, q^i_t) \to (t, q^i, \dot{t} = 1, \dot{q}^i = q^i_t) \in TQ, \\ &\lambda_{(1)} = d_t = \partial_t + q^i_t \partial_i. \end{aligned} \tag{4.6}$$

From now on, the jet manifold $J^1 Q$ is identified with its image in TQ which is an affine subbundle of TX modelled over the vertical tangent bundle VQ of a fibre bundle $Q \to \mathbb{R}$. Using the morphism (4.6), one can define the contraction

$$J^1 Q \underset{Q}{\times} T^*Q \underset{Q}{\to} Q \times \mathbb{R}, \qquad (q^i_t; \dot{\mathrm{t}}, \dot{q}_i) \to \lambda_{(1)} \rfloor (\dot{\mathrm{t}} dt + \dot{q}_i dq^i) = \dot{\mathrm{t}} + q^i_t \dot{q}_i,$$

where $(t, q^i, \dot{\mathrm{t}}, \dot{q}_i)$ are holonomic coordinates on the cotangent bundle T^*Q.

In view of the morphism $\lambda_{(1)}$ (4.6), any connection

$$\Gamma = dt \otimes (\partial_t + \Gamma^i \partial_i) \tag{4.7}$$

on a fibre bundle $Q \to \mathbb{R}$ can be identified with a nowhere vanishing *horizontal vector field*

$$\Gamma = \partial_t + \Gamma^i \partial_i \tag{4.8}$$

on Q which is the horizontal lift $\Gamma \partial_t$ (B.56) of the standard vector field ∂_t on $\mathbb{R}$ by means of the connection (4.7). Conversely, any vector field Γ on Q such that

$dt\rfloor\Gamma = 1$ defines a connection on $Q \to \mathbb{R}$. Therefore, the connections (4.7) further are identified with the vector fields (4.8). The integral curves of the vector field (4.8) coincide with the integral sections for the connection (4.7).

Connections on a fibre bundle $Q \to \mathbb{R}$ constitute an affine space modelled over a vector space of vertical vector fields on $Q \to \mathbb{R}$. Accordingly, the covariant differential (B.61), associated to a connection Γ on $Q \to \mathbb{R}$, takes its values into the vertical tangent bundle VQ of $Q \to \mathbb{R}$:

$$D_\Gamma : J^1Q \underset{Q}{\to} VQ, \qquad \overline{q}^i \circ D_\Gamma = q^i_t - \Gamma^i. \tag{4.9}$$

A connection Γ on a fibre bundle $Q \to \mathbb{R}$ is obviously flat.

Theorem 4.1 *By virtue of Theorem B.15, every connection Γ on a fibre bundle $Q \to \mathbb{R}$ defines an atlas of local constant trivializations of $Q \to \mathbb{R}$ such that the associated bundle coordinates (t, q^i_Γ) on Q possess transition functions independent of t, and*

$$\Gamma = \partial_t \tag{4.10}$$

with respect to these coordinates. Conversely, every atlas of local constant trivializations of a fibre bundle $Q \to \mathbb{R}$ determines a connection on $Q \to \mathbb{R}$ which is equal to (4.10) relative to this atlas.

A connection Γ on a fibre bundle $Q \to \mathbb{R}$ is said to be *complete* if the horizontal vector field (4.8) is complete. In accordance with Remark B.11, a connection on a fibre bundle $Q \to \mathbb{R}$ is complete if and only if it is an Ehresmann connection. The following holds [98].

Theorem 4.2 *Every trivialization of a fibre bundle $Q \to \mathbb{R}$ yields a complete connection on this fibre bundle. Conversely, every complete connection Γ on $Q \to \mathbb{R}$ defines its trivialization (4.2) such that the horizontal vector field (4.8) equals ∂_t relative to the bundle coordinates associated to this trivialization.*

It follows from Theorem 4.1 that, in mechanics unlike field theory, connections Γ (4.8) on a configuration bundle $Q \to \mathbb{R}$ fail to be dynamic variables. They characterize reference frames as follows.

From the physical viewpoint, a reference frame in mechanics determines a tangent vector at each point of a configuration space Q, which characterizes the velocity of an observer at this point. This speculation leads to the following mathematical definition of a reference frame in mechanics [62, 98, 121].

Definition 4.1 In nonrelativistic mechanics, a *reference frame* is the connection Γ (4.8) on a configuration bundle $Q \to \mathbb{R}$, i.e., a section of a velocity bundle $J^1Q \to Q$.

By virtue of this definition, one can think of the horizontal vector field (4.8) associated to a connection Γ on $Q \to \mathbb{R}$ as being a family of observers, while the corresponding covariant differential (4.9):

$$\overline{q}^i_\Gamma = D_\Gamma(q^i_t) = q^i_t - \Gamma^i, \tag{4.11}$$

determines the *relative velocity* with respect to a reference frame Γ. Accordingly, q^i_t are regarded as the *absolute velocities*.

In accordance with Theorem 4.1, any reference frame Γ on a configuration bundle $Q \to \mathbb{R}$ is associated to an atlas of local constant trivializations, and vice versa. A connection Γ takes the form $\Gamma = \partial_t$ (4.10) with respect to the corresponding coordinates (t, q^i_Γ), whose transition functions are independent of time. One can think of these coordinates as also being a reference frame, corresponding to the connection (4.10). They are called the *adapted coordinates* to a reference frame Γ. Thus, we come to the following definition, equivalent to Definition 4.1.

Definition 4.2 In mechanics, a reference frame is an atlas of local constant trivializations of a configuration bundle $Q \to \mathbb{R}$.

In particular, with respect to the coordinates q^i_Γ adapted to a reference frame Γ, the velocities relative to this reference frame (4.11) coincide with the absolute ones

$$\overline{q}^i_\Gamma = D_\Gamma(q^i_{\Gamma t}) = q^i_{\Gamma t}.$$

A reference frame is said to be *complete* if the associated connection Γ is complete. By virtue of Theorem 4.2, every complete reference frame defines a trivialization of a bundle $Q \to \mathbb{R}$, and vice versa.

4.2 Lagrangian Mechanics. Integrals of Motion

As was mentioned above, our exposition is restricted to first order Lagrangian theory on a fibre bundle $Q \to \mathbb{R}$ [62, 98, 135]. This is a standard case of Lagrangian mechanics.

In mechanics, a first order Lagrangian (3.1) is defined as a horizontal density

$$L = \mathcal{L}dt, \qquad \mathcal{L} : J^1Q \to \mathbb{R}, \tag{4.12}$$

on a velocity space J^1Q.

The corresponding second order Euler–Lagrange operator (3.2), termed the *Lagrange operator*, reads

$$\delta L = (\partial_i \mathcal{L} - d_t \partial^t_i \mathcal{L})\theta^i \wedge dt, \qquad d_t = \partial_t + q^i_t \partial_i + q^i_{tt} \partial^t_i. \tag{4.13}$$

Let us further use the notation

$$\pi_i = \partial^t_i \mathcal{L}, \qquad \pi_{ji} = \partial^t_j \partial^t_i \mathcal{L}. \tag{4.14}$$

The kernel $\mathfrak{E}_L = \operatorname{Ker}\delta L \subset J^2Q$ of the Lagrange operator (4.13) defines a second order *Lagrange equation*

$$(\partial_i - d_t\partial_i^t)\mathcal{L} = 0 \tag{4.15}$$

on Q. Its classical solutions (Definition B.13) are (local) sections c of a fibre bundle $Q \to \mathbb{R}$ whose second order jet prolongations $J^2c = \partial_{tt}c$ live in $\mathfrak{E}_L$ (4.15).

Every first order Lagrangian L (4.12) yields the Legendre map (3.5):

$$\widehat{L} : J^1Q \underset{Q}{\longrightarrow} V^*Q, \qquad p_i \circ \widehat{L} = \pi_i, \tag{4.16}$$

where the Legendre bundle $\Pi = V^*Q$ (3.8) is the vertical cotangent bundle V^*Q of $Q \to \mathbb{R}$ (Definition B.4) provided with holonomic coordinates (t, q^i, p_i). As was mentioned above, it plays the role of a *phase space* of mechanics on a configuration space $Q \to \mathbb{R}$. The corresponding Lagrangian constraint space (3.9) is

$$N_L = \widehat{L}(J^1Q). \tag{4.17}$$

Given a first order Lagrangian L, its Lepage equivalent Ξ_L (3.11) in the variational formula (1.36) is the Poincaré–Cartan form (3.12):

$$H_L = \pi_i dq^i - (\pi_i q_t^i - \mathcal{L})dt \tag{4.18}$$

(see the notation (4.14)). This form takes its values into a subbundle

$$J^1Q \underset{Q}{\times} T^*Q \subset T^*J^1Q.$$

Hence, it defines the homogeneous Legendre map (3.13):

$$\widehat{H}_L : J^1Q \to Z_Y = T^*Q, \tag{4.19}$$

whose range (3.14) is an imbedded subbundle $Z_L = \widehat{H}_L(J^1Q) \subset T^*Q$ of the homogeneous Legendre bundle $Z_Y = T^*Q$ (4.19). Let $(t, q^i, p_0 = \dot{t}, p_i = \dot{q}_i)$ denote holonomic coordinates on T^*Q which possess transition functions

$$p'_i = \frac{\partial q^j}{\partial q'^i} p_j, \qquad p'_0 = \left(p_0 + \frac{\partial q^j}{\partial t} p_j\right). \tag{4.20}$$

With respect to these coordinates, the homogeneous Legendre map $\widehat{H}_L$ (4.19) reads

$$(p_0, p_i) \circ \widehat{H}_L = (\mathcal{L} - q_t^i \pi_i, \pi_i).$$

In view of the morphism (4.19), the cotangent bundle T^*Q plays the role of a *homogeneous phase space* of mechanics.

A glance at the transition functions (4.20) shows that the canonical map (B.19):

$$\zeta : T^*Q \to V^*Q, \tag{4.21}$$

is a one-dimensional affine bundle over the vertical cotangent bundle V^*Q. Herewith, the Legendre map $\widehat{L}$ (4.16) is exactly the composition of morphisms

$$\widehat{L} = \zeta \circ H_L : J^1Q \underset{Q}{\to} V^*Q.$$

Remark 4.4 Just as in Sect. 3.2, the Poincaré–Cartan form H_L (4.18) also is the Poincaré–Cartan form $H_L = H_{\widetilde{L}}$ of a first order Lagrangian

$$\widetilde{L} = \widehat{h}_0(H_L) = (\mathcal{L} + (q^i_{(t)} - q^i_t)\pi_i)dt, \qquad \widehat{h}_0(dq^i) = q^i_{(t)}dt, \tag{4.22}$$

on the repeated jet manifold J^1J^1Q. The Lagrange operator (4.13) for $\widetilde{L}$ (4.22) reads

$$\delta\widetilde{L} = [(\partial_i\mathcal{L} - \widehat{d}_t\pi_i + \partial_i\pi_j(q^j_{(t)} - q^j_t))dq^i + \partial^t_i\pi_j(q^j_{(t)} - q^j_t)dq^i_t] \wedge dt,$$
$$\widehat{d}_t = \partial_t + \widehat{q}^i_t\partial_i + q^i_{tt}\partial^t_i.$$

Its kernel Ker $\delta\widetilde{L} \subset J^1J^1Q$ defines the Cartan equation

$$\partial^t_i\pi_j(q^j_{(t)} - q^j_t) = 0, \qquad \partial_i\mathcal{L} - \widehat{d}_t\pi_i + \partial_i\pi_j(q^j_{(t)} - q^j_t) = 0 \tag{4.23}$$

on a velocity space J^1Q.

In mechanics, the Lagrange equation (4.15) as like as the Hamilton one (4.71) is an ordinary differential equation. One can think of its classical solutions $s(t)$ as being a *motion* in a configuration space Q. In this case, the notion of integrals of motion can be introduced as follows [62, 136].

In a general setting, let an equation of motion of a mechanical system is an r-order differential equation $\mathfrak{E}$ on a fibre bundle $Y \to \mathbb{R}$ given by a closed subbundle of the jet bundle $J^rY \to \mathbb{R}$ in accordance with Definition B.12.

Definition 4.3 An *integral of motion* of this mechanical system is defined as a $(k < r)$-order differential operator Φ on Y such that $\mathfrak{E}$ belongs to the kernel of an r-order jet prolongation of a differential operator $d_t\Phi$, i.e.,

$$J^{r-k-1}(d_t\Phi)|_{\mathfrak{E}} = J^{r-k}\Phi|_{\mathfrak{E}} = 0, \qquad d_t = \partial_t + y^a_t\partial_a + y^a_{tt}\partial^t_a + \cdots. \tag{4.24}$$

It follows that an integral of motion Φ is constant on classical solutions s of a differential equation $\mathfrak{E}$, i.e., there is the *differential conservation law*

$$(J^ks)^*\Phi = \text{const.}, \qquad (J^{k+1}s)^*d_t\Phi = 0. \tag{4.25}$$

We agree to write the condition (4.24) as a *weak equality*

$$J^{r-k-1}(d_t\Phi) \approx 0,$$

which holds on-shell, i.e., on solutions of a differential equation $\mathfrak{E}$ by the formula (4.25).

In mechanics, we restrict our consideration to integrals of motion Φ which are functions on J^kY. As was mentioned above, equations of motion of mechanics mainly are either of first or second order. Accordingly, their integrals of motion are functions on $Y = J^0Y$ or J^1Y. In this case, the corresponding weak equality (4.24) takes a form

$$d_t\Phi \approx 0 \tag{4.26}$$

of a weak *conservation law* of an integral of motion.

Integrals of motion can come from symmetries. This is the case both of Lagrangian mechanics on a configuration space $Y = Q$ (Theorems 4.4 and 4.5) and Hamiltonian mechanics on a phase space $Y = V^*Q$ (Theorem 4.11).

Definition 4.4 Let an equation of motion of a mechanical system be an r-order differential equation $\mathfrak{E} \subset J^rY$. Its *infinitesimal symmetry* (or, shortly, a *symmetry*) is defined as a vector field on J^rY whose restriction to $\mathfrak{E}$ is tangent to $\mathfrak{E}$.

Following Definition 4.4, let us introduce a notion of the symmetry of differential operators in the following relevant case. Let us consider an r-order differential operator on a fibre bundle $Y \to \mathbb{R}$ which is represented by an exterior form $\mathcal{E}$ on J^rY (Definition B.15). Let its kernel Ker $\mathcal{E}$ be an r-order differential equation on $Y \to \mathbb{R}$.

Theorem 4.3 *It is readily justified that a vector field ϑ on J^rY is a symmetry of the equation* Ker $\mathcal{E}$ *in accordance with Definition 4.4 if and only if* $\mathbf{L}_\vartheta\mathcal{E} \approx 0$.

Motivated by Theorem 4.3, we come to the following notion.

Definition 4.5 Let $\mathcal{E}$ be the above mentioned differential operator. A vector field ϑ on J^rY is termed the *symmetry* of a differential operator $\mathcal{E}$ if the Lie derivative $\mathbf{L}_\vartheta\mathcal{E}$ vanishes.

By virtue of Theorem 4.3, a symmetry of a differential operator $\mathcal{E}$ also is a symmetry of a differential equation Ker $\mathcal{E}$.

Let us note that there exist integrals of motion which are not associated to symmetries of an equation of motion, e.g., the Rug–Lenz vector (5.6) in a Lagrangian Kepler system (Chap. 5).

4.3 Noether's First Theorem: Energy Conservation Laws

In Lagrangian mechanics, integrals of motion come from symmetries of a Lagrangian (Theorem 4.4) in accordance with Noether's first theorem (Theorem 2.7). However as was mentioned above, not all integrals of motion are of this type.

Given a Lagrangian system $(\mathcal{O}^*_\infty Q, L)$ (Definition 1.5) on a fibre bundle $Q \to \mathbb{R}$, its infinitesimal transformations are contact derivations ϑ of a real ring $\mathcal{O}^0_\infty Q$ (Definition 2.1). In accordance with Theorem 2.1, a contact derivation is the infinite order jet prolongation $J^\infty \upsilon$ (2.2) of the generalized vector field (2.3):

$$\upsilon = \upsilon^t \partial_t + \upsilon^i(t, q^i, q^i_\Lambda)\partial_i,$$
$$\upsilon = \upsilon_V + \upsilon_H = \upsilon^t(\partial_t + q^i_t \partial_i) + (\upsilon^i - q^i_t \upsilon^t)\partial_i,$$

on Q. Just as in first order Lagrangian field theory, we further deal with classical symmetries υ which are vector fields on a configuration space Q. Moreover, not concerned with time-reparametrization, we restrict our consideration to vector fields

$$\upsilon = \upsilon^t \partial_t + \upsilon^i(t, q^i)\partial_i, \qquad \upsilon^t = 0, 1. \tag{4.27}$$

Herewith, in first order Lagrangian mechanics, contact derivations (2.2) can be reduced to the first order jet prolongation

$$\vartheta = J^1\upsilon = \upsilon^t \partial_t + \upsilon^i \partial_i + d_t \upsilon^i \partial^t_i \tag{4.28}$$

of the vector fields υ (4.27).

Let L be the Lagrangian (4.12) on a velocity space $J^1 Q$. Its Lie derivative $\mathbf{L}_{J^1\upsilon} L$ along the contact derivation (4.28) obeys the first variational formula (3.25):

$$\mathbf{L}_{J^1\upsilon} L = \upsilon_V \rfloor \delta L + d_H(\upsilon \rfloor H_L), \tag{4.29}$$

where H_L is the Poincaré–Cartan form (4.18). Its coordinate expression reads

$$[\upsilon^t \partial_t + \upsilon^i \partial_i + d_t \upsilon^i \partial^t_i]\mathcal{L} = (\upsilon^i - q^i_t \upsilon^t)\mathcal{E}_i + d_t[\pi_i(\upsilon^i - \upsilon^t q^i_t) + \upsilon^t \mathcal{L}]. \tag{4.30}$$

In accordance with Noether's first theorem (Theorem 2.7), if the vector field υ (4.27) is a symmetry of a Lagrangian L (Definition 2.2), a corollary of the first variational formula (4.30) on-shell is the weak *Lagrangian conservation law*

$$0 \approx d_H(\upsilon \rfloor H_L - \sigma) \approx d_t(\pi_i(\upsilon^i - \upsilon^t q^i_t) + \upsilon^t \mathcal{L} - \sigma)dt \approx -d_t \mathcal{J}_\upsilon dt \tag{4.31}$$

of a symmetry current

$$\mathcal{J}_\upsilon = -(\upsilon \rfloor H_L - \sigma) = -(\pi_i(\upsilon^i - \upsilon^t q^i_t) + \upsilon^t \mathcal{L} - \sigma) \tag{4.32}$$

along υ. The symmetry current (4.32) obviously is defined with the accuracy to a constant summand.

Theorem 4.4 *It is readily observed that the conserved symmetry current* $\mathcal{J}_\upsilon$ *(4.32) along a classical symmetry is a function on a velocity space* J^1Q*, and it is an integral of motion of a Lagrangian system in accordance with Definition 4.3.*

Theorem 4.5 *If a symmetry* υ *of a Lagrangian* L *is classical, this is a symmetry of the Lagrange operator* δL *(4.13) (Remark 2.4) and, as a consequence, an infinitesimal symmetry of the Lagrange equation* $\mathfrak{E}_L$ *(4.15) (Theorem 4.3).*

Remark 4.5 Given a Lagrangian L, let $\widehat{L}$ be its partner (4.22) on the repeated jet manifold J^1J^1Q. Since H_L (4.16) is the Poincaré–Cartan form both for L and $\widetilde{L}$, a Lagrangian $\widehat{L}$ does not lead to new conserved symmetry currents.

If a symmetry υ of a Lagrangian L is exact, i.e., $\mathbf{L}_{J^1\upsilon}L = 0$, the first variational formula (4.29) takes a form

$$0 = \upsilon_V \rfloor \delta L + d_H(\upsilon \rfloor H_L). \tag{4.33}$$

It leads to the weak conservation law (4.31):

$$0 \approx -d_t \mathcal{J}_\upsilon, \tag{4.34}$$

of the symmetry current

$$\mathcal{J}_\upsilon = -\upsilon \rfloor H_L = -\pi_i(\upsilon^i - \upsilon^t q_t^i) - \upsilon^t \mathcal{L} \tag{4.35}$$

along a vector field υ.

Remark 4.6 The first variational formula (4.33) also can be utilized when a Lagrangian possesses exact symmetries, but an equation of motion is a sum

$$(\partial_i - d_t \partial_i^t)\mathcal{L} + f_i(t, q^j, q_t^j) = 0 \tag{4.36}$$

of a Lagrange equation and an additional non-Lagrangian external force. Let us substitute $\mathcal{E}_i = -f_i$ from this equality in the first variational formula (4.33). Then we have the *weak transformation law*

$$(\upsilon^i - q_t^i \upsilon^t) f_i \approx d_t \mathcal{J}_\upsilon$$

of the symmetry current $\mathcal{J}_\upsilon$ (4.35) on the shell (4.36).

It is readily observed that the first variational formula (4.30) is linear in a vector field υ. Therefore, one can consider superposition of the equalities (4.30) for different vector fields.

For instance, if υ and υ' are projectable vector fields (4.27), they are projected onto the standard vector field ∂_t on $\mathbb{R}$, and the difference of the corresponding equalities (4.30) results in the first variational formula (4.30) for a vertical vector field $\upsilon - \upsilon'$.

Conversely, every vector field υ (4.27), projected onto ∂_t, can be written as a sum

$$\upsilon = \Gamma + v \tag{4.37}$$

of some reference frame (4.8):

$$\Gamma = \partial_t + \Gamma^i \partial_i, \tag{4.38}$$

and a vertical vector field v on $Q \to \mathbb{R}$.

It follows that the first variational formula (4.30) for the vector field υ (4.27) can be represented as a superposition of those for the reference frame Γ (4.38) and a vertical vector field v.

If $\upsilon = v$ is a vertical vector field, the first variational formula (4.30) reads

$$(v^i \partial_i + d_t v^i \partial_i^t)\mathcal{L} = v^i \mathcal{E}_i + d_t(\pi_i v^i).$$

If v is an exact symmetry of L, we obtain from (4.34) the Noether conservation law $0 \approx d_t(\pi_i v^i)$ (3.31) of the Noether current

$$\mathcal{J}_v = -\pi_i v^i, \tag{4.39}$$

which is a Lagrangian integral of motion by virtue of Theorem 4.4.

Remark 4.7 Let us assume that, given a trivialization $Q = \mathbb{R} \times M$ in bundle coordinates (t, q^i), a Lagrangian L is independent of some coordinate q^a. Then a vertical vector field $v = \partial_i$ is an exact symmetry of L, and we have the conserved Noether current $\mathcal{J}_v = -\pi_i$ (4.39) which is an integral of motion.

In the case of the reference frame Γ (4.38), where $\upsilon^t = 1$, the first variational formula (4.30) reads

$$(\partial_t + \Gamma^i \partial_i + d_t \Gamma^i \partial_i^t)\mathcal{L} = (\Gamma^i - q_t^i)\mathcal{E}_i - d_t(\pi_i(q_t^i - \Gamma^i) - \mathcal{L}), \tag{4.40}$$

where

$$E_\Gamma = \mathcal{J}_\Gamma = \pi_i(q_t^i - \Gamma^i) - \mathcal{L} \tag{4.41}$$

is called the *energy function relative to a reference frame* Γ [62, 121].

With respect to the coordinates q_Γ^i adapted to a reference frame Γ, the first variational formula (4.40) takes a form

$$\partial_t \mathcal{L} = -q_{\Gamma t}^i \mathcal{E}_i - d_t(\pi_i q_{\Gamma t}^i - \mathcal{L}), \tag{4.42}$$

and the E_Γ (4.41) coincides with the *canonical energy function*

$$E_L = \pi_i q^i_{\Gamma t} - \mathcal{L}. \tag{4.43}$$

A glance at the expression (4.42) shows that the vector field Γ (4.38) is an exact symmetry of a Lagrangian L if and only if, written with respect to coordinates adapted to Γ, this Lagrangian is independent of the time t. In this case, the energy function E_Γ (4.42) relative to a reference frame Γ is conserved: $0 \approx -d_t E_\Gamma$. It is a Lagrangian integral of motion in accordance with Theorem 4.4.

Since any vector field υ (4.27) can be represented as the sum (4.37) of the reference frame Γ (4.38) and a vertical generalized vector field v, the symmetry current (4.35) along the vector field υ (4.27) is a sum $\mathcal{J}_\upsilon = E_\Gamma + \mathcal{J}_v$ of the Noether current $\mathcal{J}_v$ (4.39) along a vertical vector field v and the energy function E_Γ (4.41) relative to a reference frame Γ. Conversely, energy functions relative to different reference frames Γ and Γ' differ from each other in the Noether current (4.39) along a vertical vector field $\Gamma' - \Gamma$:

$$E_\Gamma - E_{\Gamma'} = \pi_i(\Gamma'^i - \Gamma^i) = \mathcal{J}_{\Gamma - \Gamma'}. \tag{4.44}$$

Remark 4.8 Given a configuration space Q of a mechanical system and the connection Γ (4.38) on $Q \to \mathbb{R}$, let us consider a quadratic Lagrangian

$$L = \frac{1}{2} m_{ij}(t, q^k)(q^i_t - \Gamma^i)(q^j_t - \Gamma^j)dt, \tag{4.45}$$

where m_{ij} is a nondegenerate positive-definite fibre metric in the vertical tangent bundle $VQ \to Q$. It is called the *mass tensor*. Such a Lagrangian is globally defined due to linear transformation laws of the relative velocities $\overline{q}^i_\Gamma$ (4.11). Let q^i_Γ be fibre coordinates adapted to a reference frame Γ. Then the Lagrangian (4.45) reads

$$L = \frac{1}{2} \overline{m}_{ij}(q^k) q^i_{\Gamma t} q^j_{\Gamma t} dt. \tag{4.46}$$

Since coordinates q^i_Γ possess time-independent transition functions, let us assume that a mass tensor $\overline{m}_{ij}$ is independent of time. In this case, a horizontal vector field $\Gamma\partial_t = \Gamma = \partial_t$ is an exact symmetry of the Lagrangian L (4.46) that leads to a weak conservation law of the canonical energy function (4.43):

$$E_L = \frac{1}{2} \overline{m}_{ij}(q^k) q^i_{\Gamma t} q^j_{\Gamma t} dt. \tag{4.47}$$

Relative to arbitrary bundle coordinates (t, q^i) on Q, the energy function (4.47) reads

$$E_\Gamma = \frac{1}{2} m_{ij}(t, q^k)(q^i_t - \Gamma^i)(q^j_t - \Gamma^j).$$

This is an energy function relative to a reference frame Γ.

Remark 4.9 Let us consider a one-dimensional motion of a point mass m_0 subject to friction. It is described by the dynamic equation

$$m_0 q_{tt} = -k q_t, \qquad k > 0,$$

on a configuration space $\mathbb{R}^2 \to \mathbb{R}$ coordinated by (t, q). This equation is a Lagrange equation of a Lagrangian

$$L = \frac{1}{2} m_0 \exp\left[\frac{k}{m_0} t\right] q_t^2 dt,$$

termed the *Havas Lagrangian* [62, 113]. It is readily observed that the Lie derivative of this Lagrangian along a vector field

$$\Gamma = \partial_t - \frac{1}{2}\frac{k}{m_0} q \partial_q \tag{4.48}$$

vanishes. Consequently, we have the conserved energy function (4.41) with respect to the reference frame Γ (4.48). This energy function reads

$$E_\Gamma = \frac{1}{2} m_0 \exp\left[\frac{k}{m_0} t\right] q_t \left(q_t + \frac{k}{m_0} q\right).$$

4.4 Gauge Symmetries: Noether's Second and Third Theorems

In mechanics, we follow Definition 2.4 of a gauge symmetry [62]. It takes the form (2.15):

$$u = \partial_t + \left(\sum_{0 \le |\Lambda| \le m} u_a^{i\Lambda}(t, q_\Sigma^j) \chi_\Lambda^a\right) \partial_i,$$

where χ is a section of some vector bundle $E \to \mathbb{R}$. In accordance with Theorem 2.5, we can restrict our consideration to vertical gauge symmetries

$$u = \left(\sum_{0 \le |\Lambda| \le m} u_a^{i\Lambda}(t, q_\Sigma^j) \chi_\Lambda^a\right) \partial_i. \tag{4.49}$$

Noether's direct second theorem associates to a gauge symmetry of a Lagrangian L the Noether identities (NI) of its Lagrange operator δL as follows (cf. Theorem 2.6).

Theorem 4.6 *Let u (4.49) be a gauge symmetry of a Lagrangian L, then its Lagrange operator δL obeys the NI (4.50).*

Proof The density $u\rfloor\delta L = d_t\sigma dt$ is variationally trivial and, therefore, its variational derivatives with respect to variables χ^a vanish, i.e.,

$$\mathscr{E}_a = \sum_{0\leq|\Lambda|} (-1)^{|\Lambda|} d_\Lambda (u_a^{i\Lambda}\mathscr{E}_i) = 0. \tag{4.50}$$

For instance, if the gauge symmetry u (4.49) is of second jet order in gauge parameters, i.e.,

$$u = (u_a^i\chi^a + u_a^{it}\chi_t^a + u_a^{itt}\chi_{tt}^a)\partial_i,$$

the corresponding NI (4.50) read

$$u_a^i\mathscr{E}_i - d_t(u_a^{it}\mathscr{E}_i) + d_{tt}(u_a^{itt}\mathscr{E}_i) = 0$$

(cf. the expression (2.19)).

If a Lagrangian L admits the gauge symmetry u (4.49), the weak conservation law (4.31) of the corresponding symmetry current $\mathscr{J}_u$ (4.32) holds. Because gauge symmetries depend on derivatives of gauge parameters, all gauge conservation laws in first order Lagrangian mechanics possess the following peculiarity.

Theorem 4.7 *If u (4.49) is a gauge symmetry of a first order Lagrangian L, the corresponding symmetry current $\mathscr{J}_u$ (4.32) vanishes on-shell, i.e., $\mathscr{J} \approx 0$.*

Proof Let a gauge symmetry u be at most of jet order N in gauge parameters. Then the symmetry current $\mathscr{J}_u$ is decomposed into a sum

$$\mathscr{J}_u = \sum_{1<|\Lambda|\leq N} J_a^\Lambda\chi_\Lambda^a + J_a^t\chi_t^a + J_a\chi^a. \tag{4.51}$$

The first variational formula (4.30) takes a form

$$0 = \left[\sum_{|\Lambda|=1}^{N} u_a^{i\Lambda}\chi_\Lambda^a + u_a^i\chi^a\right]\mathscr{E}_i - d_t\left(\sum_{|\Lambda|=1}^{N} J_a^\Lambda\chi_\Lambda^a + J_a\chi^a\right).$$

It falls into a set of equalities for each $\chi_{t\Lambda}^a$, χ_Λ^a, $|\Lambda| = 1, \ldots, N$, and χ^a as follows:

$$0 = J_a^\Lambda, \qquad |\Lambda| = N, \tag{4.52}$$

$$0 = -u_a^{it\Lambda}\mathscr{E}_i + J_a^\Lambda + d_t J_a^{t\Lambda}, \qquad 1 \leq |\Lambda| < N, \tag{4.53}$$

$$0 = -u_a^{it}\mathscr{E}_i + J_a + d_t J_a^t, \tag{4.54}$$

$$0 = -u_a^i\mathscr{E}_i + d_t J_a. \tag{4.55}$$

With the equalities (4.52)–(4.54), the decomposition (4.51) takes a form

$$\mathscr{J}_u = \sum_{1<|\Lambda|<N} [(u_a^{it\Lambda}\mathscr{E}_i - d_t J_a^{t\Lambda}]\chi^a_\Lambda + (u_a^{itt}\mathscr{E}_i - d_t J_a^{tt})\chi^a_t + (u_a^{it}\mathscr{E}_i + -d_t J_a^t)\chi^a.$$

A direct computation leads to the expression

$$\mathscr{J}_u = \left(\sum_{1\leq|\Lambda|<N} u_a^{it\Lambda}\chi^a_\Lambda + u_a^{it}\chi^a\right)\mathscr{E}_i - \left(\sum_{1\leq|\Lambda|<N} d_t J_a^{t\Lambda}\chi^a_\Lambda + d_t J_a^t\chi^a\right). \tag{4.56}$$

The first summand of this expression vanishes on-shell. Its second one contains the terms $d_t J_a^\Lambda$, $|\Lambda| = 1, \ldots, N$. By virtue of the equalities (4.53), every $d_t J_a^\Lambda$, $|\Lambda| < N$, is expressed in the terms vanishing on-shell and the term $d_t d_t J_a^{t\Lambda}$. Iterating the procedure and bearing in mind the equality (4.52), one can easily show that the second summand of the expression (4.56) also vanishes on-shell. Thus, a symmetry current $\mathscr{J}_u$ vanishes on-shell.

Let us note that the statement of Theorem 4.7 is a particular case of Noether's third Theorem 2.8 that symmetry currents of gauge symmetries in Lagrangian theory are reduced to a superpotential because the superpotential U (2.26) equals zero on $X = \mathbb{R}$.

4.5 Non-autonomous Hamiltonian Mechanics

As was mentioned above, a Hamiltonian formulation of non-autonomous nonrelativistic mechanics is similar to covariant Hamiltonian field theory on fibre bundles (Sect. 3.5) in the particular case of fibre bundles over $\mathbb{R}$ [62, 121, 135, 136].

In accordance with the Legendre map (4.16) and the homogeneous Legendre map (4.19), a phase space and a homogeneous phase space of mechanics on a configuration bundle $Q \to \mathbb{R}$ are the vertical cotangent bundle V^*Q and the cotangent bundle T^*Q of Q, respectively.

A key point is that a non-autonomous Hamiltonian system of k degrees of freedom on a phase space V^*Q is equivalent both to some autonomous symplectic Hamiltonian system of $k+1$ degrees of freedom on a homogeneous phase space T^*Q (Theorem 4.8) and to a particular first order Lagrangian system with the characteristic Lagrangian (4.76) on a configuration space V^*Q.

Remark 4.10 It should be emphasized that this is not the most general case of a phase space of non-autonomous nonrelativistic mechanics which is defined as a fibred manifold $\Pi \to \mathbb{R}$ provided with a Poisson structure such that the corresponding symplectic foliation belongs to the fibration $\Pi \to \mathbb{R}$ [70]. Putting $\Pi = V^*Q$, we in fact restrict our consideration to Hamiltonian systems which admit the Lagrangian counterparts on a configuration space Q.

The cotangent bundle T^*Q is endowed with holonomic coordinates (t, q^i, p_0, p_i), possessing the transition functions (4.20). It admits the canonical Liouville form (4.101):

$$\Xi_T = p_0 dt + p_i dq^i, \tag{4.57}$$

the canonical symplectic form (4.102):

$$\Omega_T = d\Xi_T = dp_0 \wedge dt + dp_i \wedge dq^i, \tag{4.58}$$

and the corresponding canonical Poisson bracket (4.103):

$$\{f, g\}_T = \partial^0 f \partial_t g - \partial^0 g \partial_t f + \partial^i f \partial_i g - \partial^i g \partial_i f, \quad f, g \in C^\infty(T^*Q). \tag{4.59}$$

There is the canonical one-dimensional affine bundle (4.21):

$$\zeta : T^*Q \to V^*Q. \tag{4.60}$$

A glance at the transformation law (4.20) shows that it is a trivial affine bundle. Indeed, given a global section h of ζ, one can equip T^*Q with a global fibre coordinate

$$I_0 = p_0 - h, \qquad I_0 \circ h = 0,$$

possessing the identity transition functions. With respect to coordinates

$$(t, q^i, I_0, p_i), \qquad i = 1, \dots, m, \tag{4.61}$$

the fibration (4.60) reads

$$\zeta : \mathbb{R} \times V^*Q \ni (t, q^i, I_0, p_i) \to (t, q^i, p_i) \in V^*Q,$$

where (t, q^i, p_i) are holonomic coordinates on the vertical cotangent bundle V^*Q possessing transition functions (B.20).

Let us consider a subring of $C^\infty(T^*Q)$ which comprises the pull-back $\zeta^* f$ onto T^*Q of functions f on the vertical cotangent bundle V^*Q by the fibration ζ (4.60). This subring is closed under the Poisson bracket (4.59). Then by virtue of the well known theorem [62, 148], there exists a degenerate coinduced *Poisson bracket*

$$\{f, g\}_V = \partial^i f \partial_i g - \partial^i g \partial_i f, \qquad f, g \in C^\infty(V^*Q), \tag{4.62}$$

on a phase space V^*Q such that

$$\zeta^*\{f, g\}_V = \{\zeta^* f, \zeta^* g\}_T.$$

Holonomic coordinates on V^*Q are canonical for the Poisson structure (4.62).

With respect to the Poisson bracket (4.62), the *Hamiltonian vector fields* of functions on V^*Q read

$$\vartheta_f = \partial^i f \partial_i - \partial_i f \partial^i, \qquad f \in C^\infty(V^*Q), \tag{4.63}$$

$$\{f, f'\}_V = \vartheta_f \rfloor df', \qquad [\vartheta_f, \vartheta_{f'}] = \vartheta_{\{f,f'\}_V}. \tag{4.64}$$

They are vertical vector fields on $V^*Q \to \mathbb{R}$. Accordingly, the characteristic distribution of the Poisson structure (4.62) is the vertical tangent bundle $VV^*Q \subset TV^*Q$ of a fibre bundle $V^*Q \to \mathbb{R}$. The corresponding symplectic foliation on a phase space V^*Q coincides with the fibration $V^*Q \to \mathbb{R}$.

However, the Poisson structure (4.62) fails to provide any dynamic equation on a phase space $V^*Q \to \mathbb{R}$ because Hamiltonian vector fields (4.63) of functions on V^*Q are vertical vector fields. Hamiltonian dynamics on V^*Q is described as a particular Hamiltonian dynamics on fibre bundles [62, 121, 136].

A *Hamiltonian* on a phase space $V^*Q \to \mathbb{R}$ is defined as a global section

$$h : V^*Q \to T^*Q, \qquad p_0 \circ h = \mathcal{H}(t, q^j, p_j), \tag{4.65}$$

of the affine bundle ζ (4.60) (Definition 3.8). Given the Liouville form Ξ_T (4.57) on T^*Q, this section yields the pull-back Hamiltonian form

$$H = (-h)^* \Xi_T = p_k dq^k - \mathcal{H} dt \tag{4.66}$$

on V^*Q. This is the well-known *invariant of Poincaré–Cartan* [62].

It should be emphasized that, in contrast with a Hamiltonian in autonomous mechanics, the Hamiltonian $\mathcal{H}$ (4.65) is not a function on V^*Q, but it obeys the transformation law

$$\mathcal{H}'(t, q'^i, p'_i) = \mathcal{H}(t, q^i, p_i) + p'_i \partial_t q'^i.$$

Remark 4.11 Any connection Γ (4.8) on a configuration bundle $Q \to \mathbb{R}$ defines the global section $h_\Gamma = p_i \Gamma^i$ (4.65) of the affine bundle ζ (4.60) and the corresponding Hamiltonian form

$$H_\Gamma = p_k dq^k - \mathcal{H}_\Gamma dt = p_k dq^k - p_i \Gamma^i dt. \tag{4.67}$$

Furthermore, given a connection Γ, any Hamiltonian form (4.66) admits a splitting

$$H = H_\Gamma - \mathcal{E}_\Gamma dt, \tag{4.68}$$

where

$$\mathcal{E}_\Gamma = \mathcal{H} - \mathcal{H}_\Gamma = \mathcal{H} - p_i \Gamma^i \tag{4.69}$$

is a function on V^*Q. It is called the *Hamiltonian function* relative to a reference frame Γ. With respect to the coordinates adapted to a reference frame Γ, we have $\mathcal{E}_\Gamma = \mathcal{H}$. Given different reference frames Γ and Γ', the decomposition (4.68) leads at once to a relation

$$\mathcal{E}_{\Gamma'} = \mathcal{E}_\Gamma + \mathcal{H}_\Gamma - \mathcal{H}_{\Gamma'} = \mathcal{E}_\Gamma + (\Gamma^i - \Gamma'^i)p_i$$

(cf. (4.44)) between the Hamiltonian functions with respect to different reference frames.

Given the Hamiltonian form H (4.66), there exists a unique Hamiltonian connection

$$\gamma_H = \partial_t + \partial^k \mathcal{H} \partial_k - \partial_k \mathcal{H} \partial^k \tag{4.70}$$

on $V^*Q \to \mathbb{R}$ such that $\gamma_H \rfloor dH = 0$. It yields a first order dynamic *Hamilton equation*

$$q_t^k = \partial^k \mathcal{H}, \qquad p_{tk} = -\partial_k \mathcal{H} \tag{4.71}$$

(cf. (3.61)) on $V^*Q \to \mathbb{R}$, where $(t, q^k, p_k, q_t^k, p_{tk})$ are adapted coordinates on the first order jet manifold J^1V^*Q of $V^*Q \to \mathbb{R}$.

A classical solution of the Hamilton equation (4.71) is an integral section r for the connection γ_H (4.70).

We agree to call (V^*Q, H) the *Hamiltonian system* of $k = \dim Q - 1$ degrees of freedom.

In order to describe evolution of a Hamiltonian system at any instant, the Hamilton connection γ_H (4.70) is assumed to be complete, i.e., it is an Ehresmann connection (Remark 4.2). In this case, the Hamilton equation (4.71) admits a unique global classical solution through each point of a phase space V^*Q. By virtue of Theorem 4.2, there exists a trivialization of a fibre bundle $V^*Q \to \mathbb{R}$ (not necessarily compatible with its fibration $V^*Q \to Q$) such that

$$\gamma_H = \partial_t, \qquad H = \widetilde{p}_i d\widetilde{q}^i$$

with respect to the associated bundle coordinates $(t, \widetilde{q}^i, \widetilde{p}_i)$. A direct computation shows that the Hamilton vector field γ_H (4.70) is an infinitesimal generator of a one-parameter group of automorphisms of a Poisson manifold $(V^*Q, \{,\}_V)$. Then one can show that $(t, \widetilde{q}^i, \widetilde{p}_i)$ are canonical coordinates for the Poisson bracket $\{,\}_V$ [98]. Since $\mathcal{H} = 0$, the Hamilton equation (4.71) in these coordinates takes a form

$$\widetilde{q}_t^i = 0, \qquad \widetilde{p}_{ti} = 0,$$

i.e., $(t, \widetilde{q}^i, \widetilde{p}_i)$ are the *initial data coordinates*.

Remark 4.12 In applications, the condition of the Hamilton connection γ_H (4.70) to be complete need not holds on the entire phase space (Chap. 5). In this case, one consider its subsets, and sometimes we have different Hamiltonian systems on different subsets of V^*Q.

As was mentioned above, one can associate to any non-autonomous Hamiltonian system on a phase space V^*Q an equivalent autonomous symplectic Hamiltonian system on the cotangent bundle T^*Q as follows (Theorem 4.8).

Given a Hamiltonian system (V^*Q, H), its Hamiltonian $\mathcal{H}$ (4.65) defines a function

$$\mathcal{H}^* = \partial_t \rfloor (\Xi_T - \zeta^*(-h)^*\Xi_T)) = p_0 + h = p_0 + \mathcal{H} \tag{4.72}$$

on T^*Q. Let us regard $\mathcal{H}^*$ (4.72) as a Hamiltonian of an autonomous Hamiltonian system on a symplectic manifold (T^*Q, Ω_T). The corresponding autonomous Hamilton equation on T^*Q takes a form

$$\dot{t} = 1, \qquad \dot{p}_0 = -\partial_t \mathcal{H}, \qquad \dot{q}^i = \partial^i \mathcal{H}, \qquad \dot{p}_i = -\partial_i \mathcal{H}. \tag{4.73}$$

Remark 4.13 Let us note that the splitting $\mathcal{H}^* = p_0 + \mathcal{H}$ (4.72) is ill defined. At the same time, any reference frame Γ yields the decomposition

$$\mathcal{H}^* = (p_0 + \mathcal{H}_\Gamma) + (\mathcal{H} - \mathcal{H}_\Gamma) = \mathcal{H}^*_\Gamma + \mathcal{E}_\Gamma,$$

where $\mathcal{H}_\Gamma$ is the Hamiltonian (4.67) and $\mathcal{E}_\Gamma$ (4.69) is the Hamiltonian function relative to a reference frame Γ.

A Hamiltonian vector field $\vartheta_{\mathcal{H}^*}$ of the function $\mathcal{H}^*$ (4.72) on T^*Q is

$$\vartheta_{\mathcal{H}^*} = \partial_t - \partial_t \mathcal{H} \partial^0 + \partial^i \mathcal{H} \partial_i - \partial_i \mathcal{H} \partial^i, \qquad \vartheta_{\mathcal{H}^*} \rfloor \Omega_T = -d\mathcal{H}^*.$$

Written relative to the coordinates (4.61), this vector field reads

$$\vartheta_{\mathcal{H}^*} = \partial_t + \partial^i \mathcal{H} \partial_i - \partial_i \mathcal{H} \partial^i. \tag{4.74}$$

It is identically projected onto the Hamiltonian connection γ_H (4.70) on V^*Q such that

$$\zeta^*(\mathbf{L}_{\gamma_H} f) = \{\mathcal{H}^*, \zeta^* f\}_T, \qquad f \in C^\infty(V^*Q). \tag{4.75}$$

Therefore, the Hamilton equation (4.71) is equivalent to the autonomous Hamilton equation (4.73).

Obviously, the Hamiltonian vector field $\vartheta_{\mathcal{H}^*}$ (4.74) is complete if the Hamilton vector field γ_H (4.70) is so.

Thus, the following has been proved [32, 62, 99].

Theorem 4.8 *A non-autonomous Hamiltonian system (V^*Q, H) of k degrees of freedom is equivalent to an autonomous Hamiltonian system $(T^*Q, \mathcal{H}^*)$ of $k+1$ degrees of freedom on a symplectic manifold (T^*Q, Ω_T) whose Hamiltonian is the function $\mathcal{H}^*$ (4.72).*

We agree to call $(T^*Q, \mathcal{H}^*)$ the *homogeneous Hamiltonian system* and $\mathcal{H}^*$ (4.72) the *homogeneous Hamiltonian*.

It is readily observed that the Hamiltonian form H (4.66) also is the Poincaré–Cartan form (4.18) of the characteristic Lagrangian

$$L_H = h_0(H) = (p_i q_t^i - \mathcal{H})dt \tag{4.76}$$

on the jet manifold J^1V^*Q of $V^*Q \to \mathbb{R}$.

Remark 4.14 In fact, the Lagrangian (4.76) is the pull-back onto J^1V^*Q of an exterior form L_H on a product $V^*Q \times_Q J^1Q$.

The Lagrange operator (4.13) associated to the characteristic Lagrangian L_H (4.76) reads

$$\mathcal{E}_H = \delta L_H = [(q_t^i - \partial^i \mathcal{H})dp_i - (p_{ti} + \partial_i \mathcal{H})dq^i] \wedge dt.$$

The corresponding Lagrange equation (4.15) is of first order, and it coincides with the Hamilton equation (4.71) on V^*Q.

Due to this fact, Hamiltonian mechanics as like as covariant Hamiltonian field theory (Sect. 3.5) can be formulated as a specific Lagrangian mechanics on a configuration space V^*Q.

In particular, let

$$u = u^t\partial_t + u^i(t, q^j)\partial_i, \qquad u^t = 0, 1, \tag{4.77}$$

be a vector field on a configuration space Q. Its functorial lift (B.30) onto the cotangent bundle T^*Q is

$$\widetilde{u} = u^t\partial_t + u^i\partial_i - p_j\partial_i u^j \partial^i. \tag{4.78}$$

This vector field is identically projected onto a vector field, also given by the expression (4.78), on a phase space V^*Q as a base of the trivial fibre bundle (4.60). Then we have the equality

$$\mathbf{L}_{\widetilde{u}} H = \mathbf{L}_{J^1\widetilde{u}} L_H = (-u^t\partial_t\mathcal{H} + p_i\partial_t u^i - u^i\partial_i\mathcal{H} + p_i\partial_j u^i \partial^j \mathcal{H})dt$$

for any Hamiltonian form H (4.66). This equality enables us to study conservation laws in Hamiltonian mechanics similarly to those in Lagrangian mechanics (Sect. 4.6).

Lagrangian and Hamiltonian formulations of mechanics as like as those of field theory fail to be equivalent, unless Lagrangians are hyperregular (Definition 3.1). The comprehensive relations between Lagrangian and Hamiltonian systems can be established in the case of almost regular Lagrangians [62, 99, 121]. This is a particular case of the relations between Lagrangian and covariant Hamiltonian theories on fibre bundles (Sect. 3.6).

If the first order Lagrangian L (4.12) is hyperregular, it admits a unique weakly associated Hamiltonian form

$$H = p_i dq^i - (p_i \widehat{L}^{-1i} - \mathscr{L}(t, q^j, \widehat{L}^{-1j}))dt \tag{4.79}$$

(cf. (3.65)) which also is L-associated. Let s be a classical solution of the Lagrange equation (4.15) for a Lagrangian L. A direct computation shows that $\widehat{L} \circ J^1 s$ is a classical solution of the Hamilton equation (4.71) for the Hamiltonian form H (4.79). Conversely, if r is a classical solution of the Hamilton equation (4.71) for the Hamiltonian form H (4.79), then $s = \pi_\Pi \circ r$ is a solution of the Lagrange equation (4.15) for L.

In the case of a regular Lagrangian L, the Lagrangian constraint space N_L (4.17) is an open subbundle of the vertical cotangent bundle $V^*Q \to Q$. If $N_L \neq V^*Q$, a weakly associated Hamiltonian form fails to be defined everywhere on V^*Q in general.

Let us restrict our consideration to almost regular Lagrangians L (Definition 3.1). Then the following are reformulations of Theorems 3.9 and 3.10.

Theorem 4.9 *Let a section r of $V^*Q \to \mathbb{R}$ be a classical solution of the Hamilton equation (4.71) for a Hamiltonian form H weakly associated to an almost regular Lagrangian L. If r lives in the Lagrangian constraint space N_L, a section $s = \pi \circ r$ of $\pi : Q \to \mathbb{R}$ satisfies the Lagrange equation (4.15), while $\overline{s} = \widehat{H} \circ r$, where*

$$\widehat{H} : V^*Q \underset{Q}{\longrightarrow} J^1Q, \qquad q^i_t \circ \widehat{H} = \partial^i \mathscr{H}$$

is the Hamiltonian map (3.62), obeys the Cartan equation (4.23).

Theorem 4.10 *Given an almost regular Lagrangian L, let a section $\overline{s}$ of the jet bundle $J^1Q \to \mathbb{R}$ be a solution of the Cartan equation (4.23). Let H be a Hamiltonian form weakly associated to L, and let H satisfy a relation*

$$\widehat{H} \circ \widehat{L} \circ \overline{s} = J^1 s, \tag{4.80}$$

where s is the projection of $\overline{s}$ onto Q. Then a section $r = \widehat{L} \circ \overline{s}$ of a fibre bundle $V^*Q \to \mathbb{R}$ is a classical solution of the Hamilton equation (4.71) for H.

A set of Hamiltonian forms H weakly associated to an almost regular Lagrangian L is said to be complete if, for each classical solution s of a Lagrange equation, there exists a classical solution r of a Hamilton equation for a Hamiltonian form H from this set such that $s = \pi_\Pi \circ r$ (Definition 3.13). By virtue of Theorem 4.10, a set of weakly associated Hamiltonian forms is complete if, for every classical solution s of a Lagrange equation for L, there exists a Hamiltonian form H from this set which fulfills the relation (4.80) where $\overline{s} = J^1 s$, i.e.,

$$\widehat{H} \circ \widehat{L} \circ J^1 s = J^1 s. \tag{4.81}$$

4.6 Hamiltonian Conservation Laws: Noether's Inverse First Theorem

As was mentioned above, integrals of motion in Lagrangian mechanics can come from Lagrangian symmetries (Theorem 4.4), but not any integral of motion is of this type. In Hamiltonian mechanics, all integrals of motion are conserved symmetry currents (Theorem 4.15). One can think of this fact as being Noether's inverse first theorem.

Definition 4.6 An *integral of motion* of a Hamiltonian system (V^*Q, H) is defined as a smooth real function Φ on V^*Q which is an integral of motion of the Hamilton equation (4.71) in accordance with Definition 4.3, i.e., it satisfies the relation (4.26).

Since the Hamilton equation (4.71) is the kernel of the covariant differential D_{γ_H}, this relation $d_t\Phi \approx 0$ is equivalent to the equality

$$\mathbf{L}_{\gamma_H}\Phi = (\partial_t + \gamma^i_H\partial_i + \gamma_{Hi}\partial^i)\Phi = \partial_t\Phi + \{\mathscr{H}, \Phi\}_V = 0, \tag{4.82}$$

i.e., the Lie derivative of Φ along the Hamilton connection γ_H (4.70) vanishes.

At the same time, it follows from Theorem 4.3 that a vector field υ on V^*Q is a symmetry of the Hamilton equation (4.71) in accordance with Definition 4.4 if and only if $[\gamma_H, \upsilon] = 0$. Given the Hamiltonian vector field ϑ_Φ (4.63) of Φ with respect to the Poisson bracket (4.62), it is easily justified that

$$[\gamma_H, \vartheta_\Phi] = \vartheta_{\mathbf{L}_{\gamma_H}\Phi}.$$

Thus, we can conclude the following.

Theorem 4.11 *The Hamiltonian vector field of an integral of motion is a symmetry of the Hamilton equation (4.71).*

Given a Hamiltonian system (V^*Q, H), let $(T^*Q, \mathscr{H}^*)$ be an equivalent homogeneous Hamiltonian system. It follows from the equality (4.75) that

$$\zeta^*(\mathbf{L}_{\gamma_H}\Phi) = \{\mathscr{H}^*, \zeta^*\Phi\}_T = \zeta^*(\partial_t\Phi + \{\mathscr{H}, \Phi\}_V) \tag{4.83}$$

for any function $\Phi \in C^\infty(V^*Q)$. This formula is equivalent to that (4.82).

Theorem 4.12 *A function $\Phi \in C^\infty(V^*Q)$ is an integral of motion of a Hamiltonian system (V^*Q, H) if and only if its pull-back $\zeta^*\Phi$ onto T^*Q is an integral of motion of a homogeneous Hamiltonian system $(T^*Q, \mathscr{H}^*)$.*

Proof The result follows from the equality (4.83):

$$\{\mathscr{H}^*, \zeta^*\Phi\}_T = \zeta^*(\mathbf{L}_{\gamma_H}\Phi) = 0. \tag{4.84}$$

Theorem 4.13 *If Φ and Φ' are integrals of motion of a Hamiltonian system, their Poisson bracket $\{\Phi, \Phi'\}_V$ also is an integral of motion.*

Proof This fact results from the equalities (4.64) and (4.84).

Consequently, integrals of motion of a Hamiltonian system (V^*Q, H) constitute a real Lie subalgebra of a Poisson algebra $C^\infty(V^*Q)$.

Let us turn to Hamiltonian conservation laws. We are based on the fact that the Hamilton equation (4.71) also is a Lagrange equation of the characteristic Lagrangian L_H (4.76). Therefore, by analogy with field theory (Sect. 3.7), one can study conservation laws in Hamiltonian mechanics on a phase space V^*Q similarly to those in Lagrangian mechanics on a configuration space V^*Q [62, 101].

Since the Hamilton equation (4.71) is of first order, we restrict our consideration to classical symmetries, i.e., vector fields on V^*Q.

Definition 4.7 A vector field on a phase space V^*Q of a Hamiltonian system (V^*Q, H) is said to be its *Hamiltonian symmetry* if it is a Lagrangian symmetry of the characteristic Lagrangian L_H.

Let

$$\upsilon = \upsilon^t \partial_t + \upsilon^i \partial_i + \upsilon_i \partial^i, \qquad \upsilon^t = 0, 1, \tag{4.85}$$

be a vector field on a phase space V^*Q. Its prolongation onto $V^*Q \times_Q J^1Q$ (Remark 4.14) reads

$$J^1\upsilon = \upsilon^t \partial_t + \upsilon^i \partial_i + \upsilon_i \partial^i + d_t \upsilon^i \partial_i^t.$$

Then the first variational formula (4.30) for the characteristic Lagrangian L_H (4.76) takes a form

$$\begin{aligned} -\upsilon^t \partial_t \mathcal{H} - \upsilon^i \partial_i \mathcal{H} + \upsilon_i (q_t^i - \partial^i \mathcal{H}) + p_i d_t \upsilon^i = \\ (q_t^i \upsilon^t - \upsilon^i)(p_{ti} + \partial_i \mathcal{H}) + (\upsilon_i - p_{ti}\upsilon^t)(q_t^i - \partial^i \mathcal{H}) + d_t(p_i \upsilon^i - \upsilon^t \mathcal{H}). \end{aligned} \tag{4.86}$$

If υ (4.85) is a symmetry of L_H, i.e.,

$$\mathbf{L}_{J^1\upsilon} L_H = d_t \sigma dt,$$

we obtain the weak *Hamiltonian conservation law* $0 \approx -d_t \mathcal{J}$ (4.31) of the *Hamiltonian symmetry current* (4.32):

$$\mathcal{J}_\upsilon = -p_i \upsilon^i + \upsilon^t \mathcal{H} + \sigma. \tag{4.87}$$

In accordance with item (iv) of Theorem 2.4, the vector field υ (4.85) on V^*Q is a symmetry of the characteristic Lagrangian L_H (4.76) if and only if

$$\upsilon^i (p_{ti} + \partial_i \mathcal{H}) - \upsilon_i (q_t^i - \partial^i \mathcal{H}) + \upsilon^t \partial_t \mathcal{H} = d_t(-\mathcal{J}_\upsilon + \upsilon^t \mathcal{H}). \tag{4.88}$$

A glance at this equality shows the following.

Theorem 4.14 *The vector field* υ *(4.85) is a Hamiltonian symmetry in accordance with Definition 4.7 only if*

$$\partial^i \upsilon_i = -\partial_i \upsilon^i. \tag{4.89}$$

Remark 4.15 It is readily observed that the Hamiltonian connection γ_H (4.70) is a symmetry of the characteristic Lagrangian L_H whose conserved Hamiltonian symmetry current (4.87) equals zero. It follows that, given a nonvertical Hamiltonian symmetry υ, $\upsilon^t = 1$, there exists a vertical Hamiltonian symmetry $\upsilon - \gamma_H$ with the same conserved Hamiltonian symmetry current as υ.

In view of Remark 2.4, any Hamiltonian symmetry, being classical symmetry of the characteristic Lagrangian L_H (4.76), also is symmetry of the Hamilton equation (4.71). In accordance with Theorem 4.4, the corresponding conserved Hamiltonian symmetry current (4.87) is an integral of motion of a Hamiltonian system which, thus, comes from its Hamiltonian symmetry.

The converse also is true.

Theorem 4.15 *Any integral of motion* Φ *of a Hamiltonian system* (V^*Q, H) *is the conserved Hamiltonian symmetry current* $\mathcal{J}_{-\vartheta_\Phi}$ *(4.87) along the Hamiltonian vector field* $-\vartheta_\Phi$ *(4.63) of* $-\Phi$.

Proof It follows from the relations (4.82) and (4.86) that

$$\mathbf{L}_{-J^1\vartheta_\Phi} = d_t(\Phi - p_i \partial^i \Phi).$$

Then the equality (4.87) results in a desired relation $\Phi = \mathcal{J}_{-\vartheta_\Phi}$.

This assertion can be regarded as above mentioned *Noether's inverse first theorem*.

For instance, if the Hamiltonian symmetry υ (4.85) is projectable onto Q (i.e., its components $\upsilon^i = u^i$ are independent of momenta p_i), then we $\upsilon_i = -p_j \partial_i u^j$ in accordance with the equality (4.89). Consequently, υ is the canonical lift $\widetilde{u}$ (4.78) onto V^*Q of the vector field u (4.77) on Q. If $\widetilde{u}$ is a symmetry of the characteristic Lagrangian L_H, it follows at once from the equality (4.88) that $\widetilde{u}$ is an exact symmetry of L_H. The corresponding conserved *Hamiltonian symmetry current* (4.87) reads

$$\widetilde{\mathcal{J}}_u = \mathcal{J}_{\widetilde{u}} = -p_i u^i + u^t \mathcal{H} \tag{4.90}$$

(cf. (3.87)).

Definition 4.8 The vector field u (4.77) on a configuration space Q is said to be the *basic Hamiltonian symmetry* if its canonical lift $\widetilde{u}$ (4.78) onto V^*Q is a Hamiltonian symmetry.

If a basic Hamiltonian symmetry u is vertical, the corresponding conserved Hamiltonian symmetry current (4.90):

$$\widetilde{\mathcal{J}}_u = -p_i u^i, \tag{4.91}$$

is a Noether current.

Now let Γ be the connection (4.8) on Q. The corresponding symmetry current (4.90) is the Hamiltonian function (4.69):

$$\widetilde{\mathcal{J}}_\Gamma = \mathcal{J}_{\widetilde{\Gamma}} = \mathcal{E}_\Gamma = \mathcal{H} - p_i \Gamma^i, \tag{4.92}$$

relative to a reference frame Γ. Given bundle coordinates adapted to Γ, we obtain the Lie derivative

$$\mathbf{L}_{J^1\widetilde{\Gamma}} L_H = -\partial_t \mathcal{H}.$$

It follows that a connection Γ is a basic Hamiltonian symmetry if and only if the Hamiltonian $\mathcal{H}$ (4.65), written with respect to the coordinates adapted to Γ, is time-independent. In this case, the Hamiltonian function (4.92) is an integral of motion of a Hamiltonian system.

As a consequence of Theorem 3.17, there is the following relation between Lagrangian symmetries and basic Hamiltonian symmetries if they are the same vector fields on a configuration space Q.

Theorem 4.16 *Let a Hamiltonian form H be associated to an almost regular Lagrangian L. Let r be a solution of the Hamilton equation (4.71) for H which lives in the Lagrangian constraint space N_L (4.17). Let $s = \pi_\Pi \circ r$ be the corresponding solution of a Lagrange equation for L so that the relation (4.81) holds. Then, for any vector field u (4.77) on a fibre bundle $Q \to \mathbb{R}$, we have*

$$\widetilde{\mathcal{J}}_u(r) = \mathcal{J}_u(\pi_\Pi \circ r), \qquad \widetilde{\mathcal{J}}_u(\widehat{L} \circ J^1 s) = \mathcal{J}_u(s), \tag{4.93}$$

where $\mathcal{J}_u$ is the symmetry current (4.35) on J^1Y and $\widetilde{\mathcal{J}}_u = \mathcal{J}_{\widetilde{u}}$ is the symmetry current (4.90) on V^*Q.

By virtue of Theorems 4.9–4.10, it follows that:

- if $\mathcal{J}_u$ in Theorem 4.16 is a conserved symmetry current, then the symmetry current $\widetilde{\mathcal{J}}_u$ (4.93) is conserved on solutions of a Hamilton equation which live in the Lagrangian constraint space;
- if $\widetilde{\mathcal{J}}_u$ in Theorem 4.16 is a conserved symmetry current, then the symmetry current $\mathcal{J}_u$ (4.93) is conserved on solutions s of a Lagrange equation which obey the condition (4.81).

In particular, let $u = \Gamma$ be a connection and E_Γ the energy function (4.41). Then the relations (4.93):

$$\mathscr{E}_\Gamma(r) = \widetilde{\mathscr{J}_\Gamma}(r) = \mathscr{J}_\Gamma(\pi_\Pi \circ r) = E_\Gamma(\pi_\Pi \circ r),$$
$$\mathscr{E}_\Gamma(\widehat{L} \circ J^1 s) = \widetilde{\mathscr{J}_\Gamma}(\widehat{L} \circ J^1 s) = \mathscr{J}_\Gamma(s) = E_\Gamma(s),$$

show that the Hamiltonian function $\mathscr{E}_\Gamma$ (4.92) can be treated as a *Hamiltonian energy function* relative to a reference frame Γ.

Remark 4.16 There exist Hamiltonian symmetries which are not basic in accordance with Definition 4.8. Therefore, they fail to be associated to Lagrangian symmetries. For instance, these are Hamiltonian symmetries (5.11) whose symmetry currents are components (5.9) of the Rung–Lenz vector in the Hamiltonian Kepler problem (Chap. 5).

4.7 Completely Integrable Hamiltonian Systems

It may happen that symmetries and the corresponding integrals of motion define a Hamiltonian system in full. This is the case of commutative and noncommutative completely integrable systems (henceforth, CISs) (Definition 4.9).

In view of Remark 4.15, we can restrict our consideration to vertical symmetries υ (4.85) where $\upsilon^t = 0$.

Definition 4.9 A non-autonomous Hamiltonian system (V^*Q, H) of $n = \dim Q - 1$ degrees of freedom is said to be *completely integrable* if it admits $n \leq k < 2n$ vertical classical symmetries υ_α which obey the following conditions.

- (i) Symmetries υ_α everywhere are linearly independent.
- (ii) They form a k-dimensional real Lie algebra $\mathfrak{g}$ of corank $m = 2n - k$ with commutation relations

$$[\upsilon_\alpha, \upsilon_\beta] = c^\nu_{\alpha\beta}\upsilon_\nu. \tag{4.94}$$

If $k = n$, then a Lie algebra $\mathfrak{g}$ is commutative, and we are in the case of a *commutative CIS*. If $n < k$, the Lie algebra (4.94) is noncommutative, and a CIS is called *noncommutative* or *superintegrable*.

The conditions of Definition 4.9 can be reformulated in terms of integrals of motion $\Phi_\alpha = -\mathscr{J}_{\upsilon_\alpha}$ corresponding to symmetries υ_α. By virtue of Noether's inverse first Theorem 4.15, $\upsilon_\alpha = \vartheta_{\Phi_\alpha}$ are the Hamiltonian vector fields (4.63) of integrals of motion Φ_α. In accordance with the relation (4.64), integrals of motion obey the commutation relations

$$\{\Phi_\alpha, \Phi_\beta\}_V = c^\nu_{\alpha\beta}\Phi_\nu. \tag{4.95}$$

Then we come to an equivalent definition of a CISs [62, 132, 136].

Definition 4.10 A non-autonomous Hamiltonian system (V^*Q, H) of $n = \dim Q - 1$ degrees of freedom is a CIS if it possesses $n \leq k < 2n$ integrals of motion $\Phi_1, \ldots, \Phi_k$, obeying the following conditions.

(i) All the functions Φ_α are independent, i.e., a k-form $d\Phi_1 \wedge \cdots \wedge d\Phi_k$ nowhere vanishes on V^*Q. It follows that a map

$$\Phi : V^*Q \to N = (\Phi_1(V^*Q), \ldots, \Phi_k(V^*Q)) \subset \mathbb{R}^k \tag{4.96}$$

is a fibred manifold (Definition B.1) over a connected open subset $N \subset \mathbb{R}^k$.

(ii) The commutation relations (4.95) are satisfied.

Given a non-autonomous CIS in accordance with Definition 4.10, the equivalent autonomous Hamiltonian system on a homogeneous phase space T^*Q (Theorem 4.8) possesses $k + 1$ integrals of motion

$$(\mathcal{H}^*, \zeta^*\Phi_1, \ldots, \zeta^*\Phi_k) \tag{4.97}$$

with the following properties (Theorem 4.12).

(i) The integrals of motion (4.97) are mutually independent, and a map

$$\begin{aligned}\widetilde{\Phi} : T^*Q \to (\mathcal{H}^*(T^*Q), \zeta^*\Phi_1(T^*Q), \ldots, \zeta^*\Phi_k(T^*Q)) = \\ (I_0, \Phi_1(V^*Q), \ldots, \Phi_k(V^*Q)) = \mathbb{R} \times N = N'\end{aligned} \tag{4.98}$$

is a fibred manifold.

(ii) The integrals of motion (4.97) obey the commutation relations

$$\{\zeta^*\Phi_\alpha, \zeta^*\Phi_\beta\} = c^\nu_{\alpha\beta}\zeta^*\Phi_\nu, \qquad \{\mathcal{H}^*, \zeta^*\Phi_\alpha\} = 0.$$

They generate a real $(k + 1)$ dimensional Lie algebra of corank $2n + 1 - k$.

As a result, integrals of motion (4.97) form an autonomous CIS on a symplectic manifold (T^*Q, Ω_T) in accordance with Definition 4.11. In order to describe it, one then can follow the Mishchenko–Fomenko theorem [18, 38, 106] extended to the case of noncompact invariant submanifolds [43, 129, 136].

Therefore, we turn to CISs (superintegrable systems) on a symplectic manifold.

Remark 4.17 Let Z be a manifold. Any exterior two-form Ω on Z yields a linear bundle morphism

$$\Omega^\flat : TZ \underset{Z}{\to} T^*Z, \qquad \Omega^\flat : v \to -v\rfloor\Omega(z), \quad v \in T_zZ, \quad z \in Z. \tag{4.99}$$

One says that a two-form Ω is nondegenerate if $\mathrm{Ker}\,\Omega^\flat = 0$. A closed nondegenerate form is called the *symplectic form*. Accordingly, a manifold Z equipped with a symplectic form is said to be the *symplectic manifold*. A symplectic manifold necessarily is even-dimensional. A closed two-form on Z is called *presymplectic* if it is not necessary degenerate. A vector field u on a symplectic manifold (Z, Ω) is said to be *Hamiltonian* if a one-form $u\rfloor\Omega$ is exact. Any smooth function $f \in C^\infty(Z)$ on

Z defines a unique Hamiltonian vector field ϑ_f, called the *Hamiltonian vector field of a function* f, such that

$$\vartheta_f \rfloor \Omega = -df, \qquad \vartheta_f = \Omega^\sharp(df), \tag{4.100}$$

where $\Omega^\sharp$ is the inverse isomorphism to $\Omega^\flat$ (4.99). Given an m-dimensional manifold M coordinated by (q^i), let T^*M be its cotangent bundle equipped with the holonomic coordinates $(q^i, \dot{q}_i)$. It is endowed with the *canonical Liouville form*

$$\Xi_T = \dot{q}_i dq^i \tag{4.101}$$

and the *canonical symplectic form*

$$\Omega_T = d\Xi_T = d\dot{q}_i \wedge dq^i. \tag{4.102}$$

The Hamiltonian vector field ϑ_f (4.100) with respect to the canonical symplectic form (4.102) reads

$$\vartheta_f = \partial^i f \partial_i - \partial_i f \partial^i.$$

A symplectic form Ω on a manifold Z defines a Poisson bracket

$$\{f, g\} = \vartheta_g \rfloor \vartheta_f \rfloor \Omega, \qquad f, g \in C^\infty(Z).$$

The canonical symplectic form Ω_T (4.102) on T^*M yields the *canonical Poisson bracket*

$$\{f, g\}_T = \frac{\partial f}{\partial \dot{q}_i}\frac{\partial g}{\partial q^i} - \frac{\partial f}{\partial q^i}\frac{\partial g}{\partial \dot{q}_i}. \tag{4.103}$$

Definition 4.11 Let (Z, Ω) be a $2n$-dimensional connected symplectic manifold, and let $(C^\infty(Z), \{,\})$ be a Poisson algebra of smooth real functions on Z. A subset

$$F = (F_1, \dots, F_k), \qquad n \leq k < 2n, \tag{4.104}$$

of a Poisson algebra $C^\infty(Z)$ is called the *CIS* or the *superintegrable system* if the following conditions hold.

(i) All the functions F_i (called the *generating functions of a CIS*) are independent, i.e., a k-form $\overset{k}{\wedge} dF_i$ nowhere vanishes on Z. It follows that a map $F : Z \to \mathbb{R}^k$ is a submersion, i.e.,

$$F : Z \to N = F(Z) \tag{4.105}$$

is a fibred manifold over a domain $N \subset \mathbb{R}^k$ endowed with the coordinates (x_i) such that $x_i \circ F = F_i$.

(ii) There exist smooth real functions s_{ij} on N such that

$$\{F_i, F_j\} = s_{ij} \circ F, \qquad i, j = 1, \dots, k. \tag{4.106}$$

(iii) The $(k \times k)$-matrix function $\mathbf{s}$ with the entries s_{ij} (4.106) is of constant corank $m = 2n - k$ at all points of N.

Remark 4.18 If $k = n$, then $\mathbf{s} = 0$, and we are in the case of commutative CISs when F_1, ..., $F - n$ are independent functions in involution.

If $k > n$, the matrix $\mathbf{s}$ is necessarily nonzero. If $k = 2n - 1$, a CIS is called *maximally integrable*.

The following two assertions clarify a structure of CISs [38, 43, 136].

Theorem 4.17 *Given a symplectic manifold (Z, Ω), let $F : Z \to N$ be a fibred manifold such that, for any two functions f, f' constant on fibres of F, their Poisson bracket $\{f, f'\}$ is so. By virtue of the well known theorem [62, 148], N is provided with an unique coinduced Poisson structure $\{,\}_N$ such that F is a Poisson morphism.*

Since any function constant on fibres of F is the pull-back of some function on N, the CIS (4.104) satisfies the condition of Theorem 4.17 due to item (ii) of Definition 4.11. Thus, a base N of the fibration (4.105) is endowed with a coinduced Poisson structure of corank m. With respect to coordinates x_i in item (i) of Definition 4.11 its bivector field reads

$$w = s_{ij}(x_k)\partial^i \wedge \partial^j. \tag{4.107}$$

Theorem 4.18 *Given a fibred manifold $F : Z \to N$ in Theorem 4.17, the following conditions are equivalent [38]:*

(i) a rank of the coinduced Poisson structure $\{,\}_N$ on N equals $2\dim N - \dim Z$,
(ii) the fibres of F are isotropic,
(iii) the fibres of F are maximal integral manifolds of the involutive distribution spanned by the Hamiltonian vector fields of the pull-back F^*C of Casimir functions C of the coinduced Poisson structure (4.107) on N.

It is readily observed that the fibred manifold F (4.105) obeys condition (i) of Theorem 4.18 due to item (iii) of Definition 4.11, namely, $k - m = 2(k - n)$.

Fibres of the fibred manifold F (4.105) are called the *invariant submanifolds*.

Remark 4.19 In practice, condition (i) of Definition 4.11 fails to hold everywhere. It can be replaced with that a subset $Z_R \subset Z$ of regular points (where $\overset{k}{\wedge} dF_i \neq 0$) is open and dense. Let M be an invariant submanifold through a regular point $z \in Z_R \subset Z$. Then it is regular, i.e., $M \subset Z_R$. Let M admit a regular open *saturated neighborhood* U_M (i.e., a fibre of F through a point of U_M belongs to U_M). For instance, any compact invariant submanifold M has such a neighborhood U_M. The restriction of functions F_i to U_M defines a CIS on U_M which obeys Definition 4.11. In this case, one says that a CIS is considered around its invariant submanifold M.

Let (Z, Ω) be a $2n$-dimensional connected symplectic manifold. Given the CIS (F_i) (4.104) on (Z, Ω), the well known Mishchenko–Fomenko theorem (Theorem 4.20) states the existence of action-angle coordinates around its connected compact invariant submanifold [18, 38, 106]. This theorem has been extended to CISs with noncompact invariant submanifolds (Theorem 4.19) [43, 129, 136]. These submanifolds are diffeomorphic to a toroidal cylinder

$$\mathbb{R}^{m-r} \times T^r, \qquad m = 2n - k, \qquad 0 \leq r \leq m. \tag{4.108}$$

Theorem 4.19 *Let the Hamiltonian vector fields ϑ_i of the functions F_i be complete, and let the fibres of the fibred manifold F (4.105) be connected and mutually diffeomorphic. Then the following hold.*

(I) The fibres of F (4.105) are diffeomorphic to the toroidal cylinder (4.108).
(II) Given a fibre M of F (4.105), there exists its open saturated neighborhood U_M which is a trivial principal bundle

$$U_M = N_M \times \mathbb{R}^{m-r} \times T^r \xrightarrow{F} N_M \tag{4.109}$$

with the structure group (4.108).
(III) A neighborhood U_M is provided with the bundle action-angle coordinates $(I_\lambda, p_s, q^s, y^\lambda)$, $\lambda = 1, \ldots, m$, $s = 1, \ldots, n - m$, such that: (i) the angle coordinates (y^λ) are those on a toroidal cylinder, i.e., fibre coordinates on the fibre bundle (4.109), (ii) (I_λ, p_s, q^s) are coordinates on its base N_M where the action coordinates (I_λ) are values of Casimir functions of the coinduced Poisson structure $\{,\}_N$ on N_M, and (iii) a symplectic form Ω on U_M reads

$$\Omega = dI_\lambda \wedge dy^\lambda + dp_s \wedge dq^s.$$

Remark 4.20 The condition of the completeness of Hamiltonian vector fields of the generating functions F_i in Theorem 4.19 is rather restrictive. One can replace this condition with that the Hamiltonian vector fields of the pull-back onto Z of Casimir functions on N are complete.

If the conditions of Theorem 4.19 are replaced with that fibres of the fibred manifold F (4.105) are compact and connected, this theorem restarts the *Mishchenko–Fomenko theorem* as follows.

Theorem 4.20 *Let the fibres of the fibred manifold F (4.105) be connected and compact. Then they are diffeomorphic to a torus T^m, and statements (II)–(III) of Theorem 4.19 hold.*

Remark 4.21 In Theorem 4.20, the Hamiltonian vector fields υ_λ are complete because fibres of the fibred manifold F (4.105) are compact. As well known, any vector field on a compact manifold is complete.

To study a CIS, one conventionally considers it with respect to action-angle coordinates. A problem is that an action-angle coordinate chart on an open subbundle

U of the fibred manifold $Z \to N$ (4.105) in Theorem 4.19 is local. The following generalizes this theorem to the case of global action-angle coordinates.

Definition 4.12 The CIS F (4.104) on a symplectic manifold (Z, Ω) in Definition 4.11 is called *globally integrable* (or, shortly, *global*) if there exist *global action-angle coordinates*

$$(I_\lambda, x^A, y^\lambda), \qquad \lambda = 1, \dots, m, \qquad A = 1, \dots, 2(n-m), \tag{4.110}$$

such that: (i) the action coordinates (I_λ) are expressed in values of some Casimir functions C_λ on a Poisson manifold $(N, \{,\}_N)$, (ii) the angle coordinates (y^λ) are coordinates on the toroidal cylinder $\mathbb{R}^{m-r} \times T^r$, $0 \le r \le m$, and (iii) a symplectic form Ω on Z reads

$$\Omega = dI_\lambda \wedge dy^\lambda + \Omega_{AB}(I_\mu, x^C) dx^A \wedge dx^B.$$

It is readily observed that the action-angle coordinates on U in Theorem 4.19 are global on U in accordance with Definition 4.12.

Forthcoming Theorem 4.21 provides the sufficient conditions of the existence of global action-angle coordinates of a CIS on a symplectic manifold (Z, Ω) [62, 101, 129, 136]. It generalizes the well-known result for the case of compact invariant submanifolds [31, 34, 38].

Theorem 4.21 *A CIS F on a symplectic manifold (Z, Ω) is globally integrable if the following conditions hold.*

(i) Hamiltonian vector fields ϑ_i of the generating functions F_i are complete.
(ii) The fibred manifold F (4.105) is a fibre bundle with connected fibres.
(iii) Its base N is simply connected and the cohomology $H^2(N; \mathbb{Z})$ with coefficients in the constant sheaf $\mathbb{Z}$ is trivial.
(iv) The coinduced Poisson structure $\{,\}_N$ on a base N admits m independent Casimir functions C_λ.

Theorem 4.21 restarts Theorem 4.19 if one considers an open subset V of N admitting the Darboux coordinates x^A on symplectic leaves of U. If invariant submanifolds of a CIS are assumed to be compact, condition (i) of Theorem 4.21 is unnecessary since vector fields ϑ_λ on compact fibres of F are complete. Condition (ii) also holds by virtue of Theorem B.6. In this case, Theorem 4.21 reproduces the well known result in [31].

Furthermore, one can show that conditions (iii) of Theorem 4.21 guarantee that fibre bundles F in conditions (ii) of these theorems are trivial [136]. Therefore, Theorem 4.21 can be reformulated as follows.

Theorem 4.22 *A CIS F on a symplectic manifold (Z, Ω) is global if and only if the following conditions hold.*

(i) The fibred manifold F (4.105) is a trivial fibre bundle.

(ii) *The coinduced Poisson structure* $\{,\}_N$ *on a base* N *admits* m *independent Casimir functions* C_λ *such that Hamiltonian vector fields of their pull-back* F^*C_λ *are complete.*

Bearing in mind the autonomous CIS (4.97), let us turn to autonomous CISs whose generating functions are integrals of motion, i.e., they are in involution with a Hamiltonian $\mathcal{H}$, and the functions $(\mathcal{H}, F_1, \dots, F_k)$ are nowhere independent, i.e.,

$$\{\mathcal{H}, F_i\} = 0, \tag{4.111}$$

$$d\mathcal{H} \wedge (\overset{k}{\wedge} dF_i) = 0. \tag{4.112}$$

Let us note that, in accordance with item (ii) of Theorem 4.22 and 4.23 below, the Hamiltonian vector field of a Hamiltonian $\mathcal{H}$ of a CIS always is complete.

Theorem 4.23 *It follows from the equality (4.112) that a Hamiltonian* $\mathcal{H}$ *is constant on invariant submanifolds. Hence, it is the pull-back of a function on* N *which is a Casimir function of the Poisson structure (4.107) because of conditions (4.111).*

Theorem 4.23 leads to the following.

Theorem 4.24 *Let* $\mathcal{H}$ *be a Hamiltonian of an autonomous global CIS provided with the action-angle coordinates* $(I_\lambda, x^A, y^\lambda)$ *(4.110). Then a Hamiltonian* $\mathcal{H}$ *depends only on the action coordinates* I_λ*. Consequently, the Hamilton equation of a global CIS takes a form*

$$\dot{y}^\lambda = \frac{\partial \mathcal{H}}{\partial I_\lambda}, \qquad I_\lambda = \text{const.}, \qquad x^A = \text{const.}$$

Remark 4.22 Given a Hamiltonian $\mathcal{H}$ of a Hamiltonian system on a symplectic manifold Z, it may happen that we have different CISs on different open subsets of Z. For instance, this is the case of the global Kepler problem (Chap. 5).

Remark 4.23 Bearing in mind again the autonomous CIS (4.97), let us also consider CISs whose generating functions $\{F_1, \dots, F_k\}$ form a k-dimensional real Lie algebra $\mathfrak{g}$ of corank m with commutation relations

$$\{F_i, F_j\} = c_{ij}^h F_h, \qquad c_{ij}^h = \text{const.} \tag{4.113}$$

Then F (4.105) is a momentum mapping of Z to the Lie coalgebra $\mathfrak{g}^*$ provided with the coordinates x_i in item (i) of Definition 4.11 [58, 68]. In this case, the coinduced Poisson structure $\{,\}_N$ coincides with the canonical Lie–Poisson structure on $\mathfrak{g}^*$. Let V be an open subset of $\mathfrak{g}^*$ such that conditions (i) and (ii) of Theorem 4.22 are satisfied. Then an open subset $F^{-1}(V) \subset Z$ is provided with the action-angle coordinates. Let Hamiltonian vector fields ϑ_i of the generating functions F_i which form a Lie algebra $\mathfrak{g}$ be complete. Then they define a locally free Hamiltonian action on Z of some simply connected Lie group G whose Lie algebra is isomorphic to

$\mathfrak{g}$ [109]. Orbits of G coincide with k-dimensional maximal integral manifolds of the regular distribution $\mathcal{V}$ on Z spanned by Hamiltonian vector fields ϑ_i [142]. Furthermore, Casimir functions of the Lie–Poisson structure on $\mathfrak{g}^*$ are exactly the coadjoint invariant functions on $\mathfrak{g}^*$. They are constant on orbits of the coadjoint action of G on $\mathfrak{g}^*$ which coincide with leaves of the symplectic foliation of $\mathfrak{g}^*$.

Now, let us return to the autonomous CIS (4.97) on homogeneous phase space of non-autonomous mechanics. There is the commutative diagram

$$\begin{array}{ccc} T^*Q & \xrightarrow{\zeta} & V^*Q \\ {\scriptstyle \widetilde{\Phi}}\downarrow & & \downarrow{\scriptstyle \Phi} \\ N' & \xrightarrow{\xi} & N \end{array}$$

where ζ (3.16) and $\xi : N' = \mathbb{R} \times N \to N$ are trivial bundles. Thus, the fibred manifold (4.98) is the pull-back $\widetilde{\Phi} = \xi^*\Phi$ of the fibred manifold Φ (4.96) onto N'.

Let the conditions of Theorem 4.19 hold. If the Hamiltonian vector fields

$$(\gamma_H, \vartheta_{\Phi_1}, \ldots, \vartheta_{\Phi_k}), \qquad \vartheta_{\Phi_\alpha} = \partial^i\Phi_\alpha\partial_i - \partial_i\Phi_\alpha\partial^i,$$

of integrals of motion Φ_α on V^*Q are complete, the Hamiltonian vector fields

$$(u_{\mathcal{H}^*}, u_{\zeta^*\Phi_1}, \ldots, u_{\zeta^*\Phi_k}), \qquad u_{\zeta^*\Phi_\alpha} = \partial^i\Phi_\alpha\partial_i - \partial_i\Phi_\alpha\partial^i,$$

on T^*Q are complete. If fibres of the fibred manifold Φ (4.96) are connected and mutually diffeomorphic, the fibres of the fibred manifold $\widetilde{\Phi}$ (4.98) also are well.

Let M be a fibre of Φ (4.96) and $h(M)$ the corresponding fibre of $\widetilde{\Phi}$ (4.98). In accordance with Theorem 4.19, there exists an open neighborhood U' of $h(M)$ which is a trivial principal bundle with the structure group

$$\mathbb{R}^{1+m-r} \times T^r \tag{4.114}$$

whose bundle coordinates are the action-angle coordinates

$$(I_0, I_\lambda, t, y^\lambda, p_A, q^A), \qquad A = 1, \ldots, n-m, \qquad \lambda = 1, \ldots, k, \tag{4.115}$$

such that:

(i) (t, y^λ) are coordinates on the toroidal cylinder (4.114),
(ii) the symplectic form Ω_T on U' reads

$$\Omega_T = dI_0 \wedge dt + dI_\alpha \wedge dy^\alpha + dp_A \wedge dq^A,$$

(iii) the action coordinates (I_0, I_α) are expressed in values of the Casimir functions $C_0 = I_0$, C_α of the coinduced Poisson structure $w = \partial^A \wedge \partial_A$ on N',

(iv) a homogeneous Hamiltonian depends on action coordinates just as $\mathcal{H}^* = I_0$,
(v) the integrals of motion $\zeta^*\Phi_1, \ldots \zeta^*\Phi_k$ are independent of coordinates (t, y^λ).

Endowed with the action-angle coordinates (4.115), the above mentioned neighborhood U' is a trivial bundle $U' = \mathbb{R} \times U_M$ where $U_M = \zeta(U')$ is an open neighborhood of a fibre M of the fibre bundle Φ (4.96). Then we come to the following.

Theorem 4.25 *Let symmetries υ_α in Definition 4.9 be complete, and let fibres of the fibred manifold Φ (4.96) defined by the corresponding conserved integrals of motion be connected and mutually diffeomorphic. Then there exists an open neighborhood U_M of a fibre M of Φ (4.96) which is a trivial principal bundle with a structure group (4.114) whose bundle coordinates are the action-angle coordinates*

$$(p_A, q^A, I_\lambda, t, y^\lambda), \qquad A = 1, \ldots, k-n, \qquad \lambda = 1, \ldots, m,$$

such that:

(i) (t, y^λ) are coordinates on the toroidal cylinder (4.114),
(ii) the Poisson bracket $\{,\}_V$ on U_M reads

$$\{f, g\}_V = \partial^A f \partial_A g - \partial^A g \partial_A f + \partial^\lambda f \partial_\lambda g - \partial^\lambda g \partial_\lambda f,$$

(iii) a Hamiltonian $\mathcal{H}$ depends only on action coordinates I_λ,
(iv) the integrals of motion $\Phi_1, \ldots \Phi_k$ are independent of coordinates (t, y^λ).

Chapter 5
Global Kepler Problem

We provide a global analysis of the Kepler problem as an example of a mechanical system which is characterized by its symmetries in full. It falls into two distinct global CISs on different open subsets of a phase space. Their integrals of motion form the Lie algebras $so(3)$ and $so(2,1)$ with compact and noncompact invariant submanifolds, respectively [62, 129, 136].

Let us consider a mechanical system of a point mass in the presence of a central potential. Its configuration space is

$$Q = \mathbb{R} \times \mathbb{R}^3 \to \mathbb{R} \tag{5.1}$$

endowed with the Cartesian coordinates (t, q^i), $i = 1, 2, 3$.

A Lagrangian of this mechanical system reads

$$\mathscr{L} = \frac{1}{2}\sum_i (q^i_t)^2 - V(r), \qquad r^2 = \sum_i (q^i)^2. \tag{5.2}$$

The vertical vector fields

$$v^a_b = q^b \partial_a - q^a \partial_b \tag{5.3}$$

on Q (5.1) are infinitesimal generators of the natural representation of a group $SO(3)$ acting on $\mathbb{R}^3$. Their jet prolongations (4.28) read

$$J^1 v^a_b = q^b \partial_a - q^a \partial_b + q^b_t \partial^t_a - q^a_t \partial^t_b.$$

It is easily justified that the vector fields (5.3) are exact symmetries of the Lagrangian (5.2). In accordance with Noether's first theorem, the corresponding conserved Noether currents (4.39) are *orbital momenta*

$$M^a_b = \mathscr{J}_{v^a_b} = (q^a \pi_b - q^b \pi_a) = q^a q^b_t - q^b q^a_t. \tag{5.4}$$

G. Sardanashvily, *Noether's Theorems*, Atlantis Studies
in Variational Geometry 3, DOI 10.2991/978-94-6239-171-0_5

They are integrals of motion, which however fail to be independent.

Let us consider the Lagrangian system (5.2) where

$$V(r) = -r^{-1} \tag{5.5}$$

is a *Kepler potential*. This Lagrangian system admits additional integrals of motion

$$A^a = \sum_b (q^a q_t^b - q^b q_t^a) q_t^b - \frac{q^a}{r}, \tag{5.6}$$

besides the orbital momenta (5.4). They are components of the *Rung–Lenz vector*. However, there are no symmetries on Q (5.1) whose symmetry currents are A^a (5.6).

Let us consider a Hamiltonian Kepler system on the configuration space Q (5.1). Its phase space is $V^*Q = \mathbb{R} \times \mathbb{R}^6$ coordinated by (t, q^i, p_i).

It is readily observed that the Lagrangian (5.2) with the Kepler potential (5.5) of a Kepler system is hyperregular. The associated Hamiltonian form reads

$$H = p_i dq^i - \left(\frac{1}{2}\sum_i (p_i)^2 - r^{-1}\right) dt. \tag{5.7}$$

The corresponding characteristic Lagrangian L_H (4.76) is

$$L_H = \left(p_i q_t^i - \frac{1}{2}\sum_i (p_i)^2 + r^{-1}\right) dt.$$

Then a Hamiltonian Kepler system possesses the following integrals of motion:

- an energy function $\mathcal{E} = \mathcal{H}$;
- orbital momenta

$$M_b^a = q^a p_b - q^b p_a; \tag{5.8}$$

- components of the Rung–Lenz vector

$$A^a = \sum_b (q^a p_b - q^b p_a) p_b - \frac{q^a}{r}. \tag{5.9}$$

By virtue of the Noether's inverse first Theorem 4.15, these integrals of motion are conserved symmetry currents of the following Hamiltonian symmetries:

- the exact symmetry ∂_t,
- the exact vertical symmetries

$$\upsilon_b^a = q^b \partial_a - q^a \partial_b - p_a \partial^b + p_b \partial^a, \tag{5.10}$$

- the vertical symmetries

$$\upsilon^a = \sum_b [p_b \upsilon^a_b + (q^b p_a - q^a p_b)\partial_b] - \partial_b \left(\frac{q^a}{r}\right) \partial^b. \tag{5.11}$$

Let us note that the Hamiltonian symmetries υ^a_b (5.10) are the canonical lift (4.78) onto V^*Q of the vector fields v^a_b (5.3) on Q, which thus are basic Hamiltonian symmetries, and integrals of motion M^a_b (5.8) are the Noether currents (4.91).

At the same time, the Hamiltonian symmetries (5.11) do not come from any vector fields on a configuration space Q. Therefore, in contrast with the Rung–Lenz vector (5.11) in Hamiltonian mechanics, the Rung–Lenz vector (5.6) in Lagrangian mechanics fails to be a conserved symmetry current of a Lagrangian symmetry.

As was mentioned above, the Hamiltonian symmetries of the Kepler problem make up CISs. Without a loss of generality, we further consider the Kepler problem on a configuration space $\mathbb{R}^2$. Its phase space is $T^*\mathbb{R}^2 = \mathbb{R}^4$ provided with the Cartesian coordinates (q_i, p_i), $i = 1, 2$, and the canonical symplectic form

$$\Omega_T = \sum_i dp_i \wedge dq_i. \tag{5.12}$$

Let us denote

$$p^2 = \sum_i (p_i)^2, \qquad r^2 = \sum_i (q^i)^2, \qquad (p, q) = \sum_i p_i q_i.$$

An autonomous Hamiltonian of a Kepler system reads

$$\mathcal{H} = \frac{1}{2} p^2 - r^{-1} \tag{5.13}$$

(cf. (5.7)). A Kepler system is a Hamiltonian system on a symplectic manifold $Z = \mathbb{R}^4 \setminus \{0\}$ endowed with the symplectic form Ω_T (5.12). Let us consider functions

$$M_{12} = -M_{21} = q_1 p_2 - q_2 p_1, \tag{5.14}$$

$$A_i = \sum_j M_{ij} p_j - \frac{q_i}{r} = q_i p^2 - p_i (p, q) - \frac{q_i}{r}, \qquad i = 1, 2, \tag{5.15}$$

on Z. They are integrals of motion of the Hamiltonian $\mathcal{H}$ (5.13) where M_{12} is an angular momentum and (A_i) is the Rung–Lenz vector. Let us denote

$$M^2 = (M_{12})^2, \qquad A^2 = (A_1)^2 + (A_a)^2 = 2M^2\mathcal{H} + 1. \tag{5.16}$$

Let $Z_0 \subset Z$ be a closed subset of points where $M_{12} = 0$. A direct computation shows that the functions (M_{12}, A_i) (5.14) and (5.15) are independent on an open submanifold $U = Z \setminus Z_0$ of Z. At the same time, the functions $(\mathcal{H}, M_{12}, A_i)$ are independent nowhere on U because it follows from the expression (5.16) that

$$\mathcal{H} = \frac{A^2 - 1}{2M^2} \tag{5.17}$$

on U. The well known dynamics of a Kepler system shows that the Hamiltonian vector field of its Hamiltonian is complete on U (but not on Z).

Poisson brackets of integrals of motion M_{12} (5.14) and A_i (5.15) read

$$\{M_{12}, A_i\} = \eta_{2i}A_1 - \eta_{1i}A_2, \qquad \{A_1, A_2\} = 2\mathcal{H}M_{12} = \frac{A^2-1}{M_{12}}, \tag{5.18}$$

where η_{ij} is an Euclidean metric on $\mathbb{R}^2$. It is readily observed that these relations take the form (4.106). However, the matrix function **s** of the relations (5.18) fails to be of constant rank at points where $\mathcal{H} = 0$. Therefore, let us consider open submanifolds $U_- \subset U$ where $\mathcal{H} < 0$ and U_+ where $\mathcal{H} > 0$. Then we observe that both a Kepler system with the Hamiltonian $\mathcal{H}$ (5.13) and the integrals of motion (M_{ij}, A_i) (5.14) and (5.15) on U_- and a Kepler system with the Hamiltonian $\mathcal{H}$ (5.13) and the integrals of motion (M_{ij}, A_i) (5.14) and (5.15) on U_+ are noncommutative CISs. Moreover, these CISs can be brought into the form (4.113) as follows.

Let us replace the integrals of motions A_i with the integrals of motion

$$L_i = A_i(-2\mathcal{H})^{-1/2} \tag{5.19}$$

on U_-, and with the integrals of motion

$$K_i = A_i(2\mathcal{H})^{-1/2} \tag{5.20}$$

on U_+. The CIS (M_{12}, L_i) on U_- obeys relations

$$\{M_{12}, L_i\} = \eta_{2i}L_1 - \eta_{1i}L_2, \qquad \{L_1, L_2\} = -M_{12}. \tag{5.21}$$

Let us denote $M_{i3} = -L_i$ and put the indexes $\mu, \nu, \alpha, \beta = 1, 2, 3$. Then the relations (5.21) are brought into a form

$$\{M_{\mu\nu}, M_{\alpha\beta}\} = \eta_{\mu\beta}M_{\nu\alpha} + \eta_{\nu\alpha}M_{\mu\beta} - \eta_{\mu\alpha}M_{\nu\beta} - \eta_{\nu\beta}M_{\mu\alpha} \tag{5.22}$$

where $\eta_{\mu\nu}$ is an Euclidean metric on $\mathbb{R}^3$. A glance at the expression (5.22) shows that the integrals of motion M_{12} (5.14) and L_i (5.19) constitute a Lie algebra $\mathfrak{g} = so(3)$. Its corank equals 1. Therefore the CIS (M_{12}, L_i) on U_- is maximally integrable. The equality (5.17) takes a form

$$M^2 + L^2 = -(2\mathcal{H})^{-1}. \tag{5.23}$$

The CIS (M_{12}, K_i) on U_+ obeys relations

$$\{M_{12}, K_i\} = \eta_{2i} K_1 - \eta_{1i} K_2, \qquad \{K_1, K_2\} = M_{12}. \tag{5.24}$$

Let us denote $M_{i3} = -K_i$ and put the indexes $\mu, \nu, \alpha, \beta = 1, 2, 3$. Then the relations (5.24) are brought into a form

$$\{M_{\mu\nu}, M_{\alpha\beta}\} = \rho_{\mu\beta} M_{\nu\alpha} + \rho_{\nu\alpha} M_{\mu\beta} - \rho_{\mu\alpha} M_{\nu\beta} - \rho_{\nu\beta} M_{\mu\alpha} \tag{5.25}$$

where $\rho_{\mu\nu}$ is a pseudo-Euclidean metric of signature $(+, +, -)$ on $\mathbb{R}^3$. A glance at the expression (5.25) shows that the integrals of motion M_{12} (5.14) and K_i (5.20) constitute a Lie algebra $so(2, 1)$. Its corank equals 1. Therefore the CIS (M_{12}, K_i) on U_+ is maximally integrable. The equality (5.17) takes a form

$$K^2 - M^2 = (2\mathscr{H})^{-1}. \tag{5.26}$$

Thus, the Kepler problem on a phase space $\mathbb{R}^4$ falls into two different maximally integrable systems on open submanifolds U_- and U_+ of $\mathbb{R}^4$. We agree to call them the Kepler CISs on U_- and U_+, respectively.

Let us study the first one, and let us put

$$\begin{aligned} &F_1 = -L_1, \qquad F_2 = -L_2, \qquad F_3 = -M_{12}, \\ &\{F_1, F_2\} = F_3, \qquad \{F_2, F_3\} = F_1, \qquad \{F_3, F_1\} = F_2. \end{aligned} \tag{5.27}$$

We have a fibred manifold

$$F : U_- \to N \subset \mathfrak{g}^*, \tag{5.28}$$

which is the momentum mapping to a Lie coalgebra $\mathfrak{g}^* = so(3)^*$, endowed with coordinates (x_i) such that integrals of motion F_i on $\mathfrak{g}^*$ read $F_i = x_i$ (Remark 4.23). A base N of the fibred manifold (5.28) is an open submanifold of $\mathfrak{g}^*$ given by a coordinate condition $x_3 \neq 0$. It is a union of two contractible components with $x_3 > 0$ and $x_3 < 0$. A coinduced Lie–Poisson structure on N is given by the bivector

$$w = x_2 \partial^3 \wedge \partial^1 + x_3 \partial^1 \wedge \partial^2 + x_1 \partial^2 \wedge \partial^3. \tag{5.29}$$

The coadjoint action of $so(3)$ on N reads

$$\varepsilon_1 = x_3 \partial^2 - x_2 \partial^3, \quad \varepsilon_2 = x_1 \partial^3 - x_3 \partial^1, \quad \varepsilon_3 = x_2 \partial^1 - x_1 \partial^2.$$

Orbits of this action are given by an equation $x_1^2 + x_2^2 + x_3^2 = \text{const}$. They are level surfaces of a Casimir function $C = x_1^2 + x_2^2 + x_3^2$ and, hence, a Casimir function

$$h = -\frac{1}{2}(x_1^2 + x_2^2 + x_3^2)^{-1}. \tag{5.30}$$

A glance at the expression (5.23) shows that the pull-back F^*h of this Casimir function (5.30) onto U_- is the Hamiltonian $\mathcal{H}$ (5.13) of a Kepler system on U_-.

As was mentioned above, the Hamiltonian vector field of F^*h is complete. Furthermore, it is known that invariant submanifolds of the Kepler CIS on U_- are compact. Therefore, the fibred manifold F (5.28) is a fibre bundle in accordance with Theorem B.6. Moreover, this fibre bundle is trivial because N is a disjoint union of two contractible manifolds. Consequently, it follows from Theorem 4.22 that the Kepler CIS on U_- is global, i.e., it admits global action-angle coordinates as follows.

The Poisson manifold N (5.28) can be endowed with the coordinates

$$(I, x_1, \gamma), \qquad I < 0, \qquad \gamma \neq \pi/2, 3\pi/2, \tag{5.31}$$

defined by the equalities

$$\begin{aligned} I &= -\frac{1}{2}(x_1^2 + x_2^2 + x_3^2)^{-1}, \\ x_2 &= (-(2I)^{-1} - x_1^2)^{1/2}\sin\gamma, \qquad x_3 = (-(2I)^{-1} - x_1^2)^{1/2}\cos\gamma. \end{aligned} \tag{5.32}$$

It is readily observed that the coordinates (5.31) are Darboux coordinates of the Lie–Poisson structure (5.29) on U_-, namely, $w = \partial_{x_1} \wedge \partial_\gamma$.

Let ϑ_I be the Hamiltonian vector field of the Casimir function I (5.32). Its flows are invariant submanifolds of the Kepler CIS on U_- (Remark 4.23). Let α be a parameter along the flow of this vector field, i.e.,

$$\vartheta_I = \partial_\alpha. \tag{5.33}$$

Then U_- is provided with the action-angle coordinates (I, x_1, γ, α) such that the Poisson bivector associated to the symplectic form Ω_T on U_- reads

$$w = \frac{\partial}{\partial I} \wedge \frac{\partial}{\partial \alpha} + \frac{\partial}{\partial x_1} \wedge \frac{\partial}{\partial \gamma}.$$

Accordingly, Hamiltonian vector fields of integrals of motion F_i (5.27) take a form

$$\begin{aligned} \vartheta_1 &= \frac{\partial}{\partial \gamma}, \\ \vartheta_2 &= \frac{1}{4I^2}\left(-\frac{1}{2I} - x_1^2\right)^{-1/2} \sin\gamma \frac{\partial}{\partial \alpha} - x_1\left(-\frac{1}{2I} - x_1^2\right)^{-1/2} \sin\gamma \frac{\partial}{\partial \gamma} - \\ &\quad \left(-\frac{1}{2I} - x_1^2\right)^{1/2} \cos\gamma \frac{\partial}{\partial x_1}, \end{aligned}$$

$$\vartheta_3 = \frac{1}{4I^2}\left(-\frac{1}{2I} - x_1^2\right)^{-1/2} \cos\gamma \frac{\partial}{\partial\alpha} - x_1\left(-\frac{1}{2I} - x_1^2\right)^{-1/2} \cos\gamma \frac{\partial}{\partial\gamma} +$$
$$\left(-\frac{1}{2I} - x_1^2\right)^{1/2} \sin\gamma \frac{\partial}{\partial x_1}.$$

A glance at these expressions shows that the vector fields ϑ_1 and ϑ_2 fail to be complete on U_- (Remark 4.20).

One can say something more about the angle coordinate α. The vector field ϑ_I (5.33) reads

$$\frac{\partial}{\partial\alpha} = \sum_i \left(\frac{\partial\mathscr{H}}{\partial p_i}\frac{\partial}{\partial q_i} - \frac{\partial\mathscr{H}}{\partial q_i}\frac{\partial}{\partial p_i}\right).$$

This equality leads to relations

$$\frac{\partial q_i}{\partial\alpha} = \frac{\partial\mathscr{H}}{\partial p_i}, \qquad \frac{\partial p_i}{\partial\alpha} = -\frac{\partial\mathscr{H}}{\partial q_i},$$

which take a form of the Hamilton equation. Therefore, the coordinate α is a cyclic time $\alpha = t \bmod 2\pi$ given by the well-known expression

$$\alpha = \phi - a^{3/2} e \sin(a^{-3/2}\phi), \qquad r = a(1 - e\cos(a^{-3/2}\phi)),$$
$$a = (2I)^{-1}, \qquad e = (1 + 2IM^2)^{1/2}.$$

Now let us turn to the Kepler CIS on U_+. It is a globally integrable system with noncompact invariant submanifolds as follows. Let us put

$$S_1 = -K_1, \qquad S_2 = -K_2, \qquad S_3 = -M_{12}, \tag{5.34}$$
$$\{S_1, S_2\} = -S_3, \qquad \{S_2, S_3\} = S_1, \qquad \{S_3, S_1\} = S_2.$$

We have a fibred manifold

$$S : U_+ \to N \subset \mathfrak{g}^*, \tag{5.35}$$

which is the momentum mapping to a Lie coalgebra $\mathfrak{g}^* = so(2,1)^*$, endowed with the coordinates (x_i) such that integrals of motion S_i on $\mathfrak{g}^*$ read $S_i = x_i$. A base N of the fibred manifold (5.35) is an open submanifold of $\mathfrak{g}^*$ given by a coordinate condition $x_3 \neq 0$. It is a union of two contractible components defined by conditions $x_3 > 0$ and $x_3 < 0$. A coinduced Lie–Poisson structure on N takes a form

$$w = x_2\partial^3 \wedge \partial^1 - x_3\partial^1 \wedge \partial^2 + x_1\partial^2 \wedge \partial^3. \tag{5.36}$$

The coadjoint action of $so(2,1)$ on N reads

$$\varepsilon_1 = -x_3\partial^2 - x_2\partial^3, \qquad \varepsilon_2 = x_1\partial^3 + x_3\partial^1, \qquad \varepsilon_3 = x_2\partial^1 - x_1\partial^2.$$

The orbits of this coadjoint action are given by an equation $x_1^2+x_2^2-x_3^2 = \text{const}$. They are the level surfaces of the Casimir function $C = x_1^2 + x_2^2 - x_3^2$ and, consequently, the Casimir function

$$h = \frac{1}{2}(x_1^2 + x_2^2 - x_3^2)^{-1}. \tag{5.37}$$

A glance at the expression (5.26) shows that the pull-back S^*h of this Casimir function (5.37) onto U_+ is the Hamiltonian $\mathscr{H}$ (5.13) of a Kepler system on U_+.

As was mentioned above, the Hamiltonian vector field of S^*h is complete. Furthermore, it is known that invariant submanifolds of the Kepler CIS on U_+ are diffeomorphic to $\mathbb{R}$. Therefore, the fibred manifold S (5.35) is a fibre bundle in accordance with Theorem B.6. Moreover, this fibre bundle is trivial because N is a disjoint union of two contractible manifolds. Consequently, it follows from Theorem 4.22 that the Kepler CIS on U_+ is globally integrable, i.e., it admits global action-angle coordinates as follows.

The Poisson manifold N (5.35) can be endowed with the coordinates (I, x_1, λ), $I > 0$, $\lambda \neq 0$, defined by the equalities

$$I = \frac{1}{2}(x_1^2 + x_2^2 - x_3^2)^{-1},$$
$$x_2 = ((2I)^{-1} - x_1^2)^{1/2} \cosh\lambda, \qquad x_3 = ((2I)^{-1} - x_1^2)^{1/2} \sinh\lambda.$$

These coordinates are Darboux coordinates of the Lie–Poisson structure (5.36) on N, namely, $w = \partial_\lambda \wedge \partial_{x_1}$.

Let ϑ_I be the Hamiltonian vector field of the Casimir function I (5.32). Its flows are invariant submanifolds of the Kepler CIS on U_+ (Remark 4.23). Let τ be a parameter along the flows of this vector field, i.e.,

$$\vartheta_I = \partial_\tau. \tag{5.38}$$

Then U_+ (5.35) is provided with the action-angle coordinates (I, x_1, λ, τ) such that the Poisson bivector associated to the symplectic form Ω_T on U_+ reads

$$W = \frac{\partial}{\partial I} \wedge \frac{\partial}{\partial \tau} + \frac{\partial}{\partial \lambda} \wedge \frac{\partial}{\partial x_1}.$$

Accordingly, Hamiltonian vector fields of integrals of motion S_i (5.34) take a form

$$\vartheta_1 = -\frac{\partial}{\partial \lambda},$$
$$\vartheta_2 = \frac{1}{4I^2}\left(\frac{1}{2I} - x_1^2\right)^{-1/2} \cosh\lambda \frac{\partial}{\partial \tau} + x_1 \left(\frac{1}{2I} - x_1^2\right)^{-1/2} \cosh\lambda \frac{\partial}{\partial \lambda} +$$
$$\left(\frac{1}{2I} - x_1^2\right)^{1/2} \sinh\lambda \frac{\partial}{\partial x_1},$$

$$\vartheta_3 = \frac{1}{4I^2}\left(\frac{1}{2I} - x_1^2\right)^{-1/2} \sinh\lambda\frac{\partial}{\partial\tau} + x_1\left(\frac{1}{2I} - x_1^2\right)^{-1/2} \sinh\lambda\frac{\partial}{\partial\lambda} +$$
$$\left(\frac{1}{2I} - x_1^2\right)^{1/2} \cosh\lambda\frac{\partial}{\partial x_1}.$$

Similarly to the angle coordinate α (5.33), the angle coordinate τ (5.38) obeys the Hamilton equation

$$\frac{\partial q_i}{\partial\tau} = \frac{\partial\mathcal{H}}{\partial p_i}, \qquad \frac{\partial p_i}{\partial\tau} = -\frac{\partial\mathcal{H}}{\partial q_i}.$$

Therefore, it is the time $\tau = t$ given by the well-known expression

$$\tau = s - a^{3/2}e\sinh(a^{-3/2}s), \qquad r = a(e\cosh(a^{-3/2}s) - 1),$$
$$a = (2I)^{-1}, \qquad e = (1 + 2IM^2)^{1/2}.$$

Chapter 6
Calculus of Variations on Graded Bundles

Throughout the book, by the Grassmann gradation is meant the $\mathbb{Z}_2$-gradation. It shortly is called the graded structure if there is no danger of confusion. The symbol [.] stands for the Grassmann parity.

From the mathematical viewpoint, we restrict our consideration to simple graded manifolds (Definition 6.4) and graded bundles over smooth manifolds (Definition 6.5). A key point is that vector fields and exterior one-forms on a simple graded manifold with a body manifold Z are represented by sections of the vector bundles (6.20) and (6.22) over Z, respectively.

We follow Definition 6.7 of graded jet manifolds of graded bundles which is compatible with the conventional Definition B.11 of jets of fibre bundles. It differs from the definition of jets of modules over graded commutative rings [58, 131] and from that of jets of fibred-graded manifolds [73, 107], but reproduces the heuristic notion of jets of odd ghosts in BRST field theory [6, 22].

6.1 Grassmann-Graded Algebraic Calculus

Let us summarize the relevant notions of the Grassmann-graded algebraic calculus [7, 29, 61, 131].

Let $\mathscr{K}$ be a commutative ring. A $\mathscr{K}$-module Q is termed *graded* if it is endowed with a *grading automorphism* γ such that $\gamma^2 = \mathrm{Id}$. A graded module falls into a direct sum $Q = Q_0 \oplus Q_1$ of $\mathscr{K}$-modules Q_0 and Q_1 of *even* and *odd* elements such that $\gamma(q) = (-1)^{[q]}q$, $q \in Q_{[q]}$. One calls Q_0 and Q_1 the even and odd parts of Q, respectively.

In particular, by a real *graded vector space* $B = B_0 \oplus B_1$ is meant a graded $\mathbb{R}$-module. A real graded vector space is said to be (n, m)-dimensional if $B_0 = \mathbb{R}^n$ and $B_1 = \mathbb{R}^m$.

A $\mathscr{K}$-algebra algebra $\mathscr{A}$ is called *graded* if it a graded $\mathscr{K}$-module such that

$$[aa'] = ([a] + [a'])\mathrm{mod}\,2, \qquad a \in \mathscr{A}_{[a]}, \qquad a' \in \mathscr{A}_{[a']}.$$

G. Sardanashvily, *Noether's Theorems*, Atlantis Studies
in Variational Geometry 3, DOI 10.2991/978-94-6239-171-0_6

Its even part $\mathscr{A}_0$ is a subalgebra of $\mathscr{A}$ and the odd one $\mathscr{A}_1$ is an $\mathscr{A}_0$-module. If $\mathscr{A}$ is a *graded ring*, then $[\mathbf{1}] = 0$.

Definition 6.1 A graded algebra $\mathscr{A}$ is called *graded commutative* if

$$aa' = (-1)^{[a][a']}a'a,$$

where a and a' are *graded-homogeneous elements* of $\mathscr{A}$, i.e., they are either even or odd.

Given a graded algebra $\mathscr{A}$, a left graded $\mathscr{A}$-module Q is defined as a left $\mathscr{A}$-module where $[aq] = ([a] + [q]) \bmod 2$. Similarly, right graded $\mathscr{A}$-modules are treated.

Definition 6.2 Let V be a real vector space, and let $\Lambda = \wedge V$ be its exterior algebra endowed with the Grassmann gradation

$$\Lambda = \Lambda_0 \oplus \Lambda_1, \qquad \Lambda_0 = \mathbb{R} \bigoplus_{k=1} \overset{2k}{\wedge} V, \qquad \Lambda_1 = \bigoplus_{k=1} \overset{2k-1}{\wedge} V. \tag{6.1}$$

It is a real graded commutative ring, termed the *Grassmann algebra*.

A Grassmann algebra, seen as an additive group, admits the decomposition

$$\Lambda = \mathbb{R} \oplus R = \mathbb{R} \oplus R_0 \oplus R_1 = \mathbb{R} \oplus (\Lambda_1)^2 \oplus \Lambda_1,$$

where R is the *ideal of nilpotents* of Λ. The corresponding projections $\sigma : \Lambda \to \mathbb{R}$ and $s : \Lambda \to R$ are called the *body* and soul maps, respectively. Let us note that there is a different definition of a Grassmann algebra [78] which is equivalent to the above mentioned one only in the case of an infinite-dimensional vector space V [29]. Hereafter, we restrict our consideration to Grassmann algebras of finite rank. Given a basis $\{c^i\}$ for a vector space V, elements of a Grassmann algebra Λ (6.1) take a form

$$a = \sum_{k=0,1,\ldots} \sum_{(i_1 \cdots i_k)} a_{i_1 \cdots i_k} c^{i_1} \cdots c^{i_k},$$

where the second sum runs through all the tuples $(i_1 \cdots i_k)$ such that no two of them are permutations of each other.

Definition 6.3 A graded (non-associative) algebra $\mathfrak{g}$ is termed a *Lie superalgebra* if its product $[.,.]$, called the *graded Lie bracket*, or *superbracket* obeys the relations

$$[\varepsilon, \varepsilon'] = -(-1)^{[\varepsilon][\varepsilon']}[\varepsilon', \varepsilon],$$
$$(-1)^{[\varepsilon][\varepsilon'']}[\varepsilon, [\varepsilon', \varepsilon'']] + (-1)^{[\varepsilon'][\varepsilon]}[\varepsilon', [\varepsilon'', \varepsilon]] + (-1)^{[\varepsilon''][\varepsilon']}[\varepsilon'', [\varepsilon, \varepsilon']] = 0.$$

Being decomposed in even and odd parts $\mathfrak{g} = \mathfrak{g}_0 \oplus \mathfrak{g}_1$, a Lie superalgebra $\mathfrak{g}$ obeys the relations

$$[\mathfrak{g}_0, \mathfrak{g}_0] \subset \mathfrak{g}_0, \qquad [\mathfrak{g}_0, \mathfrak{g}_1] \subset \mathfrak{g}_1, \qquad [\mathfrak{g}_1, \mathfrak{g}_1] \subset \mathfrak{g}_1.$$

In particular, an even part $\mathfrak{g}_0$ of a Lie superalgebra $\mathfrak{g}$ is a Lie algebra. A graded $\mathscr{K}$-module P is called the $\mathfrak{g}$-*module* if it is provided with an $\mathbb{R}$-bilinear map

$$\mathfrak{g} \times P \ni (\varepsilon, p) \to \varepsilon p \in P, \qquad [\varepsilon p] = ([\varepsilon] + [p]) \bmod 2,$$
$$[\varepsilon, \varepsilon']p = (\varepsilon \circ \varepsilon' - (-1)^{[\varepsilon][\varepsilon']}\varepsilon' \circ \varepsilon)p.$$

If $\mathscr{A}$ is graded commutative, a graded $\mathscr{K}$-module can be provided with a graded $\mathscr{A}$-*bimodule* structure by letting

$$qa = (-1)^{[a][q]}aq, \qquad a \in \mathscr{K}, \qquad q \in Q.$$

Given a graded commutative ring $\mathscr{A}$, the following are standard constructions of new graded $\mathscr{A}$-modules from old ones.

- A direct sum of graded modules and a graded quotient module are defined just as those of modules over a commutative ring.
- A *tensor product* $P \otimes Q$ of graded $\mathscr{A}$-modules P and Q is an additive group generated by elements $p \otimes q$, $p \in P$, $q \in Q$, obeying the relations

$$(p + p') \otimes q = p \otimes q + p' \otimes q,$$
$$p \otimes (q + q') = p \otimes q + p \otimes q',$$
$$ap \otimes q = (-1)^{[p][a]}pa \otimes q = (-1)^{[p][a]}p \otimes aq, \qquad a \in \mathscr{A}.$$

In particular, the tensor algebra $\otimes P$ of a graded $\mathscr{K}$-module P is defined as that (A.4) of a module over a commutative ring. Its quotient $\wedge P$ with respect to the ideal generated by elements

$$p \otimes p' - (-1)^{[p][p']}p' \otimes p, \qquad p, p' \in P,$$

is the *bigraded exterior algebra* of a graded module P with respect to the *graded exterior product*

$$p \wedge p' = (-1)^{[p][p']}p' \wedge p.$$

- A morphism $\Phi : P \to Q$ of graded $\mathscr{A}$-modules seen as additive groups is said to be *even morphism* (resp. *odd morphism*) if Φ preserves (resp. change) the Grassmann parity of all graded-homogeneous elements of P and obeys the relations

$$\Phi(ap) = (-1)^{[\Phi][a]}a\Phi(p), \qquad p \in P, \qquad a \in \mathscr{A}.$$

A morphism $\Phi : P \to Q$ of graded $\mathcal{A}$-modules as additive groups is termed a *graded $\mathcal{A}$-module morphism* if it is represented by a sum of even and odd morphisms. The set $\mathrm{Hom}_{\mathcal{K}}(P, Q)$ of graded morphisms of a graded $\mathcal{A}$-module P to a graded $\mathcal{A}$-module Q is naturally a graded $\mathcal{A}$-module. A graded $\mathcal{A}$-module $P^* = \mathrm{Hom}_{\mathcal{A}}(P, \mathcal{K})$ is called the *dual* of a graded $\mathcal{K}$-module P.

Let B be a graded real vector space. Given a Grassmann algebra Λ, it can be brought into a graded Λ-module

$$\Lambda B = (\Lambda B)_0 \oplus (\Lambda B)_1 = (\Lambda_0 \otimes B_0 \oplus \Lambda_1 \otimes B_1) \oplus (\Lambda_1 \otimes B_0 \oplus \Lambda_0 \otimes B_1),$$

called the *superspace*. A superspace

$$B^{n|m} = [(\overset{n}{\oplus} \Lambda_0) \oplus (\overset{m}{\oplus} \Lambda_1)] \oplus [(\overset{n}{\oplus} \Lambda_1) \oplus (\overset{m}{\oplus} \Lambda_0)]$$

is said to be (n, m)-dimensional. A graded Λ_0-module

$$B^{n,m} = (\overset{n}{\oplus} \Lambda_0) \oplus (\overset{m}{\oplus} \Lambda_1)$$

is termed an (n, m)-dimensional *supervector space*.

6.2 Grassmann-Graded Differential Calculus

The differential calculus over graded commutative rings (Definition 6.1) is developed similarly to that over commutative rings (Sects. A.1 and A.2) [61, 131, 134].

Let $\mathcal{K}$ be a commutative ring and $\mathcal{A}$ a graded commutative $\mathcal{K}$-ring (Definition 6.1). Let P and Q be graded $\mathcal{A}$-modules. A $\mathcal{K}$-module $\mathrm{Hom}_{\mathcal{K}}(P, Q)$ of graded $\mathcal{K}$-module homomorphisms $\Phi : P \to Q$ can be endowed with the two graded $\mathcal{A}$-module structures

$$(a\Phi)(p) = a\Phi(p), \qquad (\Phi \bullet a)(p) = \Phi(ap), \qquad a \in \mathcal{A}, \quad p \in P.$$

Let us put

$$\delta_a \Phi = a\Phi - (-1)^{[a][\Phi]} \Phi \bullet a, \qquad a \in \mathcal{A}.$$

An element $\Delta \in \mathrm{Hom}_{\mathcal{K}}(P, Q)$ is said to be a Q-valued *graded differential operator* of order s on P if $\delta_{a_0} \circ \cdots \circ \delta_{a_s} \Delta = 0$ for any tuple of $s + 1$ elements $a_0, \ldots, a_s$ of $\mathcal{A}$.

In particular, zero order graded differential operators obey a condition

$$\delta_a \Delta(p) = a\Delta(p) - (-1)^{[a][\Delta]} \Delta(ap) = 0, \qquad a \in \mathcal{A}, \qquad p \in P,$$

i.e., they coincide with graded $\mathcal{A}$-module morphisms $P \to Q$. A first order graded differential operator Δ satisfies a relation

$$\delta_a \circ \delta_b \,\Delta(p) = ab\Delta(p) - (-1)^{([b]+[\Delta])[a]} b\Delta(ap) - (-1)^{[b][\Delta]} a\Delta(bp) + (-1)^{[b][\Delta]+([\Delta]+[b])[a]} = 0, \qquad a, b \in \mathcal{A}, \quad p \in P.$$

For instance, let $P = \mathcal{A}$. Any zero order Q-valued graded differential operator Δ on $\mathcal{A}$ is defined by its value $\Delta(\mathbf{1})$. Then there is a graded $\mathcal{A}$-module isomorphism

$$\mathrm{Diff}_0(\mathcal{A}, Q) = Q, \qquad Q \ni q \to \Delta_q \in \mathrm{Diff}_0(\mathcal{A}, Q),$$

where Δ_q is given by the equality $\Delta_q(\mathbf{1}) = q$. A first order Q-valued graded differential operator Δ on $\mathcal{A}$ fulfils the condition

$$\Delta(ab) = \Delta(a)b + (-1)^{[a][\Delta]} a\Delta(b) - (-1)^{([b]+[a])[\Delta]} ab\Delta(\mathbf{1}), \qquad a, b \in \mathcal{A}.$$

It is called a Q-valued *graded derivation* of $\mathcal{A}$ if $\Delta(\mathbf{1}) = 0$, i.e., the *Grassmann-graded Leibniz rule*

$$\Delta(ab) = \Delta(a)b + (-1)^{[a][\Delta]} a\Delta(b), \quad a, b \in \mathcal{A}, \tag{6.2}$$

holds. One obtains at once that any first order graded differential operator on $\mathcal{A}$ falls into a sum

$$\Delta(a) = \Delta(\mathbf{1})a + [\Delta(a) - \Delta(\mathbf{1})a]$$

of a zero order graded differential operator $\Delta(\mathbf{1})a$ and a graded derivation $\Delta(a) - \Delta(\mathbf{1})a$. If ∂ is a graded derivation of $\mathcal{A}$, then $a\partial$ is so for any $a \in \mathcal{A}$. Hence, graded derivations of $\mathcal{A}$ constitute a graded $\mathcal{A}$-module $\mathfrak{d}(\mathcal{A}, Q)$, termed the *graded derivation module*.

If $Q = \mathcal{A}$, the graded derivation module $\mathfrak{d}\mathcal{A}$ also is a Lie superalgebra (Definition 6.3) over a commutative ring $\mathcal{K}$ with respect to a superbracket

$$[u, u'] = u \circ u' - (-1)^{[u][u']} u' \circ u, \qquad u, u' \in \mathcal{A}. \tag{6.3}$$

Since $\mathfrak{d}\mathcal{A}$ is a Lie $\mathcal{K}$-superalgebra, let us consider the Chevalley–Eilenberg complex $C^*[\mathfrak{d}\mathcal{A}; \mathcal{A}]$ where a graded commutative ring $\mathcal{A}$ is a regarded as a $\mathfrak{d}\mathcal{A}$-module [47, 131]. It is a complex

$$0 \to \mathcal{A} \xrightarrow{d} C^1[\mathfrak{d}\mathcal{A}; \mathcal{A}] \xrightarrow{d} \cdots C^k[\mathfrak{d}\mathcal{A}; \mathcal{A}] \xrightarrow{d} \cdots \tag{6.4}$$

where

$$C^k[\mathfrak{d}\mathcal{A}; \mathcal{A}] = \mathrm{Hom}_{\mathcal{K}}(\overset{k}{\wedge} \mathfrak{d}\mathcal{A}, \mathcal{A})$$

are $\mathfrak{d}\mathcal{A}$-modules of $\mathcal{K}$-linear graded morphisms of the graded exterior products $\overset{k}{\wedge}\mathfrak{d}\mathcal{A}$ of a graded $\mathcal{K}$-module $\mathfrak{d}\mathcal{A}$ to $\mathcal{A}$. Let us bring homogeneous elements of $\overset{k}{\wedge}\mathfrak{d}\mathcal{A}$ into a form

$$\varepsilon_1 \wedge \cdots \varepsilon_r \wedge \varepsilon_{r+1} \wedge \cdots \wedge \varepsilon_k, \qquad \varepsilon_i \in \mathfrak{d}\mathcal{A}_0, \quad \varepsilon_j \in \mathfrak{d}\mathcal{A}_1.$$

Then the Chevalley–Eilenberg coboundary operator d of the complex (6.4) is given by the expression

$$\begin{aligned}
&dc(\varepsilon_1 \wedge \cdots \wedge \varepsilon_r \wedge \epsilon_1 \wedge \cdots \wedge \epsilon_s) = \qquad (6.5)\\
&\sum_{i=1}^{r}(-1)^{i-1}\varepsilon_i c(\varepsilon_1 \wedge \cdots \widehat{\varepsilon_i} \cdots \wedge \varepsilon_r \wedge \epsilon_1 \wedge \cdots \epsilon_s) + \\
&\sum_{j=1}^{s}(-1)^{r}\varepsilon_i c(\varepsilon_1 \wedge \cdots \wedge \varepsilon_r \wedge \epsilon_1 \wedge \cdots \widehat{\epsilon_j} \cdots \wedge \epsilon_s) + \\
&\sum_{1\leq i<j\leq r}(-1)^{i+j}c([\varepsilon_i, \varepsilon_j] \wedge \varepsilon_1 \wedge \cdots \widehat{\varepsilon_i} \cdots \widehat{\varepsilon_j} \cdots \wedge \varepsilon_r \wedge \epsilon_1 \wedge \cdots \wedge \epsilon_s) + \\
&\sum_{1\leq i<j\leq s} c([\epsilon_i, \epsilon_j] \wedge \varepsilon_1 \wedge \cdots \wedge \varepsilon_r \wedge \epsilon_1 \wedge \cdots \widehat{\epsilon_i} \cdots \widehat{\epsilon_j} \cdots \wedge \epsilon_s) + \\
&\sum_{1\leq i<r, 1\leq j\leq s}(-1)^{i+r+1}c([\varepsilon_i, \epsilon_j] \wedge \varepsilon_1 \wedge \cdots \widehat{\varepsilon_i} \cdots \wedge \varepsilon_r \wedge \epsilon_1 \wedge \cdots \widehat{\epsilon_j} \cdots \wedge \epsilon_s),
\end{aligned}$$

where the caret $\widehat{\ }$ denotes omission. This operator is called the *graded Chevalley–Eilenberg coboundary operator*.

Let us consider the extended Chevalley–Eilenberg complex

$$0 \to \mathcal{K} \xrightarrow{\text{in}} C^*[\mathfrak{d}\mathcal{A}; \mathcal{A}].$$

It is easily justified that this complex contains a subcomplex $\mathcal{O}^*[\mathfrak{d}\mathcal{A}]$ of $\mathcal{A}$-linear graded morphisms. The $\mathbb{N}$-graded module $\mathcal{O}^*[\mathfrak{d}\mathcal{A}]$ is provided with the structure of a bigraded $\mathcal{A}$-algebra with respect to the *graded exterior product*

$$\begin{aligned}
&\phi \wedge \phi'(u_1, ..., u_{r+s}) = \qquad (6.6)\\
&\qquad \sum_{i_1<\cdots<i_r; j_1<\cdots<j_s} \mathrm{Sgn}^{i_1\cdots i_r j_1\cdots j_s}_{1\cdots r+s}\phi(u_{i_1}, \ldots, u_{i_r})\phi'(u_{j_1}, \ldots, u_{j_s}),\\
&\phi \in \mathcal{O}^r[\mathfrak{d}\mathcal{A}], \qquad \phi' \in \mathcal{O}^s[\mathfrak{d}\mathcal{A}], \qquad u_k \in \mathfrak{d}\mathcal{A},
\end{aligned}$$

where $u_1, \ldots, u_{r+s}$ are graded-homogeneous elements of $\mathfrak{d}\mathcal{A}$ and

$$u_1 \wedge \cdots \wedge u_{r+s} = \mathrm{Sgn}^{i_1\cdots i_r j_1\cdots j_s}_{1\cdots r+s} u_{i_1} \wedge \cdots \wedge u_{i_r} \wedge u_{j_1} \wedge \cdots \wedge u_{j_s}.$$

The graded Chevalley–Eilenberg coboundary operator d (6.5) and the graded exterior product $\wedge$ (6.6) bring $\mathcal{O}^*[\mathfrak{d}\mathcal{A}]$ into a *differential bigraded algebra* (henceforth, DBGA) whose elements obey relations

$$\phi \wedge \phi' = (-1)^{|\phi||\phi'|+[\phi][\phi']}\phi' \wedge \phi, \tag{6.7}$$

$$d(\phi \wedge \phi') = d\phi \wedge \phi' + (-1)^{|\phi|}\phi \wedge d\phi'. \tag{6.8}$$

It is called the *graded Chevalley–Eilenberg differential calculus* over a graded commutative $\mathcal{K}$-ring $\mathcal{A}$. In particular, we have

$$\mathcal{O}^1[\mathfrak{d}\mathcal{A}] = \mathrm{Hom}_{\mathcal{A}}(\mathfrak{d}\mathcal{A}, \mathcal{A}) = \mathfrak{d}\mathcal{A}^*. \tag{6.9}$$

One can extend this duality relation to the *graded interior product* of $u \in \mathfrak{d}\mathcal{A}$ with any element $\phi \in \mathcal{O}^*[\mathfrak{d}\mathcal{A}]$ by the rules

$$u\rfloor(bda) = (-1)^{[u][b]}u(a), \qquad a, b \in \mathcal{A},$$

$$u\rfloor(\phi \wedge \phi') = (u\rfloor\phi) \wedge \phi' + (-1)^{|\phi|+[\phi][u]}\phi \wedge (u\rfloor\phi'). \tag{6.10}$$

As a consequence, any graded derivation $u \in \mathfrak{d}\mathcal{A}$ of $\mathcal{A}$ yields a derivation

$$\mathbf{L}_u\phi = u\rfloor d\phi + d(u\rfloor\phi), \qquad \phi \in \mathcal{O}^*, \qquad u \in \mathfrak{d}\mathcal{A}, \tag{6.11}$$

$$\mathbf{L}_u(\phi \wedge \phi') = \mathbf{L}_u(\phi) \wedge \phi' + (-1)^{[u][\phi]}\phi \wedge \mathbf{L}_u(\phi'),$$

termed the *graded Lie derivative* of a DBGA $\mathcal{O}^*[\mathfrak{d}\mathcal{A}]$.

Let us note that, if $\mathcal{A}$ is a commutative ring, the graded Chevalley–Eilenberg differential calculus comes to the familiar one (Sect. A.3).

The minimal graded Chevalley–Eilenberg differential calculus $\mathcal{O}^*\mathcal{A} \subset \mathcal{O}^*[\mathfrak{d}\mathcal{A}]$ over a graded commutative ring $\mathcal{A}$ consists of monomials $a_0da_1 \wedge \cdots \wedge da_k$, $a_i \in \mathcal{A}$. The corresponding complex

$$0 \to \mathcal{K} \longrightarrow \mathcal{A} \xrightarrow{d} \mathcal{O}^1\mathcal{A} \xrightarrow{d} \cdots \mathcal{O}^k\mathcal{A} \xrightarrow{d} \cdots \tag{6.12}$$

is called the *bigraded de Rham complex* of a graded commutative $\mathcal{K}$-ring $\mathcal{A}$.

6.3 Differential Calculus on Graded Bundles

In accordance with Serre–Swan Theorem 6.2 below, if a real graded commutative algebra $\mathcal{A}$ is generated by a projective module of finite rank over the ring $C^\infty(Z)$ of smooth functions on some manifold Z, then $\mathcal{A}$ is isomorphic to the algebra of graded functions on a graded manifold with a body Z, and vice versa. Then the minimal graded Chevalley–Eilenberg differential calculus $\mathcal{O}^*\mathcal{A}$ over $\mathcal{A}$ is a DBGA of graded exterior forms on this graded manifold [61, 131, 134].

As was mentioned above, we restrict our consideration to simple graded manifolds modelled over vector bundles (Definition 6.4) and graded bundles over smooth manifolds (Definition 6.5).

Graded Manifolds

A *graded manifold* of dimension (n, m) is defined as a local-ringed space $(Z, \mathfrak{A})$ (Definition C.1) where Z is an n-dimensional smooth manifold, and $\mathfrak{A} = \mathfrak{A}_0 \oplus \mathfrak{A}_1$ is a sheaf of Grassmann algebras of rank m (Definition 6.2) such that [7, 134]:

- there is the exact sequence of sheaves

$$0 \to \mathcal{R} \to \mathfrak{A} \xrightarrow{\sigma} C^\infty_Z \to 0, \qquad \mathcal{R} = \mathfrak{A}_1 + (\mathfrak{A}_1)^2, \tag{6.13}$$

  where C^∞_Z is the sheaf of smooth real functions on Z;
- $\mathcal{R}/\mathcal{R}^2$ is a locally free sheaf of C^∞_Z-modules of finite rank (with respect to pointwise operations), and the sheaf $\mathfrak{A}$ is locally isomorphic to the exterior product $\wedge_{C^\infty_Z}(\mathcal{R}/\mathcal{R}^2)$.

A sheaf $\mathfrak{A}$ is called the *structure sheaf* of a graded manifold $(Z, \mathfrak{A})$, and a manifold Z is said to be the *body* of $(Z, \mathfrak{A})$. Sections of the sheaf $\mathfrak{A}$ are termed *graded functions* on a graded manifold $(Z, \mathfrak{A})$. They make up a graded commutative $C^\infty(Z)$-ring $\mathfrak{A}(Z)$ called the *structure ring* of $(Z, \mathfrak{A})$.

By virtue of the well-known *Batchelor theorem* [7, 14, 134], graded manifolds possess the following structure.

Theorem 6.1 *Let $(Z, \mathfrak{A})$ be a graded manifold. There exists a vector bundle $E \to Z$ with an m-dimensional typical fibre V such that the structure sheaf $\mathfrak{A}$ of $(Z, \mathfrak{A})$ is isomorphic to the structure sheaf $\mathfrak{A}_E = S_{\wedge E^*}$ of germs of sections of the exterior bundle $\wedge E^*$ (B.13), whose typical fibre is the Grassmann algebra $\wedge V^*$.*

It should be emphasized that Batchelor's isomorphism in Theorem 6.1 fails to be canonical. In applications, it however is fixed from the beginning. Therefore, we restrict our consideration to graded manifolds $(Z, \mathfrak{A}_E)$ whose structure sheaf is the sheaf of germs of sections of some exterior bundle $\wedge E^*$.

Definition 6.4 We agree to call $(Z, \mathfrak{A}_E)$ the *simple graded manifold* modelled over a vector bundle $E \to Z$, called its *characteristic vector bundle*.

Accordingly, the structure ring $\mathcal{A}_E$ of a simple graded manifold $(Z, \mathfrak{A}_E)$ is the structure module

$$\mathcal{A}_E = \mathfrak{A}_E(Z) = \wedge E^*(Z) \tag{6.14}$$

of sections of the exterior bundle $\wedge E^*$. Automorphisms of a simple graded manifold $(Z, \mathfrak{A}_E)$ are restricted to those induced by automorphisms of its characteristic vector bundles $E \to Z$ (Remark 6.3).

Combining Batchelor Theorem 6.1 and classical Serre–Swan Theorem A.10, we come to the following *Serre–Swan theorem for graded manifolds* [11, 61].

Theorem 6.2 *Let Z be a smooth manifold. A graded commutative $C^\infty(Z)$-algebra $\mathcal{A}$ is isomorphic to the structure ring of a graded manifold with a body Z if and only if it is the exterior algebra of some projective $C^\infty(Z)$-module of finite rank.*

Proof By virtue of the Batchelor theorem, any graded manifold is isomorphic to a simple graded manifold $(Z, \mathfrak{A}_E)$ modelled over some vector bundle $E \to Z$. Its structure ring $\mathcal{A}_E$ (6.14) of graded functions consists of sections of the exterior bundle $\wedge E^*$. The classical Serre–Swan theorem states that a $C^\infty(Z)$-module is isomorphic to the module of sections of a smooth vector bundle over Z if and only if it is a projective module of finite rank. □

Remark 6.1 One can treat a local-ringed space $(Z, \mathfrak{A}_0 = C^\infty_Z)$ as a *trivial graded manifold*. It is a simple graded manifold whose characteristic bundle is $E = Z \times \{0\}$. Its structure module is a ring $C^\infty(Z)$ of smooth real functions on Z.

Given a graded manifold $(Z, \mathfrak{A}_E)$, every trivialization chart $(U; z^A, y^a)$ of the vector bundle $E \to Z$ yields a splitting domain $(U; z^A, c^a)$ of $(Z, \mathfrak{A}_E)$. Graded functions on such a chart are Λ-valued functions

$$f = \sum_{k=0}^{m} \frac{1}{k!} f_{a_1\ldots a_k}(z) c^{a_1} \cdots c^{a_k}, \tag{6.15}$$

where $f_{a_1\cdots a_k}(z)$ are smooth functions on U and $\{c^a\}$ is the fibre basis for E^*. In particular, the sheaf epimorphism σ in (6.13) is induced by the body map of Λ. One calls $\{z^A, c^a\}$ the local *basis for the graded manifold* $(Z, \mathfrak{A}_E)$ [7]. Transition functions $y'^a = \rho^a_b(z^A) y^b$ of bundle coordinates on $E \to Z$ induce the corresponding transformation

$$c'^a = \rho^a_b(z^A) c^b \tag{6.16}$$

of the associated local basis for a graded manifold $(Z, \mathfrak{A}_E)$ and the according coordinate transformation law of graded functions (6.15).

Remark 6.2 Strictly speaking, elements c^a of the local basis for a graded manifold are locally constant sections c^a of $E^* \to X$ such that $y_b \circ c^a = \delta^a_b$. Therefore, graded functions are locally represented by Λ-valued functions (6.15), but they are not Λ-valued functions on a manifold Z because of the transformation law (6.16).

Remark 6.3 In general, automorphisms of a graded manifold take the form

$$c'^a = \rho^a(z^A, c^b), \tag{6.17}$$

where $\rho^a(z^A, c^b)$ are local graded functions. Considering a simple graded manifold $(Z, \mathfrak{A}_E)$, we restrict the class of graded manifold transformations (6.17) to the linear ones (6.16), compatible with given Batchelor's isomorphism.

Given a graded manifold $(Z, \mathfrak{A})$, by the *sheaf* $\mathfrak{d}\mathfrak{A}$ *of graded derivations* of $\mathfrak{A}$ is meant a subsheaf of endomorphisms of the structure sheaf $\mathfrak{A}$ such that any section $u \in \mathfrak{d}\mathfrak{A}(U)$ of $\mathfrak{d}\mathfrak{A}$ over an open subset $U \subset Z$ is a graded derivation of the real graded commutative algebra $\mathfrak{A}(U)$, i.e., $u \in \mathfrak{d}(\mathfrak{A}(U))$. Conversely, one can show that, given open sets $U' \subset U$, there is a surjection of the graded derivation modules

$$\mathfrak{d}(\mathfrak{A}(U)) \to \mathfrak{d}(\mathfrak{A}(U')).$$

It follows that any graded derivation of a local graded algebra $\mathfrak{A}(U)$ also is a local section over U of a sheaf $\mathfrak{d}\mathfrak{A}$. Global sections of $\mathfrak{d}\mathfrak{A}$ are called *graded vector fields* on the graded manifold $(Z, \mathfrak{A})$. They make up a graded derivation module $\mathfrak{d}\mathfrak{A}(Z)$ of a real graded commutative ring $\mathfrak{A}(Z)$. This module is a real Lie superalgebra with respect to the superbracket (6.3).

A key point is that graded vector fields $u \in \mathfrak{d}\mathcal{A}_E$ on a simple graded manifold $(Z, \mathfrak{A}_E)$ can be represented by sections of some vector bundle as follows [58, 100, 131].

Due to the canonical splitting $VE = E \times E$, the vertical tangent bundle VE of $E \to Z$ can be provided with the fibre bases $\{\partial/\partial c^a\}$, which are the duals of the bases $\{c^a\}$. Then graded vector fields on a splitting domain $(U; z^A, c^a)$ of $(Z, \mathfrak{A}_E)$ read

$$u = u^A \partial_A + u^a \frac{\partial}{\partial c^a}, \tag{6.18}$$

where u^λ, u^a are local graded functions on U. In particular,

$$\frac{\partial}{\partial c^a} \circ \frac{\partial}{\partial c^b} = -\frac{\partial}{\partial c^b} \circ \frac{\partial}{\partial c^a}, \qquad \partial_A \circ \frac{\partial}{\partial c^a} = \frac{\partial}{\partial c^a} \circ \partial_A.$$

The graded derivations (6.18) act on graded functions $f \in \mathfrak{A}_E(U)$ (6.15) by the rule

$$u(f_{a\ldots b} c^a \cdots c^b) = u^A \partial_A (f_{a\ldots b}) c^a \cdots c^b + u^k f_{a\ldots b} \frac{\partial}{\partial c^k} \rfloor (c^a \cdots c^b). \tag{6.19}$$

This rule implies the corresponding coordinate transformation law

$$u'^A = u^A, \qquad u'^a = \rho^a_j u^j + u^A \partial_A (\rho^a_j) c^j$$

of graded vector fields. It follows that graded vector fields (6.18) can be represented by sections of the following vector bundle $\mathcal{V}_E \to Z$. This vector bundle is locally isomorphic to a vector bundle

$$\mathcal{V}_E \approx \wedge E^* \underset{Z}{\otimes} (E \underset{Z}{\oplus} TZ), \tag{6.20}$$

and is characterized by an atlas of bundle coordinates $(z^A, z^A_{a_1\ldots a_k}, v^i_{b_1\ldots b_k})$, $k = 0, \ldots, m$, possessing the transition functions

$$z'^A_{i_1\dots i_k} = \rho^{-1}{}^{a_1}_{i_1}\cdots\rho^{-1}{}^{a_k}_{i_k} z^A_{a_1\dots a_k},$$
$$v'^i_{j_1\dots j_k} = \rho^{-1}{}^{b_1}_{j_1}\cdots\rho^{-1}{}^{b_k}_{j_k}\left[\rho^i_j v^j_{b_1\dots b_k} + \frac{k!}{(k-1)!} z^A_{b_1\dots b_{k-1}}\partial_A\rho^i_{b_k}\right],$$

which fulfil the cocycle condition (B.4). Thus, the graded derivation module $\mathfrak{d}\mathcal{A}_E$ is isomorphic to the structure module $\mathcal{V}_E(Z)$ of global sections of a vector bundle $\mathcal{V}_E \to Z$.

Given the structure ring $\mathcal{A}_E$ of graded functions on a simple graded manifold $(Z, \mathfrak{A}_E)$ and the real Lie superalgebra $\mathfrak{d}\mathcal{A}_E$ of its graded derivations, let us consider the graded Chevalley–Eilenberg differential calculus

$$\mathcal{S}^*[E; Z] = \mathcal{O}^*[\mathfrak{d}\mathcal{A}_E] \tag{6.21}$$

over $\mathcal{A}_E$ where $\mathcal{S}^0[E; Z] = \mathcal{A}_E$.

Theorem 6.3 *Since a graded derivation module $\mathfrak{d}\mathcal{A}_E$ is isomorphic to the structure module of sections of a vector bundle $\mathcal{V}_E \to Z$, elements of $\mathcal{S}^*[E; Z]$ are represented by sections of the exterior bundle $\wedge\overline{\mathcal{V}}_E$ of the $\mathcal{A}_E$-dual $\overline{\mathcal{V}}_E \to Z$ of $\mathcal{V}_E$.*

The bundle $\overline{\mathcal{V}}_E$ is locally isomorphic to a vector bundle

$$\overline{\mathcal{V}}_E \approx \wedge E^* \underset{Z}{\otimes}(E^* \underset{Z}{\oplus} T^*Z). \tag{6.22}$$

With respect to the dual fibre bases $\{dz^A\}$ for T^*Z and $\{dc^b\}$ for E^*, sections of $\overline{\mathcal{V}}_E$ take a coordinate form

$$\phi = \phi_A dz^A + \phi_a dc^a,$$

together with transition functions

$$\phi'_a = \rho^{-1}{}^b_a \phi_b, \qquad \phi'_A = \phi_A + \rho^{-1}{}^b_a \partial_A(\rho^a_j)\phi_b c^j.$$

The duality isomorphism $\mathcal{S}^1[E; Z] = \mathfrak{d}\mathcal{A}^*_E$ (6.9) is given by the *graded interior product*

$$u\rfloor\phi = u^A\phi_A + (-1)^{[\phi_a]} u^a \phi_a.$$

Elements of $\mathcal{S}^*[E; Z]$ are called *graded exterior forms* on a graded manifold $(Z, \mathfrak{A}_E)$.

Seen as an $\mathcal{A}_E$-algebra, the DBGA $\mathcal{S}^*[E; Z]$ (6.21) on a splitting domain (z^A, c^a) is locally generated by the graded one-forms dz^A, dc^i such that

$$dz^A \wedge dc^i = -dc^i \wedge dz^A, \qquad dc^i \wedge dc^j = dc^j \wedge dc^i.$$

Accordingly, the graded Chevalley–Eilenberg coboundary operator d (6.5), termed the *graded exterior differential*, reads

$$d\phi = dz^A \wedge \partial_A \phi + dc^a \wedge \frac{\partial}{\partial c^a}\phi,$$

where derivatives ∂_λ, $\partial/\partial c^a$ act on coefficients of graded exterior forms by the formula (6.19), and they are graded commutative with the graded exterior forms dz^A and dc^a. The formulas (6.7)–(6.11) hold.

Theorem 6.4 *The DBGA $\mathcal{S}^*[E;Z]$ (6.21) is a minimal differential calculus over $\mathcal{A}_E$, i.e., it is generated by elements df, $f \in \mathcal{A}_E$.*

Proof The proof follows that of Theorem 1.8. Since $\mathfrak{d}\mathcal{A}_E = \mathcal{V}_E(Z)$, it is a projective $C^\infty(Z)$- and $\mathcal{A}_E$-module of finite rank, and so is its $\mathcal{A}_E$-dual $\mathcal{S}^1[E;Z]$. Hence, $\mathfrak{d}\mathcal{A}_E$ is the $\mathcal{A}_E$-dual of $\mathcal{S}^1[E;Z]$ and, consequently, $\mathcal{S}^1[E;Z]$ is generated by elements df, $f \in \mathcal{A}_E$. □

The bigraded de Rham complex (6.12) of a minimal graded Chevalley–Eilenberg differential calculus $\mathcal{S}^*[E;Z]$ reads

$$0 \to \mathbb{R} \to \mathcal{A}_E \xrightarrow{d} \mathcal{S}^1[E;Z] \xrightarrow{d} \cdots \mathcal{S}^k[E;Z] \xrightarrow{d} \cdots. \tag{6.23}$$

Its cohomology $H^*(\mathcal{A}_E)$ is called the *de Rham cohomology of a simple graded manifold* $(Z, \mathfrak{A}_E)$.

In particular, given a DGA $\mathcal{O}^*(Z)$ of exterior forms on Z, there exist a canonical monomorphism

$$\mathcal{O}^*(Z) \to \mathcal{S}^*[E;Z] \tag{6.24}$$

and the body epimorphism $\mathcal{S}^*[E;Z] \to \mathcal{O}^*(Z)$ which are cochain morphisms of the de Rham complexes (6.23) and (A.18).

Theorem 6.5 *The de Rham cohomology of a simple graded manifold $(Z, \mathfrak{A}_E)$ equals the de Rham cohomology of its body Z.*

Proof Let $\mathfrak{A}^k_E$ denote the sheaf of germs of graded k-forms on $(Z, \mathfrak{A}_E)$. Its structure module is $\mathcal{S}^k[E;Z]$. These sheaves constitute the complex

$$0 \to \mathbb{R} \longrightarrow \mathfrak{A}_E \xrightarrow{d} \mathfrak{A}^1_E \xrightarrow{d} \cdots \mathfrak{A}^k_E \xrightarrow{d} \cdots. \tag{6.25}$$

Its members $\mathfrak{A}^k_E$ are sheaves of C^∞_Z-modules on Z and, consequently, are fine and acyclic. Furthermore, the Poincaré lemma for graded exterior forms holds [7]. It follows that the complex (6.25) is a fine resolution of the constant sheaf $\mathbb{R}$ on a manifold Z. Then, by virtue of Theorem C.5, there is an isomorphism

$$H^*(\mathcal{A}_E) = H^*(Z;\mathbb{R}) = H^*_{\rm DR}(Z) \tag{6.26}$$

of the cohomology $H^*(\mathscr{A}_E)$ to the de Rham cohomology $H^*_{\rm DR}(Z)$ of a smooth manifold Z. □

Theorem 6.6 *The cohomology isomorphism (6.26) is accompanied by the cochain monomorphism (6.24). Hence, any closed graded exterior form is decomposed into a sum $\phi = \sigma + d\xi$ where σ is a closed exterior form on Z.*

Graded Bundles

A *morphism of graded manifolds* $(Z, \mathfrak{A}) \to (Z', \mathfrak{A}')$ is defined as that of local-ringed spaces

$$\phi : Z \to Z', \qquad \widehat{\Phi} : \mathfrak{A}' \to \phi_* \mathfrak{A}, \tag{6.27}$$

where ϕ is a manifold morphism and $\widehat{\Phi}$ is a sheaf morphism of $\mathfrak{A}'$ to the direct image $\phi_*\mathfrak{A}$ of $\mathfrak{A}$ onto Z' (Sect. C.3).

The morphism (6.27) of graded manifolds is said to be:

- a monomorphism if ϕ is an injection and $\widehat{\Phi}$ is an epimorphism;
- an epimorphism if ϕ is a surjection and $\widehat{\Phi}$ is a monomorphism.

An epimorphism of graded manifolds $(Z, \mathfrak{A}) \to (Z', \mathfrak{A}')$ where $Z \to Z'$ is a fibre bundle is called the *graded bundle* [73, 140]. In this case, a sheaf monomorphism $\widehat{\Phi}$ induces a monomorphism of canonical presheaves $\overline{\mathfrak{A}'} \to \overline{\mathfrak{A}}$, which associates to each open subset $U \subset Z$ the ring of sections of $\mathfrak{A}'$ over $\phi(U)$. Accordingly, there is a pull-back monomorphism of the structure rings $\mathfrak{A}'(Z') \to \mathfrak{A}(Z)$ of graded functions on graded manifolds $(Z', \mathfrak{A}')$ and $(Z, \mathfrak{A})$.

In particular, let $(Y, \mathfrak{A})$ be a graded manifold whose body $Z = Y$ is a fibre bundle $\pi : Y \to X$. Let us consider a trivial graded manifold $(X, \mathfrak{A}_0 = C^\infty_X)$ (Remark 6.1). Then we have a graded bundle

$$(Y, \mathfrak{A}) \to (X, C^\infty_X). \tag{6.28}$$

Let us denote it by $(X, Y, \mathfrak{A})$. Given a graded bundle $(X, Y, \mathfrak{A})$, the local basis for a graded manifold $(Y, \mathfrak{A})$ can be brought into a form (x^λ, y^i, c^a) where (x^λ, y^i) are bundle coordinates of $Y \to X$.

Definition 6.5 We agree to call the graded bundle (6.28) over a trivial graded manifold (X, C^∞_X) the *graded bundle over a smooth manifold*.

If $Y \to X$ is a vector bundle, the graded bundle (6.28) is a particular case of graded fibre bundles in [73, 107] when their base is a trivial graded manifold.

Remark 6.4 Let $Y \to X$ be a fibre bundle. Then a trivial graded manifold (Y, C^∞_Y) together with a real ring monomorphism $C^\infty(X) \to C^\infty(Y)$ is a graded bundle (X, Y, C^∞_Y) of trivial graded manifolds

$$(Y, C^\infty_Y) \to (X, C^\infty_X). \tag{6.29}$$

Remark 6.5 A graded manifold $(X, \mathfrak{A})$ itself can be treated as the graded bundle $(X, X, \mathfrak{A})$ (6.28) associated to the identity smooth bundle $X \to X$.

Let $E \to Z$ and $E' \to Z'$ be vector bundles and $\Phi : E \to E'$ their bundle morphism over a morphism $\phi : Z \to Z'$. Then every section s^* of the dual bundle $E'^* \to Z'$ defines the pull-back section $\Phi^* s^*$ of the dual bundle $E^* \to Z$ by the law

$$v_z \rfloor \Phi^* s^*(z) = \Phi(v_z) \rfloor s^*(\varphi(z)), \qquad v_z \in E_z.$$

It follows that a bundle morphism (Φ, ϕ) yields a morphism of simple graded manifolds

$$(Z, \mathfrak{A}_E) \to (Z', \mathfrak{A}_{E'}). \tag{6.30}$$

This is a pair $(\phi, \widehat{\Phi} = \phi_* \circ \Phi^*)$ of a morphism ϕ of body manifolds and the composition $\phi_* \circ \Phi^*$ of the pull-back $\mathcal{A}_{E'} \ni f \to \Phi^* f \in \mathcal{A}_E$ of graded functions and the direct image ϕ_* of a sheaf $\mathfrak{A}_E$ onto Z'. Relative to local bases (z^A, c^a) and (z'^A, c'^a) for $(Z, \mathfrak{A}_E)$ and $(Z', \mathfrak{A}_{E'})$, the morphism (6.30) of simple graded manifolds reads $z' = \phi(z)$, $\widehat{\Phi}(c'^a) = \Phi^a_b(z) c^b$.

The graded manifold morphism (6.30) is a monomorphism (resp. epimorphism) if Φ is a bundle injection (resp. surjection).

In particular, the graded manifold morphism (6.30) is a graded bundle if Φ is a fibre bundle. Let $\mathcal{A}_{E'} \to \mathcal{A}_E$ be the corresponding pull-back monomorphism of the structure rings. By virtue of Theorem 6.4 it yields a monomorphism of the DBGAs

$$\mathcal{S}^*[E'; Z'] \to \mathcal{S}^*[E; Z]. \tag{6.31}$$

Let $(Y, \mathfrak{A}_F)$ be a simple graded manifold modelled over a vector bundle $F \to Y$. This is a graded bundle $(X, Y, \mathfrak{A}_F)$:

$$(Y, \mathfrak{A}_F) \to (X, C^\infty_X) \tag{6.32}$$

modelled over a composite bundle

$$F \to Y \to X. \tag{6.33}$$

The structure ring of graded functions on a simple graded manifold $(Y, \mathfrak{A}_F)$ is the graded commutative $C^\infty(X)$-ring $\mathcal{A}_F = \wedge F^*(Y)$ (6.14). Let the composite bundle (6.33) be provided with adapted bundle coordinates (x^λ, y^i, q^a) possessing transition functions

$$x'^\lambda(x^\mu), \qquad y'^i(x^\mu, y^j), \qquad q'^a = \rho^a_b(x^\mu, y^j) q^b.$$

The corresponding local basis for a simple graded manifold $(Y, \mathfrak{A}_F)$ is (x^λ, y^i, c^a) together with transition functions

$$x'^{\lambda}(x^{\mu}), \qquad y'^{i}(x^{\mu}, y^{j}), \qquad c'^{a} = \rho^{a}_{b}(x^{\mu}, j^{j})c^{b}.$$

We call it the local *basis for a graded bundle* $(X, Y, \mathfrak{A}_F)$.

Graded Jet Manifolds

As was shown above, Lagrangian theory on a smooth fibre bundle $Y \to X$ is formulated in terms of the variational bicomplex on jet manifolds J^*Y of Y. These are fibre bundles over X and, therefore, they can be regarded as trivial graded bundles $(X, J^kY, C^{\infty}_{J^kY})$. Then let us describe their partners in the case of graded bundles as follows.

Let us note that, given a graded manifold $(X, \mathfrak{A})$ and its structure ring $\mathcal{A}$, one can define the jet module $J^1\mathcal{A}$ of a $C^{\infty}(X)$-ring $\mathcal{A}$ [58, 131]. If $(X, \mathfrak{A}_E)$ is a simple graded manifold modelled over a vector bundle $E \to X$, the jet module $J^1\mathcal{A}_E$ is a module of global sections of the jet bundle $J^1(\wedge E^*)$. A problem is that $J^1\mathcal{A}_E$ fails to be a structure ring of some graded manifold. By this reason, we have suggested a different construction of jets of graded manifolds (Definition 6.6), though it is applied only to simple graded manifolds [61, 131, 134].

Let $(X, \mathcal{A}_E)$ be a simple graded manifold modelled over a vector bundle $E \to X$. Let us consider a k-order jet manifold J^kE of E. It is a vector bundle over X. Then let $(X, \mathcal{A}_{J^kE})$ be a simple graded manifold modelled over $J^kE \to X$.

Definition 6.6 We agree to call $(X, \mathcal{A}_{J^kE})$ the *graded k-order jet manifold* of a simple graded manifold $(X, \mathcal{A}_E)$.

Given a splitting domain $(U; x^{\lambda}, c^{a})$ of a graded manifold $(Z, \mathcal{A}_E)$, we have a splitting domain

$$(U; x^{\lambda}, c^{a}, c^{a}_{\lambda}, c^{a}_{\lambda_1\lambda_2}, \dots c^{a}_{\lambda_1\dots\lambda_k})$$

of a graded jet manifold $(X, \mathcal{A}_{J^kE})$.

As was mentioned above, a graded manifold is a particular graded bundle over its body (Remark 6.5). Then Definition 6.6 of graded jet manifolds is generalized to graded bundles over smooth manifolds as follows.

Let $(X, Y, \mathfrak{A}_F)$ be the graded bundle (6.32) modelled over the composite bundle (6.33). It is readily observed that the jet manifold J^rF of $F \to X$ is a vector bundle $J^rF \to J^rY$ coordinated by $(x^{\lambda}, y^{i}_{\Lambda}, q^{a}_{\Lambda})$, $0 \leq |\Lambda| \leq r$. Let $(J^rY, \mathfrak{A}_{J^rF})$ be a simple graded manifold modelled over this vector bundle. Its local generating basis is $(x^{\lambda}, y^{i}_{\Lambda}, c^{a}_{\Lambda})$, $0 \leq |\Lambda| \leq r$.

Definition 6.7 We call $(J^rY, \mathfrak{A}_{J^rF})$ the *graded r-order jet manifold of a graded bundle* $(X, Y, \mathfrak{A}_F)$.

In particular, let $Y \to X$ be a smooth bundle seen as a trivial graded bundle (X, Y, C^{∞}_{Y}) modelled over a composite bundle $Y \times \{0\} \to Y \to X$. Then its graded jet manifold is a trivial graded bundle $(X, J^rY, C^{\infty}_{J^rY})$, i.e., the jet manifold J^rY of Y.

Thus, Definition 6.7 of graded jet manifolds of graded bundles is compatible with the conventional Definition B.11 of jets of fibre bundles.

The affine bundles $J^{r+1}Y \to J^rY$ (1.1) and the corresponding fibre bundles $J^{r+1}F \to J^rF$ also yield the graded bundles

$$(J^{r+1}Y, \mathfrak{A}_{J^{r+1}F}) \to (J^rY, \mathfrak{A}_{J^rF}),$$

including the sheaf monomorphisms

$$\pi^{r+1*}_r \mathfrak{A}_{J^rF} \to \mathfrak{A}_{J^{r+1}F}, \tag{6.34}$$

where $\pi^{r+1*}_r\mathfrak{A}_r$ is the pull-back onto $J^{r+1}Y$ of the continuous fibre bundle $\mathfrak{A}_{J^rF} \to J^rY$. The sheaf monomorphism (6.34) induces a monomorphism of the canonical presheaves

$$\overline{\mathfrak{A}}_{J^rF} \to \overline{\mathfrak{A}}_{J^{r+1}F}, \tag{6.35}$$

which associates to each open subset $U \subset J^{r+1}Y$ the ring of sections of $\mathfrak{A}_{J^{r+1}F}$ over $\pi^{r+1}_r(U)$. Accordingly, there is a pull-back monomorphism of the structure rings

$$\mathcal{S}^0_r[F;Y] \to \mathcal{S}^0_{r+1}[F;Y], \qquad \mathcal{S}^0_k[F;Y] = \mathcal{S}^0[J^kF; J^kY], \tag{6.36}$$

of graded functions on graded manifolds $(J^rY, \mathfrak{A}_{J^rF})$ and $(J^{r+1}Y, \mathfrak{A}_{J^{r+1}F})$. As a consequence, we have the inverse sequence of graded manifolds

$$(Y, \mathfrak{A}_F) \longleftarrow (J^1Y, \mathfrak{A}_{J^1F}) \longleftarrow \cdots (J^{r-1}Y, \mathfrak{A}_{J^{r-1}F}) \longleftarrow (J^rY, \mathfrak{A}_{J^rF}) \longleftarrow \cdots.$$

One can think on its inverse limit $(J^\infty Y, \mathfrak{A}_{J^\infty F})$ as the *graded infinite order jet manifold* whose body is an infinite order jet manifold $J^\infty Y$ and whose structure sheaf $\mathfrak{A}_{J^\infty F}$ is a sheaf of germs of graded functions on graded manifolds $(J^*Y, \mathfrak{A}_{J^*F})$ [61, 134]. However $(J^\infty Y, \mathfrak{A}_{J^\infty F})$ fails to be a graded manifold in a strict sense because the inverse limit $J^\infty Y$ of the sequence (1.1) is a Fréche manifold, but not the smooth one.

By virtue of Theorem 6.4, the differential calculuses $\mathcal{S}^*_r[F;Y]$ of graded exterior forms on graded manifolds $(J^rY, \mathfrak{A}_{J^rF})$ are minimal. Therefore, the monomorphisms of structure rings (6.36) yields the pull-back monomorphisms (6.31) of DBGAs

$$\pi^{r+1*}_r : \mathcal{S}^*_r[F;Y] \to \mathcal{S}^*_{r+1}[F;Y], \qquad \mathcal{S}^*_k[F;Y] = \mathcal{S}^*[J^kF; J^kY]. \tag{6.37}$$

As a consequence, we have a direct system of DBGAs

$$\mathcal{S}^*[F;Y] \xrightarrow{\pi^*} \mathcal{S}^*_1[F;Y] \longrightarrow \cdots \mathcal{S}^*_{r-1}[F;Y] \xrightarrow{\pi^{r*}_{r-1}} \mathcal{S}^*_r[F;Y] \longrightarrow \cdots. \tag{6.38}$$

The DBGA $\mathcal{S}^*_\infty[F;Y]$ that we associate to a graded bundle $(X, Y, \mathfrak{A}_F)$ is defined as the direct limit

$$\mathcal{S}^*_\infty[F;Y] = \overrightarrow{\lim}\, \mathcal{S}^*_r[F;Y] \tag{6.39}$$

of the direct system (6.38). It consists of all graded exterior forms $\phi \in \mathcal{S}^*_r[F;Y]$ on graded manifolds $(J^rY, \mathfrak{A}_{J^rF})$ modulo the monomorphisms (6.37). Its elements obey the relations (6.7)–(6.8).

The cochain monomorphisms $\mathcal{O}^*_rY = \mathcal{O}^*(J^rY) \to \mathcal{S}^*_r[F;Y]$ (6.24) provide a monomorphism of the direct system (1.6) to the direct system (6.38) and, consequently, a monomorphism

$$\mathcal{O}^*_\infty Y \to \mathcal{S}^*_\infty[F;Y] \tag{6.40}$$

of their direct limits. In particular, $\mathcal{S}^*_\infty[F;Y]$ is an $\mathcal{O}^0_\infty Y$-algebra. Accordingly, the body epimorphisms $\mathcal{S}^*_r[F;Y] \to \mathcal{O}^*_rY$ yield an epimorphism of $\mathcal{O}^0_\infty Y$-algebras

$$\mathcal{S}^*_\infty[F;Y] \to \mathcal{O}^*_\infty Y. \tag{6.41}$$

It is readily observed that the morphisms (6.40) and (6.41) are cochain morphisms between the de Rham complex (1.8) of $\mathcal{O}^*_\infty Y$ and the de Rham complex

$$0 \to \mathbb{R} \longrightarrow \mathcal{S}^0_\infty[F;Y] \stackrel{d}{\longrightarrow} \mathcal{S}^1_\infty[F;Y] \cdots \stackrel{d}{\longrightarrow} \mathcal{S}^k_\infty[F;Y] \longrightarrow \cdots \tag{6.42}$$

of a DBGA $\mathcal{S}^*_\infty[F;Y]$. Moreover, the corresponding homomorphisms of cohomology groups of these complexes are isomorphisms as follows [61].

Theorem 6.7 *There is an isomorphism*

$$H^*(\mathcal{S}^*_\infty[F;Y]) = H^*_{\rm DR}(Y) \tag{6.43}$$

of the cohomology $H^*(\mathcal{S}^*_\infty[F;Y])$ of the de Rham complex (6.42) to the de Rham cohomology $H^*_{\rm DR}(Y)$ of Y.

Proof The complex (6.42) is the direct limit of the de Rham complexes of the DBGAs $\mathcal{S}^*_r[F;Y]$. Therefore, the direct limit of cohomology groups of these complexes is the cohomology of the de Rham complex (6.42) (Theorem A.8). By virtue of the cohomology isomorphism (6.26), cohomology of the de Rham complex of $\mathcal{S}^*_r[F;Y]$ equals the de Rham cohomology of J^rY and, consequently, that of Y, which is the strong deformation retract of any jet manifold J^rY. Hence, the isomorphism (6.43) holds. □

Theorem 6.8 *Any closed graded exterior form $\phi \in \mathcal{S}^*_\infty[F;Y]$ is decomposed into a sum $\phi = \sigma + d\xi$ where σ is a closed exterior form on Y.*

One can think of elements of $\mathcal{S}^*_\infty[F;Y]$ as being graded differential forms on an infinite order jet manifold $J^\infty Y$ [61, 134]. Indeed, the presheaf monomorphisms (6.35) define a direct system of presheaves

$$\overline{\mathfrak{A}}^*_F \longrightarrow \overline{\mathfrak{A}}^*_{J^1F} \longrightarrow \cdots \overline{\mathfrak{A}}^*_{J^rF} \longrightarrow \cdots,$$

whose direct limit $\overline{\mathfrak{Q}}^*_\infty[F;Y]$ is a presheaf of DBGAs of differential forms on an infinite order jet manifold $J^\infty Y$. Let $\mathfrak{Q}^*_\infty[F;Y]$ be the sheaf of DBGAs of germs of the presheaf $\overline{\mathfrak{Q}}_\infty[F;Y]$. One can think of the pair $(J^\infty Y, \mathfrak{Q}^0_\infty[F;Y])$ as being a graded Fréchet manifold, whose body is an infinite order jet manifold $J^\infty Y$ and the structure sheaf $\mathfrak{Q}^0_\infty[F;Y]$ is a sheaf of germs of graded functions on graded manifolds $(J^r Y, \mathfrak{A}_{J^r F})$. The structure module $\mathcal{Q}^*_\infty[F;Y]$ of sections of $\mathfrak{Q}^*_\infty[F;Y]$ is a DBGA such that, given an element $\phi \in \mathcal{Q}^*_\infty[F;Y]$ and a point $z \in J^\infty Y$, there exist an open neighborhood U of z and a graded exterior form $\phi^{(k)}$ on some finite order jet manifold $J^k Y$ so that $\phi|_U = \pi^{\infty *}_k \phi^{(k)}|_U$. In particular, there is a monomorphism

$$\mathcal{S}^*_\infty[F;Y] \to \mathcal{Q}^*_\infty[F;Y]. \tag{6.44}$$

Due to this monomorphism, one can restrict a DBGA $\mathcal{S}^*_\infty[F;Y]$ to the coordinate chart (1.3) of $J^\infty Y$ with the local basis

$$\{c^a_\Lambda, dx^\lambda, \theta^a_\Lambda = dc^a_\Lambda - c^a_{\lambda+\Lambda}dx^\lambda, \theta^i_\Lambda = dy^i_\Lambda - y^i_{\lambda+\Lambda}dx^\lambda\}, \qquad 0 \le |\Lambda|,$$

where c^a_Λ, θ^a_Λ are odd and dx^λ, θ^i_Λ are even. Let the collective symbol s^A stand for its elements. Accordingly, the notation s^A_Λ for their jets and the notation

$$\theta^A_\Lambda = ds^A_\Lambda - s^A_{\lambda+\Lambda}dx^\lambda \tag{6.45}$$

for the contact forms are introduced. For the sake of simplicity, we further denote $[A] = [s^A]$.

Remark 6.6 Let $(X, Y, \mathfrak{A}_F)$ and $(X, Y', \mathfrak{A}_{F'})$ be graded bundles modelled over composite bundles $F \to Y \to X$ and $F' \to Y' \to X$, respectively. Let $F \to F'$ be a fibre bundle over a fibre bundle $Y \to Y'$ over X. Then we have a graded bundle

$$(X, Y, \mathfrak{A}_F) \to (X, Y', \mathfrak{A}_{F'})$$

together with the pull-back monomorphism (6.31) of DBGAs

$$\mathcal{S}^*[F';Y'] \to \mathcal{S}^*[F;Y]. \tag{6.46}$$

Let $(X, J^r Y, \mathfrak{A}_{J^r F})$ and $(X, J^r Y', \mathfrak{A}_{J^r F'})$ be graded bundles modelled over composite bundles $J^r F \to J^r Y \to X$ and $J^r F' \to J^r Y' \to X$, respectively. Since $J^r F \to J^r F'$ is a fibre bundle over a fibre bundle $J^r Y \to J^r Y'$ over X [53], we also get a graded bundle

$$(X, J^r Y, \mathfrak{A}_{J^r F}) \to (X, J^r Y', \mathfrak{A}_{J^r F'})$$

together with the pull-back monomorphism of DBGAs

$$\mathcal{S}^*_r[F';Y'] \to \mathcal{S}^*_r[F;Y]. \tag{6.47}$$

The monomorphisms (6.46) and (6.47), $r = 1, 2, \ldots$, provide a monomorphism of their direct limits (6.39):

$$\mathscr{S}^*_\infty[F'; Y'] \to \mathscr{S}^*_\infty[F; Y]. \tag{6.48}$$

Remark 6.7 Let $(X, Y, \mathfrak{A}_F)$ and $(X, Y', \mathfrak{A}_{F'})$ be graded bundles modelled over composite bundles $F \to Y \to X$ and $F' \to Y' \to X$, respectively. We define their product over X as the graded bundle

$$(X, Y, \mathfrak{A}_F) \underset{X}{\times} (X, Y', \mathfrak{A}_{F'}) = (X, Y \underset{X}{\times} Y', \mathfrak{A}_{F \underset{X}{\times} F'}) \tag{6.49}$$

modelled over a composite bundle

$$F \underset{X}{\times} F' = F \underset{Y \times Y'}{\times} F' \to Y \underset{X}{\times} Y' \to X. \tag{6.50}$$

Let us consider the corresponding DBGA (6.39):

$$\mathscr{S}^*_\infty[F \underset{X}{\times} F'; Y \underset{X}{\times} Y']. \tag{6.51}$$

Then in accordance with Remark 6.6, there are the monomorphisms (6.48):

$$\mathscr{S}^*_\infty[F; Y] \to \mathscr{S}^*_\infty[F \underset{X}{\times} F; Y \underset{X}{\times} Y'], \qquad \mathscr{S}^*_\infty[F'; Y'] \to \mathscr{S}^*_\infty[F \underset{X}{\times} F; Y \underset{X}{\times} Y']. \tag{6.52}$$

6.4 Grassmann-Graded Variational Bicomplex

Let $(X, Y, \mathfrak{A}_F)$ be the graded bundle (6.32) modelled over the composite bundle (6.33) over an n-dimensional smooth manifold X, and let $\mathscr{S}^*_\infty[F; Y]$ be the associated DBGA (6.39) of graded exterior forms on graded jet manifolds of $(X, Y, \mathfrak{A}_F)$. As was mentioned above Grassmann-graded Lagrangian theory of even and odd variables on a graded bundle is formulated in terms of the variational bicomplex which the DBGA $\mathscr{S}^*_\infty[F; Y]$ is split in [12, 59, 61, 134].

A DBGA $\mathscr{S}^*_\infty[F; Y]$ is decomposed into $\mathscr{S}^0_\infty[F; Y]$-modules $\mathscr{S}^{k,r}_\infty[F; Y]$ of *k-contact and r-horizontal graded forms* together with the corresponding projections

$$h_k : \mathscr{S}^*_\infty[F; Y] \to \mathscr{S}^{k,*}_\infty[F; Y], \qquad h^m : \mathscr{S}^*_\infty[F; Y] \to \mathscr{S}^{*,m}_\infty[F; Y].$$

In particular, the horizontal projection h_0 takes a form

$$h_0 : \mathscr{S}^*_\infty[F; Y] \to \mathscr{S}^{0,*}_\infty[F, Y], \qquad h_0(dx^\lambda) = dx^\lambda, \qquad h_0(\theta^i_\Lambda) = 0.$$

Accordingly, the graded exterior differential d on $\mathcal{S}^*_\infty[F;Y]$ falls into a sum $d = d_V + d_H$ of the *vertical graded differential*

$$d_V \circ h^m = h^m \circ d \circ h^m, \qquad d_V(\phi) = \theta^A_\Lambda \wedge \partial^\Lambda_A \phi, \qquad \phi \in \mathcal{S}^*_\infty[F;Y],$$

and the *total graded differential*

$$d_H \circ h_k = h_k \circ d \circ h_k, \qquad d_H \circ h_0 = h_0 \circ d, \qquad d_H(\phi) = dx^\lambda \wedge d_\lambda(\phi),$$

where

$$d_\lambda = \partial_\lambda + \sum_{0 \leq |\Lambda|} s^A_{\lambda+\Lambda} \partial^\Lambda_A$$

are the *graded total derivatives*. These differentials obey the nilpotent relations (1.15).

Similarly to the DGA $\mathcal{O}^*_\infty Y$ (1.7), a DBGA $\mathcal{S}^*_\infty[F;Y]$ is provided with the graded projection endomorphism

$$\rho = \sum_{k>0} \frac{1}{k} \overline{\rho} \circ h_k \circ h^n : \mathcal{S}^{*>0,n}_\infty[F;Y] \to \mathcal{S}^{*>0,n}_\infty[F;Y], \tag{6.53}$$

$$\overline{\rho}(\phi) = \sum_{0 \leq |\Lambda|} (-1)^{|\Lambda|} \theta^A \wedge [d_\Lambda(\partial^\Lambda_A \rfloor \phi)], \qquad \phi \in \mathcal{S}^{>0,n}_\infty[F;Y],$$

such that $\rho \circ d_H = 0$, and with the nilpotent *graded variational operator*

$$\delta = \rho \circ d \mathcal{S}^{*,n}_\infty[F;Y] \to \mathcal{S}^{*+1,n}_\infty[F;Y].$$

With these operators a DBGA $\mathcal{S}^*_\infty[F;Y]$ is decomposed into the *Grassmann-graded variational bicomplex*

$$\begin{array}{ccccccccc}
 & & \vdots & & \vdots & & \vdots & & \vdots \\
 & & {\scriptstyle d_V}\uparrow & & {\scriptstyle d_V}\uparrow & & {\scriptstyle d_V}\uparrow & & {\scriptstyle -\delta}\uparrow \\
0 & \to & \mathcal{S}^{1,0}_\infty & \overset{d_H}{\longrightarrow} & \mathcal{S}^{1,1}_\infty & \overset{d_H}{\longrightarrow} \cdots & \mathcal{S}^{1,n}_\infty & \overset{\rho}{\longrightarrow} & \mathbf{E}_1 \to 0 \\
 & & {\scriptstyle d_V}\uparrow & & {\scriptstyle d_V}\uparrow & & {\scriptstyle d_V}\uparrow & & {\scriptstyle -\delta}\uparrow \\
0 \to \mathbb{R} & \to & \mathcal{S}^0_\infty & \overset{d_H}{\longrightarrow} & \mathcal{S}^{0,1}_\infty & \overset{d_H}{\longrightarrow} \cdots & \mathcal{S}^{0,n}_\infty & \equiv & \mathcal{S}^{0,n}_\infty \\
 & & \uparrow & & \uparrow & & \uparrow & & \\
0 \to \mathbb{R} & \to & \mathcal{O}^0(X) & \overset{d}{\longrightarrow} & \mathcal{O}^1(X) & \overset{d}{\longrightarrow} \cdots & \mathcal{O}^n(X) & \overset{d}{\longrightarrow} & 0 \\
 & & \uparrow & & \uparrow & & \uparrow & & \\
 & & 0 & & 0 & & 0 & &
\end{array} \tag{6.54}$$

where $\mathcal{S}^*_\infty = \mathcal{S}^*_\infty[F;Y]$ and $\mathbf{E}_k = \rho(\mathcal{S}^{k,n}_\infty[F;Y])$ (cf. (1.21)).

We restrict our consideration to its short *variational subcomplex*

$$0 \to \mathbb{R} \to \mathcal{S}^0_\infty[F;Y] \xrightarrow{d_H} \mathcal{S}^{0,1}_\infty[F;Y] \cdots \xrightarrow{d_H} \mathcal{S}^{0,n}_\infty[F;Y] \xrightarrow{\delta} \mathbf{E}_1, \tag{6.55}$$

and the subcomplex of one-contact graded forms

$$0 \to \mathcal{S}^{1,0}_\infty[F;Y] \xrightarrow{d_H} \mathcal{S}^{1,1}_\infty[F;Y] \cdots \xrightarrow{d_H} \mathcal{S}^{1,n}_\infty[F;Y] \xrightarrow{\varrho} \mathbf{E}_1 \to 0. \tag{6.56}$$

They possess the following cohomology [59, 61, 126, 134].

Theorem 6.9 *Cohomology of the complex (6.55) equals the de Rham cohomology $H^*_{DR}(Y)$ of Y.*

Proof See the proof below. □

Theorem 6.10 *The complex (6.56) is exact.*

Proof See the proof below. □

Cohomology of the Grassmann-Graded Variational Bicomplex

The proof of Theorem 6.9 follows the scheme of the proof of Theorem 1.12 [59, 61, 126, 134]. It falls into the three steps.

(I) We start with showing that the complexes (6.55) and (6.56) are locally exact.

Lemma 6.1 *If $Y = \mathbb{R}^{n+k} \to \mathbb{R}^n$, the complex (6.55) is acyclic.*

Proof Referring to [6] for the proof, we summarize a few formulas. Any horizontal graded form $\phi \in \mathcal{S}^{0,*}_\infty[F;Y]$ admits a decomposition

$$\phi = \phi_0 + \widetilde{\phi}, \qquad \widetilde{\phi} = \int_0^1 \frac{d\lambda}{\lambda} \sum_{0\leq|\Lambda|} s^A_\Lambda \partial^\Lambda_A \phi, \tag{6.57}$$

where ϕ_0 is an exterior form on $\mathbb{R}^{n+k}$. Let $\phi \in \mathcal{S}^{0,m<n}_\infty[F;Y]$ be d_H-closed. Then its component ϕ_0 (6.57) is an exact exterior form on $\mathbb{R}^{n+k}$ and $\widetilde{\phi} = d_H\xi$, where ξ is given by the following expressions. Let us introduce the operator

$$D^{+\nu}\widetilde{\phi} = \int_0^1 \frac{d\lambda}{\lambda} \sum_{0\leq k} k\delta^\nu_{(\mu_1}\delta^{\alpha_1}_{\mu_2} \cdots \delta^{\alpha_{k-1}}_{\mu_k)} \lambda s^A_{(\alpha_1\ldots\alpha_{k-1})} \partial^{\mu_1\ldots\mu_k}_A \widetilde{\phi}(x^\mu, \lambda s^A_\Lambda, dx^\mu).$$

The relation $[D^{+\nu}, d_\mu]\widetilde{\phi} = \delta^\nu_\mu \widetilde{\phi}$ holds, and it leads to the desired expression

$$\xi = \sum_{k=0} \frac{(n-m-1)!}{(n-m+k)!} D^{+\nu} P_k \partial_\nu \rfloor \widetilde{\phi}, \tag{6.58}$$

$$P_0 = 1, \qquad P_k = d_{\nu_1} \cdots d_{\nu_k} D^{+\nu_1} \cdots D^{+\nu_k}.$$

Now let $\phi \in \mathcal{S}^{0,n}_\infty[F;Y]$ be a graded density such that $\delta\phi = 0$. Then its component ϕ_0 (6.57) is an exact n-form on $\mathbb{R}^{n+k}$ and $\widetilde{\phi} = d_H\xi$, where ξ is given by the expression

$$\xi = \sum_{|\Lambda|\geq 0}\sum_{\Sigma+\Xi=\Lambda}(-1)^{|\Sigma|}s^A_\Xi d_\Sigma \partial^{\mu+\Lambda}_A\widetilde{\phi}\omega_\mu. \tag{6.59}$$

We also quote the homotopy operator (5.107) in [108] which leads to the expression

$$\xi = \int_0^1 I(\phi)(x^\mu, \lambda s^A_\Lambda, dx^\mu)\frac{d\lambda}{\lambda}, \tag{6.60}$$

$$I(\phi) = \sum_{0\leq|\Lambda|}\sum_\mu \frac{\Lambda_\mu + 1}{n-m+|\Lambda|+1}\cdot$$

$$d_\Lambda\left[\sum_{0\leq|\Xi|}(-1)^\Xi\frac{(\mu+\Lambda+\Xi)!}{(\mu+\Lambda)!\Xi!}s^A d_\Xi \partial^{\mu+\Lambda+\Xi}_A(\partial_\mu\rfloor\phi)\right],$$

where $\Lambda! = \Lambda_{\mu_1}!\cdots\Lambda_{\mu_n}!$, and Λ_μ denotes the number of occurrences of the index μ in Λ [108]. The graded forms (6.59) and (6.60) differ in a d_H-exact graded form. □

Lemma 6.2 *If $Y = \mathbb{R}^{n+k} \to \mathbb{R}^n$, the complex (6.56) is exact.*

Proof The fact that a d_H-closed graded $(1,m)$-form $\phi \in \mathcal{S}^{1,m<n}_\infty[F;Y]$ is d_H-exact is derived from Lemma 6.1 as follows. We write

$$\phi = \sum \phi^\Lambda_A \wedge \theta^A_\Lambda, \tag{6.61}$$

where $\phi^\Lambda_A \in \mathcal{S}^{0,m}_\infty[F;Y]$ are horizontal graded m-forms. Let us introduce additional variables $\overline{s}^A_\Lambda$ of the same Grassmann parity as s^A_Λ. Then one can associate to each graded $(1,m)$-form ϕ (6.61) a unique horizontal graded m-form

$$\overline{\phi} = \sum \phi^\Lambda_A \overline{s}^A_\Lambda, \tag{6.62}$$

whose coefficients are linear in the variables $\overline{s}^A_\Lambda$, and vice versa. Let us put a modified total differential

$$\overline{d}_H = d_H + dx^\lambda \wedge \sum_{0<|\Lambda|}\overline{s}^A_{\lambda+\Lambda}\overline{\partial}^\Lambda_A,$$

acting on graded forms (6.62), where $\overline{\partial}^\Lambda_A$ is the dual of $d\overline{s}^A_\Lambda$. Comparing the equalities

$$\overline{d}_H\overline{s}^A_\Lambda = dx^\lambda s^A_{\lambda+\Lambda}, \qquad d_H\theta^A_\Lambda = dx^\lambda \wedge \theta^A_{\lambda+\Lambda},$$

one can easily justify that $\overline{d_H\phi} = \overline{d}_H\overline{\phi}$. Let the graded $(1, m)$-form ϕ (6.61) be d_H-closed. Then the associated horizontal graded m-form $\overline{\phi}$ (6.62) is $\overline{d}_H$-closed and, by virtue of Lemma 6.1, it is $\overline{d}_H$-exact, i.e., $\overline{\phi} = \overline{d}_H\overline{\xi}$, where $\overline{\xi}$ is a horizontal graded $(m-1)$-form given by the expression (6.58) depending on additional variables $\overline{s}^A_\Lambda$. A glance at this expression shows that, since $\overline{\phi}$ is linear in the variables $\overline{s}^A_\Lambda$, so is

$$\overline{\xi} = \sum \xi^\Lambda_A \overline{s}^A_\Lambda.$$

It follows that $\phi = d_H\xi$ where

$$\xi = \sum \xi^\Lambda_A \wedge \theta^A_\Lambda.$$

It remains to prove the exactness of the complex (6.56) at the last term $\mathbf{E}_1$. If

$$\rho(\sigma) = \sum_{0\leq|\Lambda|} (-1)^{|\Lambda|}\theta^A \wedge [d_\Lambda(\partial^\Lambda_A \rfloor \sigma)] =$$
$$\sum_{0\leq|\Lambda|} (-1)^{|\Lambda|}\theta^A \wedge [d_\Lambda \sigma^\Lambda_A]\omega = 0, \qquad \sigma \in \mathscr{S}^{1,n}_\infty[F;Y],$$

a direct computation gives

$$\sigma = d_H\xi, \qquad \xi = -\sum_{0\leq|\Lambda|}\sum_{\Sigma+\Xi=\Lambda} (-1)^{|\Sigma|}\theta^A_\Xi \wedge d_\Sigma\sigma^{\mu+\Lambda}_A \omega_\mu.$$

□

Remark 6.8 The proof of Lemma 6.2 fails to be extended to complexes of higher contact forms because the products $\theta^A_\Lambda \wedge \theta^B_\Sigma$ and $s^A_\Lambda s^B_\Sigma$ obey different commutation rules.

(II) Let us now prove Theorem 6.9 for a DBGA $\mathscr{Q}^*_\infty[F;Y]$. Similarly to $\mathscr{S}^*_\infty[F;Y]$, the sheaf $\mathfrak{Q}^*_\infty[F;Y]$ and the DBGA $\mathscr{Q}^*_\infty[F;Y]$ are decomposed into the Grassmann-graded variational bicomplexes. We consider their subcomplexes

$$0 \to \mathbb{R} \to \mathfrak{Q}^0_\infty[F;Y] \xrightarrow{d_H} \mathfrak{Q}^{0,1}_\infty[F;Y] \cdots \xrightarrow{d_H} \mathfrak{Q}^{0,n}_\infty[F;Y] \xrightarrow{\delta} \mathbf{E}_1, \tag{6.63}$$

$$0 \to \mathfrak{Q}^{1,0}_\infty[F;Y] \xrightarrow{d_H} \mathfrak{Q}^{1,1}_\infty[F;Y] \cdots \xrightarrow{d_H} \mathfrak{Q}^{1,n}_\infty[F;Y] \xrightarrow{\rho} \mathbf{E}_1 \to 0, \tag{6.64}$$

$$0 \to \mathbb{R} \to \mathscr{Q}^0_\infty[F;Y] \xrightarrow{d_H} \mathscr{Q}^{0,1}_\infty[F;Y] \cdots \xrightarrow{d_H} \mathscr{Q}^{0,n}_\infty[F;Y] \xrightarrow{\delta} \Gamma(\mathbf{E}_1), \tag{6.65}$$

$$0 \to \mathscr{Q}^{1,0}_\infty[F;Y] \xrightarrow{d_H} \mathscr{Q}^{1,1}_\infty[F;Y] \cdots \xrightarrow{d_H} \mathscr{Q}^{1,n}_\infty[F;Y] \xrightarrow{\rho} \Gamma(\mathbf{E}_1) \to 0, \tag{6.66}$$

where $\mathbf{E}_1 = \rho(\mathfrak{Q}^{1,n}_\infty[F;Y])$. By virtue of Lemmas 6.1 and 6.2, the complexes (6.63) and (6.64) are acyclic. The terms $\mathfrak{Q}^{*,*}_\infty[F;Y]$ of the complexes (6.63) and (6.64) are sheaves of $\mathscr{Q}^0_\infty$-modules (see (1.10)). Since $J^\infty Y$ admits the partition of unity just by elements of $\mathscr{Q}^0_\infty$, these sheaves are fine and, consequently, acyclic. By virtue of abstract de Rham Theorem C.5, cohomology of the complex (6.65) equals the cohomology of $J^\infty Y$ with coefficients in the constant sheaf $\mathbb{R}$ and, consequently, the de Rham cohomology of Y in accordance with the isomorphisms (B.79). Similarly, the complex (6.66) is proved to be exact.

Due to monomorphisms $\mathcal{O}^*_\infty Y \to \mathcal{S}^*_\infty[F;Y] \to \mathcal{Q}^*_\infty[F;Y]$ (6.40) and (6.44), this proof gives something more.

Lemma 6.3 *Every d_H-closed graded form $\phi \in \mathcal{Q}^{0,m<n}_\infty[F;Y]$ falls into a sum*

$$\phi = h_0\sigma + d_H\xi, \qquad \xi \in \mathcal{Q}^{0,m-1}_\infty[F;Y], \tag{6.67}$$

where σ is a closed m-form on Y. Any δ-closed $\phi \in \mathcal{Q}^{0,n}_\infty[F;Y]$ is a sum

$$\phi = h_0\sigma + d_H\xi, \qquad \xi \in \mathcal{Q}^{0,n-1}_\infty[F;Y], \tag{6.68}$$

where σ is a closed n-form on Y.

(III) It remains to prove that cohomology of the complexes (6.55) and (6.56) equals that of the complexes (6.65) and (6.66). The proof of this fact straightforwardly follows the proof of Theorem 1.12 [134].

Let the common symbol D stand for d_H and δ. Bearing in mind the decompositions (6.67) and (6.68), it suffices to show that, if an element $\phi \in \mathcal{S}^*_\infty[F;Y]$ is D-exact in an algebra $\mathcal{Q}^*_\infty[F;Y]$, then it is so in an algebra $\mathcal{S}^*_\infty[F;Y]$.

Lemma 6.1 states that, if Y is a contractible bundle and a D-exact graded form ϕ on $J^\infty Y$ is of finite jet order $[\phi]$ (i.e., $\phi \in \mathcal{S}^*_\infty[F;Y]$), there exists a graded form $\varphi \in \mathcal{S}^*_\infty[F;Y]$ on $J^\infty Y$ such that $\phi = D\varphi$. Moreover, a glance at the expressions (6.58) and (6.59) shows that a jet order $[\varphi]$ of φ is bounded by an integer $N([\phi])$, depending only on a jet order of ϕ. Let us call this fact the finite exactness of an operator D. Lemma 6.1 shows that the finite exactness takes place on $J^\infty Y|_U$ over any domain $U \subset Y$. Let us prove the following.

Lemma 6.4 *Given a family $\{U_\alpha\}$ of disjoint open subsets of Y, let us suppose that the finite exactness takes place on $J^\infty Y|_{U_\alpha}$ over every subset U_α from this family. Then it is true on $J^\infty Y$ over the union $\bigcup_\alpha U_\alpha$ of these subsets.*

Proof Let $\phi \in \mathcal{S}^*_\infty[F;Y]$ be a D-exact graded form on $J^\infty Y$. The finite exactness on $(\pi^\infty_0)^{-1}(\cup U_\alpha)$ holds since $\phi = D\varphi_\alpha$ on every $(\pi^\infty_0)^{-1}(U_\alpha)$ and $[\varphi_\alpha] < N([\phi])$. □

Lemma 6.5 *If the finite exactness of an operator D takes place on $J^\infty Y$ over open subsets $U, V \subset Y$ and their non-empty overlap $U \cap V$, then it also is true on $J^\infty Y|_{U\cup V}$.*

Proof Let $\phi = D\varphi \in \mathcal{S}^*_\infty[F;Y]$ be a D-exact form on $J^\infty Y$. By assumption, it can be brought into the form $D\varphi_U$ on $(\pi^\infty_0)^{-1}(U)$ and $D\varphi_V$ on $(\pi^\infty_0)^{-1}(V)$, where φ_U and φ_V are graded forms of bounded jet order. Let us consider their difference $\varphi_U - \varphi_V$ on $(\pi^\infty_0)^{-1}(U \cap V)$. It is a D-exact graded form of bounded jet order $[\varphi_U - \varphi_V] < N([\phi])$ which, by assumption, can be written as $\varphi_U - \varphi_V = D\sigma$

where σ also is of bounded jet order $[\sigma] < N(N([\phi]))$. Lemma 6.6 below shows that $\sigma = \sigma_U + \sigma_V$ where σ_U and σ_V are graded forms of bounded jet order on $(\pi_0^\infty)^{-1}(U)$ and $(\pi_0^\infty)^{-1}(V)$, respectively. Then, putting

$$\varphi'|_U = \varphi_U - D\sigma_U, \qquad \varphi'|_V = \varphi_V + D\sigma_V,$$

we have a graded form ϕ, equal to $D\varphi'_U$ on $(\pi_0^\infty)^{-1}(U)$ and $D\varphi'_V$ on $(\pi_0^\infty)^{-1}(V)$, respectively. Since the difference $\varphi'_U - \varphi'_V$ on $(\pi_0^\infty)^{-1}(U \cap V)$ vanishes, we obtain $\phi = D\varphi'$ on $(\pi_0^\infty)^{-1}(U \cup V)$ where

$$\varphi' = \begin{cases} \varphi'|_U = \varphi'_U \\ \varphi'|_V = \varphi'_V \end{cases}$$

is of bounded jet order $[\varphi'] < N(N([\phi]))$. □

Lemma 6.6 *Let U and V be open subsets of a bundle Y and $\sigma \in \mathfrak{G}^*_\infty$ a graded form of bounded jet order on $(\pi_0^\infty)^{-1}(U \cap V) \subset J^\infty Y$. Then σ is decomposed into a sum $\sigma_U + \sigma_V$ of graded forms σ_U and σ_V of bounded jet order on $(\pi_0^\infty)^{-1}(U)$ and $(\pi_0^\infty)^{-1}(V)$, respectively.*

Proof By taking a smooth partition of unity on $U \cup V$ subordinate to a cover $\{U, V\}$ and passing to a function with support in V, one gets a smooth real function f on $U \cup V$ which equals 0 on a neighborhood of $U \setminus V$ and 1 on a neighborhood of $V \setminus U$ in $U \cup V$. Let $(\pi_0^\infty)^* f$ be the pull-back of f onto $(\pi_0^\infty)^{-1}(U \cup V)$. A graded form $((\pi_0^\infty)^* f)\sigma$ equals 0 on a neighborhood of $(\pi_0^\infty)^{-1}(U)$ and, therefore, can be extended by 0 to $(\pi_0^\infty)^{-1}(U)$. Let us denote it σ_U. Accordingly, a graded form $(1 - (\pi_0^\infty)^* f)\sigma$ has an extension σ_V by 0 to $(\pi_0^\infty)^{-1}(V)$. Then $\sigma = \sigma_U + \sigma_V$ is a desired decomposition because σ_U and σ_V are of jet order which does not exceed that of σ. □

To prove the finite exactness of D on $J^\infty Y$, it remains to choose an appropriate cover of Y. A smooth manifold Y admits a countable cover $\{U_\xi\}$ by domains U_ξ, $\xi \in \mathbb{N}$, and its refinement $\{U_{ij}\}$, where $j \in \mathbb{N}$ and i runs through a finite set, such that $U_{ij} \cap U_{ik} = \emptyset$, $j \neq k$ [67]. Then Y has a finite cover $\{U_i = \cup_j U_{ij}\}$. Since the finite exactness of an operator D takes place over any domain U_ξ, it also holds over any member U_{ij} of the refinement $\{U_{ij}\}$ of $\{U_\xi\}$ and, in accordance with Lemma 6.4, over any member of a finite cover $\{U_i\}$ of Y. Then by virtue of Lemma 6.5, the finite exactness of D takes place on $J^\infty Y$ over Y.

Similarly, one can show that, restricted to $\mathcal{S}^{k,n}_\infty[F; Y]$, an operator ρ is exact.

6.5 Grassmann-Graded Lagrangian Theory

Decomposed into the Grassmann-graded variational bicomplex (6.54), the DBGA $\mathcal{S}^*_\infty[F; Y]$ (6.39) describes Grassmann-graded Lagrangian theory on a graded bundle $(X, Y, \mathfrak{A}_F)$ modelled over the composite bundle (6.33) [61, 134].

Its *graded Lagrangian* is defined as an element

$$L = \mathcal{L}\omega \in \mathcal{S}_\infty^{0,n}[F;Y] \tag{6.69}$$

of the graded variational complex (6.55). Accordingly, a graded exterior form

$$\delta L = \theta^A \wedge \mathcal{E}_A\omega = \sum_{0\leq|\Lambda|} (-1)^{|\Lambda|}\theta^A \wedge d_\Lambda(\partial_A^\Lambda L)\omega \in \mathbf{E}_1 \tag{6.70}$$

is said to be its *graded Euler–Lagrange operator*. Its kernel defines a *graded Euler–Lagrange equation*

$$\delta L = 0, \qquad \mathcal{E}_A = \sum_{0\leq|\Lambda|} (-1)^{|\Lambda|} d_\Lambda(\partial_A^\Lambda L) = 0. \tag{6.71}$$

The following is a corollary of Theorem 6.9.

Theorem 6.11 *(i) Every d_H-closed graded form $\phi \in \mathcal{S}_\infty^{0,m<n}[F;Y]$ falls into a sum*

$$\phi = h_0\sigma + d_H\xi, \qquad \xi \in \mathcal{S}_\infty^{0,m-1}[F;Y],$$

where σ is a closed m-form on Y.

(ii) Any δ-closed graded Lagrangian $L \in \mathcal{S}_\infty^{0,n}[F;Y]$ is a sum

$$L = h_0\sigma + d_H\xi, \qquad \xi \in \mathcal{S}_\infty^{0,n-1}[F;Y],$$

where σ is a closed n-form on Y.

Proof The complex (6.55) possesses the same cohomology as the short variational subcomplex

$$0 \to \mathbb{R} \to \mathcal{O}_\infty^0 Y \xrightarrow{d_H} \mathcal{O}_\infty^{0,1} Y \cdots \xrightarrow{d_H} \mathcal{O}_\infty^{0,n} Y \xrightarrow{\delta} \mathbf{E}_1 \tag{6.72}$$

of the variational complex (1.22) of the DGA $\mathcal{O}_\infty^* Y$. The monomorphism (6.40) and the body epimorphism (6.41) yield the corresponding cochain morphisms of the complexes (6.55) and (6.72). Therefore, cohomology of the complex (6.55) is the image of the cohomology of $\mathcal{O}_\infty^* Y$. □

Item (ii) of Theorem 6.11 leads to the following.

Theorem 6.12 *Any variationally trivial odd Lagrangian is d_H-exact.*

The exactness of the complex (6.56) by virtue of Theorem 6.10 results in the following [59, 134].

Theorem 6.13 *Given a graded Lagrangian L, there is the decomposition*

$$dL = \delta L - d_H \Xi_L, \qquad \Xi \in \mathcal{S}_\infty^{n-1}[F;Y], \tag{6.73}$$

$$\Xi_L = L + \sum_{s=0} \theta^A_{\nu_s \ldots \nu_1} \wedge F_A^{\lambda\nu_s\ldots\nu_1}\omega_\lambda, \tag{6.74}$$

$$F_A^{\nu_k\ldots\nu_1} = \partial_A^{\nu_k\ldots\nu_1}\mathcal{L} - d_\lambda F_A^{\lambda\nu_k\ldots\nu_1} + \psi_A^{\nu_k\ldots\nu_1}, \qquad k = 1, 2, \ldots,$$

where local graded functions σ *obey the relations*

$$\psi_A^\nu = 0, \qquad \psi_A^{(\nu_k\nu_{k-1})\ldots\nu_1} = 0.$$

Proof The decomposition (6.73) is a straightforward consequence of the exactness of the complex (6.56) at a term $\mathcal{S}_\infty^{1,n}[F,Y]$ and the fact that the endomorphism ρ (6.53) is a projector. The expression (6.74) results from a direct computation

$$\begin{aligned}
-d_H\Xi = {} & -d_H[\theta^A F_A^\lambda + \theta_\nu^A F_A^{\lambda\nu} + \cdots + \theta^A_{\nu_s\ldots\nu_1} F_A^{\lambda\nu_s\ldots\nu_1} \\
& +\theta^A_{\nu_{s+1}\nu_s\ldots\nu_1} \wedge F_A^{\lambda\nu_{s+1}\nu_s\ldots\nu_1} + \cdots] \wedge \omega_\lambda = [\theta^A d_\lambda F_A^\lambda + \theta_\nu^A (F_A^\nu + d_\lambda F_A^{\lambda\nu}) + \cdots \\
& +\theta^A_{\nu_{s+1}\nu_s\ldots\nu_1}(F_A^{\nu_{s+1}\nu_s\ldots\nu_1} + d_\lambda F_A^{\lambda\nu_{s+1}\nu_s\ldots\nu_1}) + \cdots] \wedge \omega = \\
& [\theta^A d_\lambda F_A^\lambda + \theta_\nu^A(\partial_A^\nu \mathcal{L}) + \cdots + \theta^A_{\nu_{s+1}\nu_s\ldots\nu_1}(\partial_A^{\nu_{s+1}\nu_s\ldots\nu_1}\mathcal{L}) + \cdots] \wedge \omega = \\
& \theta^A(d_\lambda F_A^\lambda - \partial_A\mathcal{L}) \wedge \omega + dL = -\delta L + dL.
\end{aligned}$$

□

The graded exterior form Ξ_L (6.74) provides a global *graded Lepage equivalent* of a graded Lagrangian L. In particular, one can locally choose Ξ_L (6.74) where all graded functions ψ vanish.

By analogy with the variational formula (1.36), the decomposition (6.73) is called the *graded variational formula*.

Definition 6.8 We agree to call a pair $(\mathcal{S}_\infty^*[F;Y], L)$ the *Grassmann-graded Lagrangian system* on a graded bundle $(X, Y, \mathfrak{A}_F)$.

6.6 Noether's First Theorem: Supersymmetries

Given a Grassmann-graded Lagrangian system $(\mathcal{S}_\infty^*[F;Y], L)$, by its infinitesimal transformations are meant contact graded derivations of a real graded commutative ring $\mathcal{S}_\infty^0[F;Y]$. They constitute a $\mathcal{S}_\infty^0[F;Y]$-module $\mathfrak{d}\mathcal{S}_\infty^0[F;Y]$ which is a real Lie superalgebra relative to the Lie superbracket (6.3). The following holds [59, 61].

Theorem 6.14 *The derivation module* $\mathfrak{d}\mathcal{S}_\infty^0[F;Y]$ *is isomorphic to the* $\mathcal{S}_\infty^0[F;Y]$-*dual* $(\mathcal{S}_\infty^1[F;Y])^*$ *of the module of graded one-forms* $\mathcal{S}_\infty^1[F;Y]$.

Proof At first, let us show that $\mathcal{S}^*_\infty[F;Y]$ is generated by elements df, $f \in \mathcal{S}^0_\infty[F;Y]$. It suffices to justify that any element of $\mathcal{S}^1_\infty[F;Y]$ is a finite $\mathcal{S}^0_\infty[F;Y]$-linear combination of elements df, $f \in \mathcal{S}^0_\infty[F;Y]$. Indeed, every $\phi \in \mathcal{S}^1_\infty[F;Y]$ is a graded exterior form on some finite order jet manifold J^rY, i.e., a section of a vector bundle $\overline{\mathcal{V}}_{J^rF} \to J^rY$ in accordance with Theorem 6.3. Then by virtue of the classical Serre–Swan Theorem A.10, a $C^\infty(J^rY)$-module $\mathcal{S}^1_r[F;Y]$ of graded one-forms on J^rY is a projective module of finite rank, i.e., ϕ is represented by a finite $C^\infty(J^rY)$-linear combination of elements df, $f \in \mathcal{S}^0_r[F;Y] \subset \mathcal{S}^0_\infty[F;Y]$. Any element $\Phi \in (\mathcal{S}^1_\infty[F;Y])^*$ yields a derivation $\vartheta_\Phi(f) = \Phi(df)$ of a real ring $\mathcal{S}^0_\infty[F;Y]$. Since a module $\mathcal{S}^1_\infty[F;Y]$ is generated by elements df, $f \in \mathcal{S}^0_\infty[F;Y]$, different elements of $(\mathcal{S}^1_\infty[F;Y])^*$ provide different derivations of $\mathcal{S}^0_\infty[F;Y]$, i.e., there is a monomorphism $(\mathcal{S}^0_\infty[F;Y])^* \to \mathfrak{d}\mathcal{S}^0_\infty[F;Y]$. By the same formula, any derivation $\vartheta \in \mathfrak{d}\mathcal{S}^0_\infty[F;Y]$ sends $df \to \vartheta(f)$ and, since $\mathcal{S}^0_\infty[F;Y]$ is generated by elements df, it defines a morphism $\Phi_\vartheta : \mathcal{S}^1_\infty[F;Y] \to \mathcal{S}^0_\infty[F;Y]$. Moreover, different derivations ϑ provide different morphisms Φ_ϑ. Thus, we have a monomorphism $\mathfrak{d}\mathcal{S}^0_\infty[F;Y] \to (\mathcal{S}^1_\infty[F;Y])^*$ and, consequently, isomorphism $\mathfrak{d}\mathcal{S}^0_\infty[F;Y] = (\mathcal{S}^0_\infty[F;Y])^*$. □

The proof of Theorem 6.14 gives something more. It follows that a DBGA $\mathcal{S}^*_\infty[F;Y]$ is the Chevalley–Eilenberg minimal differential calculus over a real graded commutative ring $\mathcal{S}^0_\infty[F;Y]$.

Let $\vartheta\rfloor\phi$, $\vartheta \in \mathfrak{d}\mathcal{S}^0_\infty[F;Y]$, $\phi \in \mathcal{S}^1_\infty[F;Y]$, denote the corresponding interior product. Extended to a DBGA $\mathcal{S}^*_\infty[F;Y]$, it obeys the rule

$$\vartheta\rfloor(\phi\wedge\sigma) = (\vartheta\rfloor\phi)\wedge\sigma + (-1)^{|\phi|+[\phi][\vartheta]}\phi\wedge(\vartheta\rfloor\sigma), \qquad \phi,\sigma \in \mathcal{S}^*_\infty[F;Y].$$

Restricted to a coordinate chart (1.3) of $J^\infty Y$, an algebra $\mathcal{S}^*_\infty[F;Y]$ is a free $\mathcal{S}^0_\infty[F;Y]$-module generated by graded one-forms dx^λ and θ^A_Λ, $[\theta^A_\Lambda] = [A]$. Due to the isomorphism stated in Theorem 6.14, any graded derivation $\vartheta \in \mathfrak{d}\mathcal{S}^0_\infty[F;Y]$reads

$$\vartheta = \vartheta^\lambda\partial_\lambda + \vartheta^A\partial_A + \sum_{0<|\Lambda|}\vartheta^A_\Lambda\partial^\Lambda_A, \tag{6.75}$$

where the graded derivations ∂^Λ_A, $[\partial^\Lambda_A] = [A]$, obey the relations

$$\partial^\Lambda_A(s^B_\Sigma) = \partial^\Lambda_A\rfloor ds^B_\Sigma = \delta^B_A\delta^\Lambda_\Sigma$$

up to permutations of multi-indices Λ and Σ. Its coefficients ϑ^λ, ϑ^A, ϑ^A_Λ are local graded functions of finite jet order possessing the transformation law

$$\vartheta'^\lambda = \frac{\partial x'^\lambda}{\partial x^\mu}\vartheta^\mu, \qquad \vartheta'^A = \frac{\partial s'^A}{\partial s^B}\vartheta^B + \frac{\partial s'^A}{\partial x^\mu}\vartheta^\mu,$$
$$\vartheta'^A_\Lambda = \sum_{|\Sigma|\leq|\Lambda|}\frac{\partial s'^A_\Lambda}{\partial s^B_\Sigma}\vartheta^B_\Sigma + \frac{\partial s'^A_\Lambda}{\partial x^\mu}\vartheta^\mu.$$

Every graded derivation ϑ (6.75) of a graded commutative ring ring $\mathcal{S}^0_\infty[F;Y]$ yields a graded derivation (called the *graded Lie derivative*) $\mathbf{L}_\vartheta$ of a DBGA $\mathcal{S}^*_\infty[F;Y]$ given by the relations

$$\mathbf{L}_\vartheta\phi = \vartheta\rfloor d\phi + d(\vartheta\rfloor\phi), \qquad \phi\in\mathcal{S}^*_\infty[F;Y],$$
$$\mathbf{L}_\vartheta(\phi\wedge\sigma) = \mathbf{L}_\vartheta(\phi)\wedge\sigma + (-1)^{[\vartheta][\phi]}\phi\wedge\mathbf{L}_\vartheta(\sigma).$$

The graded derivation ϑ (6.75) is called contact if the graded Lie derivative $\mathbf{L}_\vartheta$ preserves the ideal of contact graded forms of the DBGA $\mathcal{S}^*_\infty[F;Y]$ generated by contact graded one-forms (6.45).

Theorem 6.15 *With respect to a local generating basis (s^A) for a DBGA $\mathcal{S}^*_\infty[F;Y]$, any its contact graded derivation takes a form*

$$\vartheta = \vartheta_H + \vartheta_V = \upsilon^\lambda d_\lambda + [\upsilon^A\partial_A + \sum_{|\Lambda|>0} d_\Lambda(\upsilon^A - s^A_\mu\upsilon^\mu)\partial^\Lambda_A], \tag{6.76}$$

where ϑ_H and ϑ_V denotes the horizontal and vertical parts of ϑ.

Proof The expression (6.76) results from a direct computation similar to that of the expression (2.1). □

A glance at the expression (6.76) shows that a contact graded derivation ϑ is the infinite order jet prolongation

$$\vartheta = J^\infty\upsilon \tag{6.77}$$

of its restriction

$$\upsilon = \upsilon^\lambda\partial_\lambda + \upsilon^A\partial_A = \upsilon_H + \upsilon_V = \upsilon^\lambda d_\lambda + (u^A\partial_A - s^A_\lambda\partial^\lambda_A) \tag{6.78}$$

to a graded commutative ring $\mathcal{S}^0[F;Y]$. By analogy with υ (2.3), we call υ (6.78) the *generalized graded vector field* on a simple graded manifold $(Y,\mathfrak{A}_F)$. This fails to be a graded vector field on $(Y,\mathfrak{A}_F)$ because its component depends on jets of elements of the local generating basis for $(Y,\mathfrak{A}_F)$ in general. At the same tine, any graded vector field u on $(Y,\mathfrak{A}_F)$ is the generalized graded vector field (6.78) generating the contact graded derivation $J^\infty u$ (6.77).

In particular, the *vertical contact graded derivation* (6.78) reads

$$\vartheta = \upsilon^A\partial_A + \sum_{|\Lambda|>0} d_\Lambda\upsilon^A\partial^\Lambda_A. \tag{6.79}$$

Theorem 6.16 *Any vertical contact graded derivation (6.79) obeys the relations*

$$\vartheta\rfloor d_H\phi = -d_H(\vartheta\rfloor\phi), \tag{6.80}$$
$$\mathbf{L}_\vartheta(d_H\phi) = d_H(\mathbf{L}_\vartheta\phi), \qquad \phi\in\mathcal{S}^*_\infty[F;Y]. \tag{6.81}$$

Proof The proof is similar to that of the equalities (2.8) and (2.9). □

The vertical contact graded derivation ϑ (6.79) is said to be *nilpotent* if

$$\mathbf{L}_\vartheta(\mathbf{L}_\vartheta\phi) = \sum_{0\leq|\Sigma|,0\leq|\Lambda|} (\upsilon^B_\Sigma \partial^\Sigma_B(\upsilon^A_\Lambda)\partial^\Lambda_A + (-1)^{[s^B][\upsilon^A]}\upsilon^B_\Sigma\upsilon^A_\Lambda\partial^\Sigma_B\partial^\Lambda_A)\phi = 0$$

for any horizontal graded form $\phi \in \mathcal{S}^{0,*}_\infty[F, Y]$.

Theorem 6.17 *The vertical contact graded derivation (6.79) is nilpotent only if it is odd.*

Remark 6.9 If there is no danger of confusion, the common symbol υ further stands for a generalized graded vector field υ (6.78), the contact graded derivation ϑ (6.77) determined by υ, and the Lie derivative $\mathbf{L}_\vartheta$. We call all these operators, in brief, a graded derivation of the structure algebra of a graded Lagrangian system.

Remark 6.10 For the sake of convenience, *right graded derivations* $\overleftarrow{\upsilon} = \partial_A \upsilon^A$ also are considered. They act on graded differential forms ϕ on the right by the rules

$$\overleftarrow{\upsilon}(\phi) = d\phi\lfloor\overleftarrow{\upsilon} + d(\phi\lfloor\overleftarrow{\upsilon}), \qquad \theta_{\Lambda B}\lfloor\partial^{\Sigma A} = \delta^A_B\delta^\Sigma_\Lambda,$$
$$\overleftarrow{\upsilon}(\phi\wedge\phi') = (-1)^{[\phi']}\,\overleftarrow{\upsilon}(\phi)\wedge\phi' + \phi\wedge\overleftarrow{\upsilon}(\phi').$$

Definition 6.9 Let $(\mathcal{S}^*_\infty[F; Y], L)$ be a Grassmann-graded Lagrangian system (Definition 6.8). The generalized graded vector field υ (6.78) is called the *supersymmetry* (henceforth, *SUSY*) of a graded Lagrangian L if a graded Lie derivative $\mathbf{L}_\vartheta L$ of L along the contact graded derivation ϑ (6.77) is d_H-exact, i.e., $\mathbf{L}_\vartheta L = d_H\sigma$.

A corollary of the graded variational formula (6.73) is the *graded first variational formula* for a graded Lagrangian [9, 59, 61].

Theorem 6.18 *The graded Lie derivative of a graded Lagrangian along any contact graded derivation (6.76) fulfils the graded first variational formula*

$$\mathbf{L}_\vartheta L = \upsilon_V \rfloor \delta L + d_H(h_0(\vartheta\rfloor\Xi_L)) + d_V(\upsilon_H\rfloor\omega)\mathcal{L}, \tag{6.82}$$

where Ξ_L is the graded Lepage equivalent (6.74) of a graded Lagrangian L.

Proof The formula (6.82) is derived from the decomposition (6.73) and the relations (6.80) and (6.81) similarly to the proof of Theorem 2.3. □

A glance at the expression (6.82) shows the following.

Theorem 6.19 *(i) A generalized graded vector field υ is SUSY only if it is projected onto X.*

(ii) Any projectable generalized graded vector field is SUSY of a variationally trivial graded Lagrangian.

(iii) A generalized graded vector field υ is SUSY if and only if its vertical part υ_V (6.78) is well.

(iv) It is SUSY if and only if the graded density $\upsilon_V \rfloor \delta L$ is d_H-exact.

SUSY of a graded Lagrangian L constitute a real vector subspace $\mathcal{SG}_L$ of the graded derivation module $\mathfrak{d}\mathcal{S}^0_\infty[F; Y]$. By virtue of item (ii) of Theorem 6.19, the Lie superbracket

$$\mathbf{L}_{[\vartheta,\vartheta']} = [\mathbf{L}_\vartheta, \mathbf{L}_{\vartheta'}]$$

of SUSY is SUSY, and their vector space $\mathcal{SG}_L$ is a real Lie superalgebra.

An immediate corollary of the graded first variational formula (6.82) is *Noether's first theorem in a very general setting.*

Theorem 6.20 *If the generalized graded vector field υ (6.78) is SUSY of a graded Lagrangian L, the first variational formula (6.82) leads to a weak conservation law*

$$0 \approx -d_H(-h_0(\vartheta \rfloor \Xi_L) + \sigma) \tag{6.83}$$

of the SUSY current

$$\mathcal{J}_\upsilon = \mathcal{J}^\mu_\vartheta \omega_\mu = -h_0(\vartheta \rfloor \Xi_L) + \sigma \tag{6.84}$$

on the shell (6.71).

Chapter 7
Noether's Second Theorems

As was mentioned above, any Euler–Lagrange operator satisfies Noether identities (henceforth, NI) which, therefore, should be separated into the trivial and notrivial ones. These NI obey first-stage NI, which in turn are subject to the second-stage NI, and so on.

We follow the general notion of NI of differential operators on fibre bundles (Definition D.1). They are represented by one-cycles of a certain chain complex (Theorem 7.1). Its boundaries are trivial NI, and nontrivial NI modulo the trivial ones are given by first homology of this complex. To describe $(k+1)$-stage NI, let us assume that nontrivial k-stage NI are generated by a projective $C^\infty(X)$-module $\mathscr{C}_{(k)}$ of finite rank and that a certain homology regularity condition (Definition 7.2) holds. In this case, $(k+1)$-stage NI are represented by $(k+2)$-cycles of some chain complex of modules of antifields isomorphic to $\mathscr{C}_{(i)}$, $i \leq k$, by virtue of Serre–Swan Theorem 6.2. Accordingly, trivial $(k+1)$-stage NI are defined as its boundaries (Sect. 7.1). Iterating the arguments, we come to the exact Koszul–Tate (henceforth, KT) complex (7.28) with the boundary KT operator (7.26) whose nilpotentness is equivalent to all nontrivial NI (7.12) and higher-stage NI (7.29) (Theorem 7.5).

Noether's inverse second Theorem 7.9 associates to the KT complex (7.28) the cochain sequence (7.38) with the ascent operator (7.39), called the gauge operator. Its components (7.44) and (7.47) are nontrivial gauge and higher-stage gauge supersymmetries of Grassmann-graded Lagrangian theory which are indexed by Grassmann-graded ghosts, but not gauge parameters (Remark 7.7). Herewith, k-stage gauge supersymmetries act on $(k-1)$-stage ghosts.

Conversely, given these gauge and higher-stage supersymmetries, Noether's direct second theorem extended to higher-gauge supersymmetries (Theorem 7.10) states that the corresponding NI and higher-stage NI hold.

As was mentioned above, gauge symmetries fail to form an algebra in general. We say that gauge and higher-stage gauge symmetries are algebraically closed if the gauge operator (7.39) admits the nilpotent BRST extension (7.66) where k-stage gauge symmetries are generalized to k-stage BRST transformations acting both on $(k-1)$-stage and k-stage ghosts (Sect. 7.5). The BRST operator (7.66) brings the cochain sequence (7.38) into the BRST complex.

G. Sardanashvily, *Noether's Theorems*, Atlantis Studies
in Variational Geometry 3, DOI 10.2991/978-94-6239-171-0_7

The KT and BRST complexes provide the above mentioned BRST extension of original Lagrangian field theory. This extension exemplifies so called field-antifield theory whose Lagrangians are required to satisfy the particular condition (7.82) called the classical master equation. We show that an original Lagrangian is extended to a proper solution of the master equation if the gauge operator (7.39) admits a nilpotent BRST extension (Theorem 7.16) [12].

Given the BRST operator (7.66), a desired proper solution of the master equation is constructed by the formula (7.86).

7.1 Noether Identities: Reducible Degenerate Lagrangian Systems

Let $(\mathcal{S}^*_\infty[F;Y], L)$ be a Grassmann-graded Lagrangian system (Definition 6.8). Without a lose of generality, let its graded Lagrangian L (6.69) be even, and let the Euler–Lagrange operator δL (6.70) be at least of first order.

Remark 7.1 Let us introduce the following notation. If $E \to X$ is a vector bundle, we call

$$\overline{E} = E^* \underset{X}{\otimes} \overset{n}{\wedge} T^*X$$

the *density-dual* of E. Given a fibre bundle $Y \to X$, by the density-dual of its vertical tangent bundle VY is meant a fibre bundle

$$\overline{VY} = V^*Y \underset{Y}{\otimes} \overset{n}{\wedge} T^*X. \tag{7.1}$$

If $Y = E$ is a vector bundle, then $\overline{VE} = \overline{E} \times_X E$. Let

$$E = E^0 \oplus_X E^1 \tag{7.2}$$

be a *graded vector bundle* over X possessing an even part $E^0 \to X$ and the odd one $E^1 \to X$. Its *graded density-dual* is defined to be

$$\overline{E} = \overline{E}^1 \underset{X}{\oplus} \overline{E}^0, \tag{7.3}$$

with an even part $\overline{E}^1 \to X$ and the odd one $\overline{E}^0 \to X$. Given a graded vector bundle E, we consider the product $(X, E^0 \times_X Y, \mathfrak{A}_{E\times_X F})$ (6.49) of graded bundles over the product (6.50) of composite bundles F (6.33) and $E \to E^0 \to X$. The corresponding DBGA (6.51) reads

$$\mathcal{S}^*_\infty[F \underset{X}{\times} E; Y \underset{X}{\times} E^0]. \tag{7.4}$$

In particular, we treat the composite bundle F (6.33) as a graded vector bundle over Y only with an odd part. The density-dual (7.1) of the vertical tangent bundle VF of $F \to X$ is

$$\overline{VF} = V^*F \underset{F}{\otimes} \overset{n}{\wedge} T^*X \to F,$$

where V^*F is the vertical cotangent bundle of $F \to X$. It however is not a vector bundle over Y. Therefore, we restrict our consideration to the case of a pull-back composite bundle F (6.33), that is,

$$F = Y \underset{X}{\times} F^1 \to Y \to X,$$

where $F^1 \to X$ is a vector bundle. Then

$$\overline{VF} = \overline{F}^1 \underset{Y}{\oplus} ((V^*Y \underset{Y}{\otimes} \overset{n}{\wedge} T^*X) \underset{Y}{\oplus} F^1) \tag{7.5}$$

is a graded vector bundle over Y. It can be seen as a product of composite bundles

$$\overline{VF^1} = \overline{F}^1 \underset{X}{\oplus} F^1 \to \overline{F}^1 \to X, \qquad \overline{VY} \to Y \to X.$$

Let we consider the corresponding graded bundle (6.49) and DBGA (6.51):

$$\mathscr{P}^*_\infty[\overline{VF}; Y] = \mathscr{S}^*_\infty[\overline{VF}; Y \underset{X}{\times} \overline{F}^1] = \mathscr{S}^*_\infty[\overline{VF}^1 \underset{X}{\times} \overline{VY}; Y \underset{X}{\times} \overline{F}^1]. \tag{7.6}$$

Theorem 7.1 *One can associate to any Grassmann-graded Lagrangian system* $(\mathscr{S}^*_\infty[F; Y], L)$ *the chain complex (7.7) whose one-boundaries vanish on-shell.*

Proof Let us consider the density-dual $\overline{VF}$ (7.5) of the vertical tangent bundle $VF \to F$, and let us enlarge an original DBGA $\mathscr{S}^*_\infty[F; Y]$ to the DBGA $\mathscr{P}^*_\infty[\overline{VF}; Y]$ (7.6) with the local generating basis $(s^A, \overline{s}_A)$, $[\overline{s}_A] = ([A] + 1)\text{mod}\, 2$. Following the terminology of Lagrangian BRST theory [6, 63], we agree to call its elements $\overline{s}_A$ the *antifields* of antifield number $\text{Ant}[\overline{s}_A] = 1$. A DBGA $\mathscr{P}^*_\infty[\overline{VF}; Y]$ is endowed with the nilpotent right graded derivation $\overline{\delta} = \overleftarrow{\partial}^A \mathscr{E}_A$, where $\mathscr{E}_A$ are the variational derivatives (6.70). Then we have the chain complex

$$0 \leftarrow \text{Im}\, \overline{\delta} \overset{\overline{\delta}}{\longleftarrow} \mathscr{P}^{0,n}_\infty[\overline{VF}; Y]_1 \overset{\overline{\delta}}{\longleftarrow} \mathscr{P}^{0,n}_\infty[\overline{VF}; Y]_2 \tag{7.7}$$

of densities of antifield number ≤ 2. Its one-boundaries $\overline{\delta}\Phi$, $\Phi \in \mathscr{P}^{0,n}_\infty[\overline{VF}; Y]_2$, by very definition, vanish on-shell.

Any one-cycle

$$\Phi = \sum_{0 \leq |\Lambda|} \Phi^{A,\Lambda} \overline{s}_{\Lambda A} \omega \in \mathscr{P}^{0,n}_\infty[\overline{VF}; Y]_1 \tag{7.8}$$

of the complex (7.7) is a differential operator on a bundle $\overline{VF}$ such that it is linear on fibres of $\overline{VF} \to F$ and its kernel contains a graded Euler–Lagrange operator δL (6.70), i.e.,

$$\overline{\delta}\Phi = 0, \qquad \sum_{0\leq|\Lambda|} \Phi^{A,\Lambda} d_\Lambda \mathcal{E}_A \omega = 0. \tag{7.9}$$

In accordance with Definition D.1, the one-cycles (7.8) define the NI (7.9) of an Euler–Lagrange operator δL, which we agree to call *Noether identities* of a Grassmann-graded Lagrangian system $(\mathcal{S}^*_\infty[F; Y], L)$.

In particular, one-chains Φ (7.8) are necessarily NI if they are boundaries. Therefore, these NI are called *trivial*. They are of the form

$$\Phi = \sum_{0\leq|\Lambda|,|\Sigma|} T^{(A\Lambda)(B\Sigma)} d_\Sigma \mathcal{E}_B \overline{s}_{\Lambda A}\omega, \qquad T^{(A\Lambda)(B\Sigma)} = -(-1)^{[A][B]} T^{(B\Sigma)(A\Lambda)}.$$

Accordingly, *nontrivial* NI modulo the trivial ones are associated to elements of the first homology $H_1(\overline{\delta})$ of the complex (7.7). A graded Lagrangian L is called *degenerate* if there exist nontrivial NI.

Nontrivial NI can obey first-stage NI. In order to describe them, let us assume that the module $H_1(\overline{\delta})$ is finitely generated. Namely, there exists a graded projective $C^\infty(X)$-module $\mathcal{C}_{(0)} \subset H_1(\overline{\delta})$ of finite rank possessing a local basis $\{\Delta_r\omega\}$:

$$\Delta_r\omega = \sum_{0\leq|\Lambda|} \Delta_r^{A,\Lambda} \overline{s}_{\Lambda A}\omega, \qquad \Delta_r^{A,\Lambda} \in \mathcal{S}^0_\infty[F; Y], \tag{7.10}$$

such that any element $\Phi \in H_1(\overline{\delta})$ factorizes as

$$\Phi = \sum_{0\leq|\Xi|} \Phi^{r,\Xi} d_\Xi \Delta_r\omega, \qquad \Phi^{r,\Xi} \in \mathcal{S}^0_\infty[F; Y], \tag{7.11}$$

through elements (7.10) of $\mathcal{C}_{(0)}$. Thus, all nontrivial NI (7.9) result from NI

$$\overline{\delta}\Delta_r = \sum_{0\leq|\Lambda|} \Delta_r^{A,\Lambda} d_\Lambda \mathcal{E}_A = 0, \tag{7.12}$$

called the *complete* NI. Certainly, the factorization (7.11) is independent of specification of a local basis $\{\Delta_r\omega\}$. Note that, being representatives of $H_1(\overline{\delta})$, the graded densities $\Delta_r\omega$ (7.10) are not $\overline{\delta}$-exact.

A Lagrangian system whose nontrivial NI are finitely generated is called *finitely degenerate*. Hereafter, degenerate Lagrangian systems only of this type are considered.

Theorem 7.2 *If the homology $H_1(\overline{\delta})$ of the complex (7.7) is finitely generated in the above mentioned sense, this complex can be extended to the one-exact chain complex (7.15) with a boundary operator whose nilpotency conditions are equivalent to the complete NI (7.12).*

Proof By virtue of Serre–Swan Theorem 6.2, the graded module $\mathcal{C}_{(0)}$ is isomorphic to a module of sections of the density-dual $\overline{E}_0$ (7.3) of some graded vector bundle $E_0 \to X$. Let us enlarge a DBGA $\mathcal{P}^*_\infty[\overline{VF};Y]$ to a DBGA

$$\overline{\mathcal{P}}^*_\infty\{0\} = \mathcal{P}^*_\infty[\overline{VF} \underset{Y}{\oplus} \overline{E}_0; Y] = \mathcal{S}^*_\infty[\overline{VF} \underset{Y}{\oplus} \overline{E}_0; Y \underset{X}{\times} (\overline{F}^1 \underset{X}{\oplus} \overline{E}^1_0)] \tag{7.13}$$

possessing the local generating basis $(s^A, \overline{s}_A, \overline{c}_r)$ where $\overline{c}_r$ are *Noether antifields* of Grassmann parity $[\overline{c}_r] = ([\Delta_r]+1)\bmod 2$ and antifield number $\mathrm{Ant}[\overline{c}_r] = 2$. The DBGA (7.13) is provided with an odd right derivation

$$\delta_0 = \overline{\delta} + \overleftarrow{\partial}^r \Delta_r \tag{7.14}$$

which is nilpotent if and only if the complete NI (7.12) hold. Then δ_0 (7.14) is a boundary operator of the chain complex

$$0 \leftarrow \mathrm{Im}\,\overline{\delta} \overset{\overline{\delta}}{\longleftarrow} \mathcal{P}^{0,n}_\infty[\overline{VF};Y]_1 \overset{\delta_0}{\longleftarrow} \overline{\mathcal{P}}^{0,n}_\infty\{0\}_2 \overset{\delta_0}{\longleftarrow} \overline{\mathcal{P}}^{0,n}_\infty\{0\}_3 \tag{7.15}$$

of graded densities of antifield number ≤ 3. Let $H_*(\delta_0)$ denote its homology. We have

$$H_0(\delta_0) = H_0(\overline{\delta}) = 0.$$

Furthermore, any one-cycle Φ up to a boundary takes the form (7.11) and, therefore, it is a δ_0-boundary

$$\Phi = \sum_{0\leq|\Sigma|} \Phi^{r,\Xi} d_\Xi \Delta_r \omega = \delta_0 \left(\sum_{0\leq|\Sigma|} \Phi^{r,\Xi} \overline{c}_{\Xi r}\omega \right).$$

Hence, $H_1(\delta_0) = 0$, i.e., the complex (7.15) is one-exact.

Let us consider the second homology $H_2(\delta_0)$ of the complex (7.15). Its two-chains read

$$\Phi = G + H = \sum_{0\leq|\Lambda|} G^{r,\Lambda}\overline{c}_{\Lambda r}\omega + \sum_{0\leq|\Lambda|,|\Sigma|} H^{(A,\Lambda)(B,\Sigma)}\overline{s}_{\Lambda A}\overline{s}_{\Sigma B}\omega. \tag{7.16}$$

Its two-cycles define the *first-stage* NI

$$\delta_0 \Phi = 0, \qquad \sum_{0\leq|\Lambda|} G^{r,\Lambda} d_\Lambda \Delta_r \omega = -\overline{\delta} H. \tag{7.17}$$

Conversely, let the equality (7.17) hold. Then it is a cycle condition of the two-chain (7.16).

Remark 7.2 It should be emphasized that this definition of first-stage NI is independent on specification of a generating module $\mathcal{C}_{(0)}$. Given a different one, there exists a chain isomorphism between the corresponding complexes (7.15).

The first-stage NI (7.17) are *trivial* either if the two-cycle Φ (7.16) is a δ_0-boundary or its summand G vanishes on-shell. Therefore, nontrivial first-stage NI fail to exhaust the second homology $H_2(\delta_0)$ of the complex (7.15) in general.

Theorem 7.3 *Nontrivial first-stage NI modulo the trivial ones are identified with elements of the homology* $H_2(\delta_0)$ *if and only if any* $\overline{\delta}$*-cycle* $\phi \in \overline{\mathcal{P}}^{0,n}_\infty\{0\}_2$ *is a* δ_0*-boundary.*

Proof It suffices to show that, if a summand G of the two-cycle Φ (7.16) is $\overline{\delta}$-exact, then Φ is a boundary. If $G = \overline{\delta}\Psi$, let us write

$$\Phi = \delta_0 \Psi + (\overline{\delta} - \delta_0)\Psi + H. \tag{7.18}$$

Hence, the cycle condition (7.17) reads

$$\delta_0 \Phi = \overline{\delta}((\overline{\delta} - \delta_0)\Psi + H) = 0.$$

Since any $\overline{\delta}$-cycle $\phi \in \overline{\mathcal{P}}^{0,n}_\infty\{0\}_2$, by assumption, is δ_0-exact, then $(\overline{\delta} - \delta_0)\Psi + H$ is a δ_0-boundary. Consequently, Φ (7.18) is δ_0-exact. Conversely, let $\Phi \in \overline{\mathcal{P}}^{0,n}_\infty\{0\}_2$ be a $\overline{\delta}$-cycle, i.e.,

$$\overline{\delta}\Phi = 2\Phi^{(A,\Lambda)(B,\Sigma)} \overline{s}_{\Lambda A} \overline{\delta} \overline{s}_{\Sigma B} \omega = 2\Phi^{(A,\Lambda)(B,\Sigma)} \overline{s}_{\Lambda A} d_\Sigma \mathcal{E}_B \omega = 0.$$

It follows that $\Phi^{(A,\Lambda)(B,\Sigma)} \overline{\delta} \overline{s}_{\Sigma B} = 0$ for all indices (A, Λ). Omitting a $\overline{\delta}$-boundary term, we obtain

$$\Phi^{(A,\Lambda)(B,\Sigma)} \overline{s}_{\Sigma B} = G^{(A,\Lambda)(r,\Xi)} d_\Xi \Delta_r.$$

Hence, Φ takes a form

$$\Phi = G'^{(A,\Lambda)(r,\Xi)} d_\Xi \Delta_r \overline{s}_{\Lambda A} \omega.$$

Then there exists a three-chain

$$\Psi = G'^{(A,\Lambda)(r,\Xi)}\overline{c}_{\Xi r}\overline{s}_{\Lambda A}\omega$$

such that

$$\delta_0\Psi = \Phi + \sigma = \Phi + G''^{(A,\Lambda)(r,\Xi)}d_\Lambda \mathscr{E}_A \overline{c}_{\Xi r}\omega. \tag{7.19}$$

Owing to an equality $\overline{\delta}\Phi = 0$, we have $\delta_0\sigma = 0$. Thus, σ in the expression (7.19) is $\overline{\delta}$-exact δ_0-cycle. By assumption, it is a δ_0-exact, i.e., $\sigma = \delta_0\psi$. Consequently, a $\overline{\delta}$-cycle Φ is a δ_0-boundary $\Phi = \delta_0\Psi - \delta_0\psi$.

Remark 7.3 It is easily justified that the two-cycle Φ (7.16) is δ_0-exact if and only if Φ up to a $\overline{\delta}$-boundary takes a form

$$\Phi = \sum_{0\leq|\Lambda|,|\Sigma|} G'^{(r,\Sigma)(r',\Lambda)}d_\Sigma \Delta_r d_\Lambda \Delta_{r'}\omega.$$

NI of a Lagrangian system are called *irreducible* if there are no nontrivial first-stage NI. Accordingly, a degenerate Lagrangian system is said to be *reducible* (resp. *irreducible*) if it admits (resp. does not admit) nontrivial first-stage NI.

If the condition of Theorem 7.3 is satisfied, let us assume that nontrivial first-stage NI are finitely generated as follows. There exists a graded projective $C^\infty(X)$-module $\mathscr{C}_{(1)} \subset H_2(\delta_0)$ of finite rank possessing a local basis $\{\Delta_{r_1}\omega\}$:

$$\Delta_{r_1}\omega = \sum_{0\leq|\Lambda|} \Delta_{r_1}^{r,\Lambda}\overline{c}_{\Lambda r}\omega + h_{r_1}\omega, \tag{7.20}$$

such that any element $\Phi \in H_2(\delta_0)$ factorizes as

$$\Phi = \sum_{0\leq|\Xi|} \Phi^{r_1,\Xi}d_\Xi \Delta_{r_1}\omega, \qquad \Phi^{r_1,\Xi} \in \mathscr{S}^0_\infty[F;Y], \tag{7.21}$$

through elements (7.20) of $\mathscr{C}_{(1)}$. Thus, all nontrivial first-stage NI (7.17) result from the equalities

$$\sum_{0\leq|\Lambda|} \Delta_{r_1}^{r,\Lambda}d_\Lambda \Delta_r + \overline{\delta}h_{r_1} = 0, \tag{7.22}$$

called the *complete first-stage* NI. Note that, by virtue of the condition of Theorem 7.3, first summands of the graded densities $\Delta_{r_1}\omega$ (7.20) are not $\overline{\delta}$-exact.

A degenerate Lagrangian system is called *finitely reducible* if admits finitely generated nontrivial first-stage NI.

Theorem 7.4 *The one-exact complex (7.15) of a finitely reducible Lagrangian system is extended to the two-exact one (7.24) with a boundary operator whose nilpotency conditions are equivalent to the complete NI (7.12) and the complete first-stage NI (7.22).*

Proof By virtue of Serre–Swan Theorem 6.2, a graded module $\mathcal{C}_{(1)}$ is isomorphic to a module of sections of the density-dual $\overline{E}_1$ (7.3) of some graded vector bundle $E_1 \to X$. Let us enlarge the DBGA $\overline{\mathcal{P}}^*_\infty\{0\}$ (7.13) to a DBGA

$$\overline{\mathcal{P}}^*_\infty\{1\} = \mathcal{P}^*_\infty[\overline{VF} \underset{Y}{\oplus} \overline{E}_0 \underset{Y}{\oplus} \overline{E}_1; Y]$$

possessing the local generating basis $\{s^A, \overline{s}_A, \overline{c}_r, \overline{c}_{r_1}\}$ where $\overline{c}_{r_1}$ are *first-stage Noether antifields* of Grassmann parity $[\overline{c}_{r_1}] = ([\Delta_{r_1}] + 1)\text{mod}\,2$ and of antifield number $\text{Ant}[\overline{c}_{r_1}] = 3$. This DBGA is provided with an odd right derivation

$$\delta_1 = \delta_0 + \partial^{r_1}\Delta_{r_1} \tag{7.23}$$

which is nilpotent if and only if the complete NI (7.12) and the complete first-stage NI (D.13) hold. Then δ_1 (7.23) is a boundary operator of the chain complex

$$0 \leftarrow \text{Im}\,\overline{\delta} \overset{\overline{\delta}}{\leftarrow} \mathcal{P}^{0,n}_\infty[\overline{VF}; Y]_1 \overset{\delta_0}{\leftarrow} \overline{\mathcal{P}}^{0,n}_\infty\{0\}_2 \overset{\delta_1}{\leftarrow} \overline{\mathcal{P}}^{0,n}_\infty\{1\}_3 \overset{\delta_1}{\leftarrow} \overline{\mathcal{P}}^{0,n}_\infty\{1\}_4 \tag{7.24}$$

of graded densities of antifield number ≤ 4. Let $H_*(\delta_1)$ denote its homology. It is readily observed that

$$H_0(\delta_1) = H_0(\overline{\delta}), \qquad H_1(\delta_1) = H_1(\delta_0) = 0.$$

By virtue of the expression (7.21), any two-cycle of the complex (7.24) is a boundary

$$\Phi = \sum_{0\leq|\Xi|} \Phi^{r_1,\Xi} d_\Xi \Delta_{r_1}\omega = \delta_1\left(\sum_{0\leq|\Xi|} \Phi^{r_1,\Xi}\overline{c}_{\Xi r_1}\omega\right).$$

It follows that $H_2(\delta_1) = 0$, i.e., the complex (7.24) is two-exact.

If the third homology $H_3(\delta_1)$ of the complex (7.24) is not trivial, its elements correspond to second-stage NI which the complete first-stage ones satisfy, and so on. Iterating the arguments, one comes to the following.

Definition 7.1 A degenerate Grassmann-graded Lagrangian system $(\mathcal{S}^*_\infty[F; Y], L)$ is called *N-stage reducible* if it admits finitely generated nontrivial N-stage NI, but no nontrivial $(N + 1)$-stage ones.

It is characterized as follows [11].

Theorem 7.5 *One can associate to degenerate N-stage reducible Lagrangian theory the exact KT complex (7.28) with the boundary operator (7.26) whose nilpotency property restarts all NI and higher-stage NI (7.12) and (7.29) if these identities are finitely generated and if and only if this complex obeys the homology regularity condition (Definition 7.2).*

Namely, one can show the following.

- There are graded vector bundles $E_0, \ldots, E_N$ over X, and a DBGA $\mathcal{P}^*_\infty[\overline{VF}; Y]$ is enlarged to a DBGA

$$\overline{\mathcal{P}}^*_\infty\{N\} = \mathcal{P}^*_\infty[\overline{VF} \underset{Y}{\oplus} \overline{E}_0 \underset{Y}{\oplus} \cdots \underset{Y}{\oplus} \overline{E}_N; Y] \tag{7.25}$$

with the local generating basis $(s^A, \overline{s}_A, \overline{c}_r, \overline{c}_{r_1}, \ldots, \overline{c}_{r_N})$, where $\overline{c}_{r_k}$ are *Noether k-stage antifields* of antifield number $\mathrm{Ant}[\overline{c}_{r_k}] = k + 2$.
- The DBGA (7.25) is provided with the nilpotent right graded derivation

$$\delta_{\mathrm{KT}} = \delta_N = \overline{\delta} + \sum_{0\leq|\Lambda|} \overleftarrow{\partial}^r \Delta_r^{A,\Lambda} \overline{s}_{\Lambda A} + \sum_{1\leq k\leq N} \overleftarrow{\partial}^{r_k} \Delta_{r_k}, \tag{7.26}$$

$$\Delta_{r_k}\omega = \sum_{0\leq|\Lambda|} \Delta_{r_k}^{r_{k-1},\Lambda} \overline{c}_{\Lambda r_{k-1}}\omega + \tag{7.27}$$

$$\sum_{0\leq|\Sigma|,|\Xi|} (h_{r_k}^{(r_{k-2},\Sigma)(A,\Xi)} \overline{c}_{\Sigma r_{k-2}} \overline{s}_{\Xi A} + \cdots)\omega \in \overline{\mathcal{P}}^{0,n}_\infty\{k-1\}_{k+1},$$

of antifield number -1. The index $k = -1$ here stands for $\overline{s}_A$. The nilpotent derivation δ_{KT} (7.26) is called the *KT operator*.
- With this graded derivation, a module $\overline{\mathcal{P}}^{0,n}_\infty\{N\}_{\leq N+3}$ of densities of antifield number $\leq (N+3)$ is decomposed into an exact *KT chain complex*

$$0 \leftarrow \mathrm{Im}\,\overline{\delta} \xleftarrow{\overline{\delta}} \mathcal{P}^{0,n}_\infty[\overline{VF}; Y]_1 \xleftarrow{\delta_0} \overline{\mathcal{P}}^{0,n}_\infty\{0\}_2 \xleftarrow{\delta_1} \overline{\mathcal{P}}^{0,n}_\infty\{1\}_3 \cdots \tag{7.28}$$
$$\xleftarrow{\delta_{N-1}} \overline{\mathcal{P}}^{0,n}_\infty\{N-1\}_{N+1} \xleftarrow{\delta_{\mathrm{KT}}} \overline{\mathcal{P}}^{0,n}_\infty\{N\}_{N+2} \xleftarrow{\delta_{\mathrm{KT}}} \overline{\mathcal{P}}^{0,n}_\infty\{N\}_{N+3}$$

which satisfies the following condition.

Definition 7.2 One says that the *homology regularity condition* holds if any $\delta_{k<N}$-cycle

$$\phi \in \overline{\mathcal{P}}^{0,n}_\infty\{k\}_{k+3} \subset \overline{\mathcal{P}}^{0,n}_\infty\{k+1\}_{k+3}$$

is a δ_{k+1}-boundary.

Remark 7.4 The exactness of the complex (7.28) means that any $\delta_{k<N}$-cycle $\phi \in \overline{\mathcal{P}}^{0,n}_\infty\{k\}_{k+3}$, is a δ_{k+2}-boundary, but not necessary a δ_{k+1}-one.

- The nilpotentness $\delta^2_{\mathrm{KT}} = 0$ of the KT operator (7.26) is equivalent to the complete nontrivial NI (7.12) and the *complete nontrivial* $(k \leq N)$*-stage NI*

$$\sum_{0\le|\Lambda|} \Delta_{r_k}^{r_{k-1},\Lambda} d_\Lambda \left(\sum_{0\le|\Sigma|} \Delta_{r_{k-1}}^{r_{k-2},\Sigma} \overline{c}_{\Sigma r_{k-2}}\right) = \tag{7.29}$$
$$-\overline{\delta}\left(\sum_{0\le|\Sigma|,|\Xi|} h_{r_k}^{(r_{k-2},\Sigma)(A,\Xi)} \overline{c}_{\Sigma r_{k-2}} \overline{s}_{\Xi A}\right).$$

This item means the following.

Theorem 7.6 *Any δ_k-cocycle $\Phi \in \overline{\mathscr{P}}_\infty^{0,n}\{k\}_{k+2}$ is a k-stage NI, and* vice versa.

Proof Any $(k+2)$-chain $\Phi \in \overline{\mathscr{P}}_\infty^{0,n}\{k\}_{k+2}$ takes a form

$$\Phi = G + H = \sum_{0\le|\Lambda|} G^{r_k,\Lambda} \overline{c}_{\Lambda r_k} \omega + \sum_{0\le\Sigma,\Xi} (H^{(A,\Xi)(r_{k-1},\Sigma)} \overline{s}_{\Xi A} \overline{c}_{\Sigma r_{k-1}} + \cdots)\omega. \tag{7.30}$$

If it is a δ_k-cycle, then the equality

$$\sum_{0\le|\Lambda|} G^{r_k,\Lambda} d_\Lambda \left(\sum_{0\le|\Sigma|} \Delta_{r_k}^{r_{k-1},\Sigma} \overline{c}_{\Sigma r_{k-1}}\right) + \tag{7.31}$$
$$\overline{\delta}\left(\sum_{0\le\Sigma,\Xi} H^{(A,\Xi)(r_{k-1},\Sigma)} \overline{s}_{\Xi A} \overline{c}_{\Sigma r_{k-1}}\right) = 0$$

is a k-stage NI. Conversely, let the conditions (7.31) hold. Then it can be extended to a cycle condition as follows. It is brought into a form

$$\delta_k \left(\sum_{0\le|\Lambda|} G^{r_k,\Lambda} \overline{c}_{\Lambda r_k} + \sum_{0\le\Sigma,0\le\Xi} H^{(A,\Xi)(r_{k-1},\Sigma)} \overline{s}_{\Xi A} \overline{c}_{\Sigma r_{k-1}}\right) =$$
$$-\sum_{0\le|\Lambda|} G^{r_k,\Lambda} d_\Lambda h_{r_k} + \sum_{0\le\Sigma,\Xi} H^{(A,\Xi)(r_{k-1},\Sigma)} \overline{s}_{\Xi A} d_\Sigma \Delta_{r_{k-1}}.$$

A glance at the expression (7.27) shows that the term in the right-hand side of this equality belongs to $\overline{\mathscr{P}}_\infty^{0,n}\{k-2\}_{k+1}$. It is a δ_{k-2}-cycle and, consequently, a δ_{k-1}-boundary $\delta_{k-1}\Psi$ in accordance with the homology regularity condition (Definition 7.2). Then the equality (7.31) is a $\overline{c}_{\Sigma r_{k-1}}$-dependent part of a cycle condition

$$\delta_k \left(\sum_{0\le|\Lambda|} G^{r_k,\Lambda} \overline{c}_{\Lambda r_k} + \sum_{0\le\Sigma,\Xi} H^{(A,\Xi)(r_{k-1},\Sigma)} \overline{s}_{\Xi A} \overline{c}_{\Sigma r_{k-1}} - \Psi\right) = 0,$$

but $\delta_k\Psi$ does not make a contribution to this condition.

Theorem 7.7 *Any trivial k-stage NI is a δ_k-boundary $\Phi \in \overline{\mathscr{P}}^{0,n}_\infty\{k\}_{k+2}$.*

Proof The k-stage NI (7.31) are trivial either if a δ_k-cycle Φ (7.30) is a δ_k-boundary or its summand G vanishes on-shell. Let us show that, if a summand G of Φ (7.30) is $\overline{\delta}$-exact, then Φ is a δ_k-boundary. If $G = \overline{\delta}\Psi$, one can write

$$\Phi = \delta_k\Psi + (\overline{\delta} - \delta_k)\Psi + H.$$

Hence, the δ_k-cycle condition reads

$$\delta_k\Phi = \delta_{k-1}((\overline{\delta} - \delta_k)\Psi + H) = 0.$$

By virtue of the homology regularity condition, any δ_{k-1}-cycle $\phi \in \overline{\mathscr{P}}^{0,n}_\infty\{k-1\}_{k+2}$ is δ_k-exact. Then $(\overline{\delta}-\delta_k)\Psi + H$ is a δ_k-boundary. Consequently, Φ (7.30) is δ_k-exact.

Theorem 7.8 *All nontrivial k-stage NI (7.31), by assumption, factorize as*

$$\Phi = \sum_{0\leq|\Xi|} \Phi^{r_k,\Xi} d_\Xi \Delta_{r_k}\omega, \qquad \Phi^{r_1,\Xi} \in \mathscr{S}^0_\infty[F;Y],$$

through the complete ones (7.29).

It may happen that a Grassmann-graded Lagrangian field system possesses nontrivial NI of any stage. However, we restrict our consideration to N-reducible Lagrangian systems for a finite integer N. In this case, the KT operator (7.26) and the gauge operator (7.39) below contain finite terms.

7.2 Noether's Inverse Second Theorem

Different variants of the second Noether theorem have been suggested in order to relate reducible NI and gauge symmetries [6, 9, 10, 49]. The *inverse second Noether theorem* (Theorem 7.9), that we formulate in homology terms, associates to the KT complex (7.28) of nontrivial NI the cochain sequence (7.38) with the ascent operator **u** (7.39) whose components are nontrivial complete gauge and higher-stage gauge SUSY of a Grassmann-graded Lagrangian system (Definitions 7.4–7.5). Let us start with the following notation.

Remark 7.5 Given the DBGA $\overline{\mathscr{P}}^*_\infty\{N\}$ (7.25), we consider a DBGA

$$\mathscr{P}^*_\infty\{N\} = \mathscr{P}^*_\infty[F \underset{Y}{\oplus} E_0 \underset{Y}{\oplus} \cdots \underset{Y}{\oplus} E_N; Y], \tag{7.32}$$

possessing the local generating basis $(s^A, c^r, c^{r_1}, \dots, c^{r_N})$, $[c^{r_k}] = ([\overline{c}_{r_k}]+1)\mathrm{mod}\,2$, and a DBGA

$$P^*_\infty\{N\} = \mathcal{P}^*_\infty[\overline{VF} \underset{Y}{\oplus} E_0 \oplus \cdots \underset{Y}{\oplus} E_N \underset{Y}{\oplus} \overline{E}_0 \underset{Y}{\oplus} \cdots \underset{Y}{\oplus} \overline{E}_N; Y] \tag{7.33}$$

with the local generating basis $(s^A, \overline{s}_A, c^r, c^{r_1}, \dots, c^{r_N}, \overline{c}_r, \overline{c}_{r_1}, \dots, \overline{c}_{r_N})$. Their elements c^{r_k} are called k-stage *ghosts* of ghost number $\mathrm{gh}[c^{r_k}] = k+1$ and antifield number $\mathrm{Ant}[c^{r_k}] = -(k+1)$. A $C^\infty(X)$-module $\mathcal{C}^{(k)}$ of k-stage ghosts is the density-dual of the module $\mathcal{C}_k$ of k-stage antifields. The DBGAs $\overline{\mathcal{P}}^*_\infty\{N\}$ (7.25) and $\mathcal{P}^*_\infty\{N\}$ (7.32) are subalgebras of $P^*_\infty\{N\}$ (7.33). The KT operator δ_{KT} (7.26) naturally is extended to a graded derivation of a DBGA $P^*_\infty\{N\}$.

Remark 7.6 Any graded differential form $\phi \in \mathcal{S}^*_\infty[F;Y]$ and any finite tuple (f^Λ), $0 \le |\Lambda| \le k$, of local graded functions $f^\Lambda \in \mathcal{S}^0_\infty[F;Y]$ obey the following relations [10]:

$$\sum_{0\le|\Lambda|\le k} f^\Lambda d_\Lambda \phi \wedge \omega = \sum_{0\le|\Lambda|} (-1)^{|\Lambda|} d_\Lambda(f^\Lambda)\phi \wedge \omega + d_H\sigma, \tag{7.34}$$

$$\sum_{0\le|\Lambda|\le k} (-1)^{|\Lambda|} d_\Lambda(f^\Lambda \phi) = \sum_{0\le|\Lambda|\le k} \eta(f)^\Lambda d_\Lambda \phi, \tag{7.35}$$

$$\eta(f)^\Lambda = \sum_{0\le|\Sigma|\le k-|\Lambda|} (-1)^{|\Sigma+\Lambda|} \frac{(|\Sigma+\Lambda|)!}{|\Sigma|!|\Lambda|!} d_\Sigma f^{\Sigma+\Lambda}, \tag{7.36}$$

$$\eta(\eta(f))^\Lambda = f^\Lambda. \tag{7.37}$$

Theorem 7.9 *Given the KT complex (7.28), a module of graded densities* $\mathcal{P}^{0,n}_\infty\{N\}$ *is decomposed into a cochain sequence*

$$0 \to \mathcal{S}^{0,n}_\infty[F;Y] \xrightarrow{\mathbf{u}} \mathcal{P}^{0,n}_\infty\{N\}^1 \xrightarrow{\mathbf{u}} \mathcal{P}^{0,n}_\infty\{N\}^2 \xrightarrow{\mathbf{u}} \cdots, \tag{7.38}$$

$$\mathbf{u} = u + u^{(1)} + \cdots + u^{(N)} = u^A \frac{\partial}{\partial s^A} + u^r \frac{\partial}{\partial c^r} + \cdots + u^{r_{N-1}} \frac{\partial}{\partial c^{r_{N-1}}}, \tag{7.39}$$

graded in ghost number. Its ascent operator $\mathbf{u}$ *(7.39) is an odd derivation of ghost number 1 where* u *(7.44) is a SUSY of a graded Lagrangian* L *and the graded derivations* $u_{(k)}$ *(7.47),* $k = 1, \dots, N$*, obey the relations (7.46).*

Proof Given the KT operator (7.26), let us extend an original graded Lagrangian L to a graded Lagrangian

$$L_e = L + L_1 = L + \sum_{0\le k\le N} c^{r_k} \Delta_{r_k}\omega = L + \delta_{\mathrm{KT}}(\sum_{0\le k\le N} c^{r_k}\overline{c}_{r_k}\omega) \tag{7.40}$$

of zero antifield number. It is readily observed that the KT operator δ_{KT} (7.26) is an exact symmetry of the extended Lagrangian $L_e \in P^{0,n}_\infty\{N\}$ (7.40). Since a graded derivation δ_{KT} is vertical, it follows from the first variational formula (6.82) that

$$\left[\frac{\overleftarrow{\delta}\mathcal{L}_e}{\delta\overline{s}_A}\mathcal{E}_A + \sum_{0\le k\le N}\frac{\overleftarrow{\delta}\mathcal{L}_e}{\delta\overline{c}_{r_k}}\Delta_{r_k}\right]\omega = \left[\upsilon^A\mathcal{E}_A + \sum_{0\le k\le N}\upsilon^{r_k}\frac{\delta\mathcal{L}_e}{\delta c^{r_k}}\right]\omega = d_H\sigma, \quad (7.41)$$

$$\upsilon^A = \frac{\overleftarrow{\delta}\mathcal{L}_e}{\delta\overline{s}_A} = u^A + w^A = \sum_{0\le|\Lambda|} c^r_\Lambda\eta(\Delta^A_r)^\Lambda + \sum_{1\le i\le N}\sum_{0\le|\Lambda|} c^{r_i}_\Lambda\eta(\overleftarrow{\partial}^A(h_{r_i}))^\Lambda,$$

$$\upsilon^{r_k} = \frac{\overleftarrow{\delta}\mathcal{L}_e}{\delta\overline{c}_{r_k}} = u^{r_k} + w^{r_k} = \sum_{0\le|\Lambda|} c^{r_{k+1}}_\Lambda\eta(\Delta^{r_k}_{r_{k+1}})^\Lambda + \sum_{k+1<i\le N}\sum_{0\le|\Lambda|} c^{r_i}_\Lambda\eta(\overleftarrow{\partial}^{r_k}(h_{r_i}))^\Lambda.$$

The equality (7.41) is split into a set of equalities

$$\frac{\overleftarrow{\delta}(c^r\Delta_r)}{\delta\overline{s}_A}\mathcal{E}_A\omega = u^A\mathcal{E}_A\omega = d_H\sigma_0, \quad (7.42)$$

$$\left[\frac{\overleftarrow{\delta}(c^{r_k}\Delta_{r_k})}{\delta\overline{s}_A}\mathcal{E}_A + \sum_{0\le i<k}\frac{\overleftarrow{\delta}(c^{r_k}\Delta_{r_k})}{\delta\overline{c}_{r_i}}\Delta_{r_i}\right]\omega = d_H\sigma_k, \quad (7.43)$$

where $k = 1, \ldots, N$. A glance at the equality (7.42) shows that, by virtue of item (iv) of Theorem 6.19, an odd derivation

$$u = u^A\frac{\partial}{\partial s^A}, \qquad u^A = \sum_{0\le|\Lambda|} c^r_\Lambda\eta(\Delta^A_r)^\Lambda, \quad (7.44)$$

of $\mathcal{P}^0_\infty\{0\}$ is SUSY of a graded Lagrangian L. Every equality (7.43) falls into a set of equalities graded by the polynomial degree in antifields. Let us consider that of them which is linear in antifields $\overline{c}_{r_{k-2}}$. We have

$$\frac{\overleftarrow{\delta}}{\delta\overline{s}_A}\left(c^{r_k}\sum_{0\le|\Sigma|,|\Xi|} h^{(r_{k-2},\Sigma)(A,\Xi)}_{r_k}\overline{c}_{\Sigma r_{k-2}}\overline{s}_{\Xi A}\right)\mathcal{E}_A\omega +$$
$$\frac{\overleftarrow{\delta}}{\delta\overline{c}_{r_{k-1}}}\left(c^{r_k}\sum_{0\le|\Sigma|}\Delta^{r'_{k-1},\Sigma}_{r_k}\overline{c}_{\Sigma r'_{k-1}}\right)\sum_{0\le|\Xi|}\Delta^{r_{k-2},\Xi}_{r_{k-1}}\overline{c}_{\Xi r_{k-2}}\omega = d_H\sigma_k.$$

This equality is brought into a form

$$\sum_{0\le|\Xi|}(-1)^{|\Xi|}d_\Xi\left(c^{r_k}\sum_{0\le|\Sigma|} h^{(r_{k-2},\Sigma)(A,\Xi)}_{r_k}\overline{c}_{\Sigma r_{k-2}}\right)\mathcal{E}_A\omega +$$
$$u^{r_{k-1}}\sum_{0\le|\Xi|}\Delta^{r_{k-2},\Xi}_{r_{k-1}}\overline{c}_{\Xi r_{k-2}}\omega = d_H\sigma_k.$$

Using the relation (7.34), we obtain an equality

$$\sum_{0\leq|\Xi|} c^{r_k} \sum_{0\leq|\Sigma|} h_{r_k}^{(r_{k-2},\Sigma)(A,\Xi)} \overline{c}_{\Sigma r_{k-2}} d_\Xi \mathscr{E}_A \omega + \\ u^{r_{k-1}} \sum_{0\leq|\Xi|} \Delta_{r_{k-1}}^{r_{k-2},\Xi} \overline{c}_{\Xi r_{k-2}} \omega = d_H \sigma'_k. \tag{7.45}$$

The variational derivative of both its sides with respect to $\overline{c}_{r_{k-2}}$ leads to a relation

$$\sum_{0\leq|\Sigma|} d_\Sigma u^{r_{k-1}} \frac{\partial}{\partial c_\Sigma^{r_{k-1}}} u^{r_{k-2}} = \overline{\delta}(\alpha^{r_{k-2}}), \tag{7.46}$$

$$\alpha^{r_{k-2}} = -\sum_{0\leq|\Sigma|} \eta(h_{r_k}^{(r_{k-2})(A,\Xi)})^\Sigma d_\Sigma (c^{r_k} \overline{s}_{\Xi A}),$$

which the odd derivation

$$u^{(k)} = u^{r_{k-1}} \frac{\partial}{\partial c^{r_{k-1}}} = \sum_{0\leq|\Lambda|} c_\Lambda^{r_k} \eta(\Delta_{r_k}^{r_{k-1}})^\Lambda \frac{\partial}{\partial c^{r_{k-1}}}, \quad k = 1, \ldots, N, \tag{7.47}$$

satisfies. Graded derivations u (7.44) and $u^{(k)}$ (7.47) are assembled into the ascent operator $\mathbf{u}$ (7.39) of the cochain sequence (7.38).

7.3 Gauge Supersymmetries: Noether's Direct Second Theorem

A glance at the expression (7.44) shows that SUSY u is a linear differential operator on the $C^\infty(X)$-module $\mathscr{C}^{(0)}$ of ghosts with values into a real vector space $\mathscr{S}\mathscr{G}_L$ of SUSY of L. Bearing in mind Definition 2.4 of gauge symmetries, we come to the following generalization of this notion to Grassmann-graded Lagrangian theory on graded bundles in a general setting (Definition 7.3).

Let $(\mathscr{S}_\infty^*[F;Y], L)$ be a Grassmann-graded Lagrangian system on a graded bundle $(X, Y, \mathfrak{A}_F)$ with the local generating basis (s^A). Let E be the graded vector bundle (7.2). Let us consider the corresponding DBGA $\mathscr{S}_\infty^*[F \times_X E; Y \times_X E^0]$ (7.4) together with the monomorphisms (6.52) of DBGAs

$$\mathscr{S}_\infty^*[F;Y] \to \mathscr{S}_\infty^*[F \underset{X}{\times} E; Y \underset{X}{\times} E^0], \qquad \mathscr{S}_\infty^*[E;E^0] \to \mathscr{S}_\infty^*[F \underset{X}{\times} E; Y \underset{X}{\times} E^0].$$

Given a graded Lagrangian $L \in \mathscr{S}_\infty^{0,n}[F;Y]$, let us define its pull-back

$$L \in \mathscr{S}_\infty^{0,n}[F;Y] \subset \mathscr{S}_\infty^{0,n}[F \underset{X}{\times} E; Y \underset{X}{\times} E^0], \tag{7.48}$$

and consider an extended Grassmann-graded Lagrangian system

$$(\mathcal{S}^*_\infty[F \underset{X}{\times} E; Y \underset{X}{\times} E^0], L) \tag{7.49}$$

provided with the local generating basis (s^A, c^r).

Definition 7.3 *Gauge* SUSY of a graded Lagrangian L is defined to be SUSY u of its pull-back (7.48) (Definition 6.9) such that the contact graded derivation J^*u (6.77) of the ring $\mathcal{S}^0_\infty[F \times_X E; Y \times_X E^0]$ equals zero on a subring

$$\mathcal{S}^0_\infty[E; E^0] \subset \mathcal{S}^0_\infty[F \underset{X}{\times} E; Y \underset{X}{\times} E^0].$$

In view of the second condition in Definition 7.3, the graded variables c^r of the extended Lagrangian system (7.49) can be treated as graded gauge parameters of gauge SUSY u.

Bearing in mind Remark 2.8, we further assume that gauge SUSY u is linear in graded gauge parameters c^r and their jets c^r_Λ. Then it reads

$$u = \left(\sum_{0\leq|\Lambda|\leq m} u_r^{\lambda\Lambda}(x^\mu) c^r_\Lambda \right) \partial_\lambda + \left(\sum_{0\leq|\Lambda|\leq m} u_r^{A\Lambda}(x^\mu, s^B_\Sigma) c^r_\Lambda \right) \partial_A. \tag{7.50}$$

By virtue of item (iii) of Theorem 6.19, the generalized graded vector field υ (7.50) is gauge SUSY if and only if its vertical part is so. Therefore, we can restrict our consideration to vertical gauge SUSY.

In accordance with Definition 7.3, the graded derivation u (7.44) in Noether's inverse second Theorem 7.9) is gauge SUSY of a graded Lagrangian L. Therefore, graded gauge parameters c^r of gauge SUSY (7.50) are called the ghosts.

Remark 7.7 In contrast with Definitions 2.3 and 2.4, if a graded Lagrangian L depends only on even variables, gauge SUSY in Definition 7.3 are parameterized by odd ghosts, but not even gauge parameters. Given the gauge symmetry u (2.15) defined as a derivation of the real ring $\mathcal{O}^0_\infty[Y \times E]$, one can associate to it gauge SUSY

$$u = \left(\sum_{0\leq|\Lambda|\leq m} u_a^{\lambda\Lambda}(x^\mu) c^a_\Lambda \right) \partial_\lambda + \left(\sum_{0\leq|\Lambda|\leq m} u_a^{i\Lambda}(x^\mu, y^j_\Sigma) c^a_\Lambda \right) \partial_i,$$

which is an odd derivation of a real ring $\mathcal{S}^0_\infty[E; Y]$, and vice versa.

The gauge SUSY u (7.44) is complete in the following sense. Let

$$\sum_{0\leq|\Xi|} C^R G_R^{r,\Xi} d_\Xi \Delta_r \omega$$

be some projective $C^\infty(X)$-module of finite rank of the nontrivial NI (7.11) parameterized by the corresponding ghosts C^R. We have the equalities

$$0 = \sum_{0\leq|\Xi|} C^R G_R^{r,\Xi} d_\Xi \left(\sum_{0\leq|\Lambda|} \Delta_r^{A,\Lambda} d_\Lambda \mathscr{E}_A \right) \omega =$$

$$\sum_{0\leq|\Lambda|} \left(\sum_{0\leq|\Xi|} \eta(G_R^r)^\Xi C_\Xi^R \right) \Delta_r^{A,\Lambda} d_\Lambda \mathscr{E}_A \omega + d_H(\sigma) =$$

$$\sum_{0\leq|\Lambda|} (-1)^{|\Lambda|} d_\Lambda \left(\Delta_r^{A,\Lambda} \sum_{0\leq|\Xi|} \eta(G_R^r)^\Xi C_\Xi^R \right) \mathscr{E}_A \omega + d_H \sigma =$$

$$\sum_{0\leq|\Lambda|} \eta(\Delta_r^A)^\Lambda d_\Lambda \left(\sum_{0\leq|\Xi|} \eta(G_R^r)^\Xi C_\Xi^R \right) \mathscr{E}_A \omega + d_H \sigma =$$

$$\sum_{0\leq|\Lambda|} u_r^{A,\Lambda} d_\Lambda \left(\sum_{0\leq|\Xi|} \eta(G_R^r)^\Xi C_\Xi^R \right) \mathscr{E}_A \omega + d_H \sigma.$$

It follows that the graded derivation

$$d_\Lambda \left(\sum_{0\leq|\Xi|} \eta(G_R^r)^\Xi C_\Xi^R \right) u_r^{A,\Lambda} \frac{\partial}{\partial s^A}$$

is SUSY of a graded Lagrangian L and, consequently, its gauge SUSY parameterized by ghosts C^R. It factorizes through the gauge SUSY (7.44) by putting ghosts

$$c^r = \sum_{0\leq|\Xi|} \eta(G_R^r)^\Xi C_\Xi^R.$$

Definition 7.4 The odd derivation u (7.44) is said to be a *complete nontrivial gauge* SUSY of a graded Lagrangian L associated to the complete NI (7.12).

Turn now to the relation (7.46). For $k = 1$, it takes a form

$$\sum_{0\leq|\Sigma|} d_\Sigma u^r \frac{\partial}{\partial c_\Sigma^r} u^A = \overline{\delta}(\alpha^A)$$

of a first-stage gauge SUSY condition on-shell which the nontrivial gauge SUSY u (7.44) satisfies. Therefore, one can treat the odd derivation

$$u^{(1)} = u^r \frac{\partial}{\partial c^r}, \qquad u^r = \sum_{0\leq|\Lambda|} c_\Lambda^{r_1} \eta(\Delta_{r_1}^r)^\Lambda,$$

as *first-stage gauge* SUSY associated to the complete first-stage NI

$$\sum_{0\leq|\Lambda|} \Delta_{r_1}^{r,\Lambda} d_\Lambda \left(\sum_{0\leq|\Sigma|} \Delta_r^{A,\Sigma} \overline{s}_{\Sigma A}\right) = -\overline{\delta}\left(\sum_{0\leq|\Sigma|,|\Xi|} h_{r_1}^{(B,\Sigma)(A,\Xi)} \overline{s}_{\Sigma B} \overline{s}_{\Xi A}\right).$$

Iterating the arguments, one comes to the relation (7.46) which provides a k-stage gauge SUSY condition which is associated to the complete k-stage NI (7.29).

The odd derivation $u_{(k)}$ (7.47) is called the k*-stage gauge* SUSY. It is complete as follows. Let

$$\sum_{0\leq|\Xi|} C^{R_k} G_{R_k}^{r_k,\Xi} d_\Xi \Delta_{r_k}\omega$$

be a projective $C^\infty(X)$-module of finite rank of nontrivial k-stage NI (7.11) factorizing through the complete ones (7.27) and parameterized by the corresponding ghosts C^{R_k}. One can show that it defines k-stage gauge SUSY factorizing through $u^{(k)}$ (7.47) by putting k-stage ghosts

$$c^{r_k} = \sum_{0\leq|\Xi|} \eta(G_{R_k}^{r_k})^\Xi C_\Xi^{R_k}.$$

Definition 7.5 The odd derivation $u_{(k)}$ (7.47) is said to be a *complete nontrivial* k*-stage gauge* SUSY of a graded Lagrangian L.

Gauge SUSY of a graded Lagrangian are called *irreducible* if they do not admit the nontrivial first-stage ones.

In accordance with Definitions 7.4–7.5, components of the ascent operator $\mathbf{u}$ (7.39) are complete nontrivial gauge and higher-stage gauge SUSY. Therefore, we agree to call this operator the *gauge operator*.

Remark 7.8 With the gauge operator (7.39), the extended Lagrangian L_e (7.40) takes a form

$$L_e = L + \mathbf{u}(\sum_{0\leq k\leq N} c^{r_{k-1}}\overline{c}_{r_{k-1}})\omega + L_1^* + d_H\sigma, \tag{7.51}$$

where L_1^* is a term of polynomial degree in antifields exceeding 1.

The correspondence of complete nontrivial gauge and higher-stage gauge SUSY to complete nontrivial Noether and higher-stage Noether identities is unique due to forthcoming *Noether's direct second theorem.*

Theorem 7.10 *(i) If* u *(7.44) is gauge SUSY, the variational derivative of a* d_H*-exact density* $u^A\mathcal{E}_A\omega$ *(7.42) with respect to ghosts* c^r *leads to the equality*

$$\delta_r(u^A\mathcal{E}_A\omega) = \sum_{0\leq|\Lambda|} (-1)^{|\Lambda|} d_\Lambda[u_r^{A\Lambda}\mathcal{E}_A] = \tag{7.52}$$

$$\sum_{0\leq|\Lambda|}(-1)^{|\Lambda|}d_\Lambda(\eta(\Delta_r^A)^\Lambda\mathcal{E}_A)=\sum_{0\leq|\Lambda|}(-1)^{|\Lambda|}\eta(\eta(\Delta_r^A))^\Lambda d_\Lambda\mathcal{E}_A=0,$$

which reproduces the complete Noether identities (7.12) by means of the relation (7.37).

(ii) Given the k-stage gauge symmetry condition (7.46), the variational derivative of the equality (7.45) with respect to ghosts c^{r_k} leads to the equality, reproducing the k-stage Noether identities (7.29) by means of the relations (7.35)–(7.37).

Remark 7.9 If the gauge SUSY u (4.49) is of second jet order in ghosts, i.e.,

$$u=(u_r^Ac^r+u_r^{A\mu}c_\mu^r+u_r^{A\nu\mu}c_{\nu\mu}^r)\partial_A, \tag{7.53}$$

the corresponding NI (7.52) take a form

$$u_r^A\mathcal{E}_A-d_\mu(u_r^{A\mu}\mathcal{E}_A)+d_{\nu\mu}(u_r^{A\nu\mu}\mathcal{E}_A)=0, \tag{7.54}$$

and vice versa (cf. Remark 2.6).

Remark 7.10 A glance at the expression (7.54) shows that, if the gauge SUSY u (7.53) is independent of jets of ghosts, then all variational derivatives of a graded Lagrangian equal zero, i.e., this Lagrangian is variationally trivial. Therefore, such gauge SUSY usually are not considered. At the same time, let a graded Lagrangian L be variationally trivial. Its variational derivatives $\mathcal{E}_A\equiv 0$ obey irreducible complete NI

$$\overline{\delta}\Delta_A=0,\qquad \Delta_A=\overline{s}_A. \tag{7.55}$$

By the formula (7.39), the associated irreducible gauge SUSY is given by the gauge operator

$$\mathbf{u}=c^A\frac{\partial}{\partial s^A}. \tag{7.56}$$

7.4 Noether's Third Theorem: Superpotential

Being SUSY, the gauge SUSY u (7.50) defines the weak conservation law (6.83) in accordance with Noether's first Theorem 6.20. The peculiarity of this conservation law is that the SUSY current $\mathcal{J}_u$ (6.84) is the graded total differential (7.57), i.e., it is reduced to a superpotential (Definition 2.6).

Theorem 7.11 *If u (7.50) is gauge SUSY of a graded Lagrangian L, the corresponding SUSY current $\mathcal{J}_u$ (6.84) along u is linear in ghosts and their jets (up to a d_H-closed term), and it takes a form*

$$\mathcal{J}_u=W+d_HU=(W^\mu+d_\nu U^{\nu\mu})\omega_\mu, \tag{7.57}$$

where a term W vanishes on-shell and $U^{\nu\mu} = -U^{\mu\nu}$ is a superpotential which takes the form (7.64).

Proof Let the gauge SUSY u (7.50) be at most of jet order N in ghosts. Then the SUSY current $\mathcal{J}_u$ (6.84), being linear in ghosts and their jets, is decomposed into a sum

$$\mathcal{J}_u^\mu = J_r^{\mu\mu_1\ldots\mu_M} c^r_{\mu_1\ldots\mu_M} + \sum_{1<k<M} J_r^{\mu\mu_k\ldots\mu_M} c^r_{\mu_k\ldots\mu_M} + J_r^{\mu\mu_M} c^r_{\mu_M} + J_r^\mu c^r, \quad (7.58)$$

where $N \leq M$, and the first variational formula (6.82) takes a form

$$0 = [\sum_{k=1}^N u_{V\,r}^{A\,\mu_k\ldots\mu_N} c^r_{\mu_k\ldots\mu_N} + u_{V\,r}^A c^r]\mathcal{E}_A - d_\mu(\sum_{k=1}^M J_r^{\mu\mu_k\ldots\mu_M} c^r_{\mu_k\ldots\mu_M} + J_r^\mu c^r).$$

This equality provides the following set of equalities for each $c^r_{\mu\mu_1\ldots\mu_M}$, $c^r_{\mu_k\ldots\mu_M}$ $(k = 1, \ldots, M-N-1)$, $c^r_{\mu_k\ldots\mu_N}$ $(k = 1, \ldots, N-1)$, c^r_μ and c^r:

$$0 = J_r^{(\mu\mu_1)\ldots\mu_M}, \quad (7.59)$$
$$0 = J_r^{(\mu_k\mu_{k+1})\ldots\mu_M} + d_\nu J_r^{\nu\mu_k\ldots\mu_M}, \qquad 1 \leq k < M-N, \quad (7.60)$$
$$0 = u_{V\,r}^{A\,\mu_k\ldots\mu_N}\mathcal{E}_A - J_r^{(\mu_k\mu_{k+1})\ldots\mu_N} - d_\nu J_r^{\nu\mu_k\ldots\mu_N}, \qquad 1 \leq k < N, \quad (7.61)$$
$$0 = u_{V\,r}^{A\,\mu}\mathcal{E}_A - J_r^\mu - d_\nu J_r^{\nu\mu}, \quad (7.62)$$

where $(\mu\nu)$ means symmetrization of indices in accordance with the splitting

$$J_r^{\mu_k\mu_{k+1}\ldots\mu_N} = J_r^{(\mu_k\mu_{k+1})\ldots\mu_N} + J_r^{[\mu_k\mu_{k+1}]\ldots\mu_N}.$$

We also have the equality

$$0 = u_{V\,r}^A\mathcal{E}_A - d_\mu J_r^\mu.$$

With the equalities (7.59)–(7.62), the decomposition (7.58) takes a form

$$\mathcal{J}_u^\mu = J_r^{[\mu\mu_1]\ldots\mu_M} c^r_{\mu_1\ldots\mu_M} + \sum_{1<k\leq M-N} [(J_r^{[\mu\mu_k]\ldots\mu_M} - d_\nu J_r^{\nu\mu\mu_k\ldots\mu_M})c^r_{\mu_k\ldots\mu_M}] +$$
$$\sum_{1<k<N} [(u_V^i A_r^{\mu\mu_k\ldots\mu_N}\mathcal{E}_A - d_\nu J_r^{\nu\mu\mu_k\ldots\mu_N} + J_r^{[\mu\mu_k]\ldots\mu_N})c^r_{\mu_k\ldots\mu_N}] +$$
$$(u_{V\,r}^{A\,\mu\mu_N}\mathcal{E}_A - d_\nu J_r^{\nu\mu\mu_N} + J_r^{[\mu\mu_N]})c^r_{\mu_N} + (u_{V\,r}^{A\,\mu}\mathcal{E}_A - d_\nu J_r^{\nu\mu})c^r.$$

A direct computation

$$
\begin{aligned}
\mathcal{J}^\mu_u &= d_\nu(J_r^{[\mu\nu]\mu_2\ldots\mu_M} c^r_{\mu_2\ldots\mu_M}) - d_\nu J_r^{[\mu\nu]\mu_2\ldots\mu_M} c^r_{\mu_2\ldots\mu_M} + \\
&\quad \sum_{1<k\leq M-N} [d_\nu(J_r^{[\mu\nu]\mu_{k+1}\ldots\mu_M} c^r_{\mu_{k+1}\ldots\mu_M}) - \\
&\quad d_\nu J_r^{[\mu\nu]\mu_{k+1}\ldots\mu_M} c^r_{\mu_{k+1}\ldots\mu_M} - d_\nu J_r^{\nu\mu\mu_k\ldots\mu_M} c^r_{\mu_k\ldots\mu_M}] + \\
&\quad \sum_{1<k<N} [(u_{Vr}^{A\,\mu\mu_k\ldots\mu_N} \mathcal{E}_A - d_\nu J_r^{\nu\mu\mu_k\ldots\mu_N}) c^r_{\mu_k\ldots\mu_N} + \\
&\quad d_\nu(J_r^{[\mu\nu]\mu_{k+1}\ldots\mu_N} c^r_{\mu_{k+1}\ldots\mu_N}) - d_\nu J_r^{[\mu\nu]\mu_{k+1}\ldots\mu_N} c^r_{\mu_{k+1}\ldots\mu_N}] + \\
&\quad [(u_{Vr}^{A\,\mu\mu_N} \mathcal{E}_A - d_\nu J_r^{\nu\mu\mu_N}) c^r_{\mu_N} + d_\nu(J_r^{[\mu\nu]} c^r) - d_\nu J_r^{[\mu\nu]} c^r] + (u_{Vr}^{A\,\mu} \mathcal{E}_A - d_\nu J_r^{\nu\mu}) c^r \\
&= d_\nu(J_r^{[\mu\nu]\mu_2\ldots\mu_M} c^r_{\mu_2\ldots\mu_M}) + \\
&\quad \sum_{1<k\leq M-N} [d_\nu(J_r^{[\mu\nu]\mu_{k+1}\ldots\mu_M} c^r_{\mu_{k+1}\ldots\mu_M}) - d_\nu J_r^{(\nu\mu)\mu_k\ldots\mu_M} c^r_{\mu_k\ldots\mu_M}] + \\
&\quad \sum_{1<k<N} [(u_{Vr}^{A\,\mu\mu_k\ldots\mu_N} \mathcal{E}_A - d_\nu J_r^{(\nu\mu)\mu_k\ldots\mu_N}) c^r_{\mu_k\ldots\mu_N} + d_\nu(J_r^{[\mu\nu]\mu_{k+1}\ldots\mu_N} c^r_{\mu_{k+1}\ldots\mu_N})] + \\
&\quad [(u_{Vr}^{A\,\mu\mu_N} \mathcal{E}_A - d_\nu J_r^{(\nu\mu)\mu_N}) c^r_{\mu_N} + d_\nu(J_r^{[\mu\nu]} c^r)] + (u_{Vr}^{A\,\mu} \mathcal{E}_A - d_\nu J_r^{(\nu\mu)}) c^r
\end{aligned}
$$

leads to the expression

$$
\begin{aligned}
\mathcal{J}^\mu_u &= (\sum_{1<k\leq N} u_{Vr}^{i\,\mu\mu_k\ldots\mu_N} c^r_{\mu_k\ldots\mu_N} + u_{Vr}^{A\,\mu} c^r)\mathcal{E}_A - \\
&\quad (\sum_{1<k\leq M} d_\nu J^{(\nu\mu)\mu_k\ldots\mu_M} c^r_{\mu_k\ldots\mu_M} + d_\nu J_r^{(\nu\mu)} c^r) - \\
&\quad d_\nu(\sum_{1<k\leq M} J^{[\nu\mu]\mu_k\ldots\mu_M} c^r_{\mu_k\ldots\mu_M} + J_r^{[\nu\mu]} c^r).
\end{aligned}
\tag{7.63}
$$

The first summand of this expression vanishes on-shell. Its second one contains terms $d_\nu J^{(\nu\mu_k)\mu_{k+1}\ldots\mu_M}$, $k = 1, \ldots, M$. By virtue of the equalities (7.60)–(7.61), every $d_\nu J^{(\nu\mu_k)\mu_{k+1}\ldots\mu_M}$ is expressed into the terms vanishing on-shell and the term $d_\nu J^{(\nu\mu_{k-1})\mu_k\ldots\mu_M}$. Iterating the procedure and bearing in mind the equality (7.59), one can easily show that a second summand of the expression (7.63) also vanishes on-shell. Thus a SUSY current takes the form (7.57) where

$$
U^{\nu\mu} = -\sum_{1<k\leq M} J^{[\nu\mu]\mu_k\ldots\mu_M} c^r_{\mu_k\ldots\mu_M} - J_r^{[\nu\mu]} c^r. \tag{7.64}
$$

Remark 7.11 If an exact gauge SUSY

$$
u = (u_r^\lambda c^r + u_r^{\lambda\mu} c^r_\mu)\partial_\lambda + (u_r^A c^r + u_r^{A\mu} c^r_\mu)\partial_A
$$

of a graded Lagrangian L depends at most on first jets of ghosts, then the decomposition (7.63) takes a form

$$\mathcal{J}_u^\mu = u_{Vr}^{A\mu}\mathcal{E}_A c^r - d_\nu(J_r^{[\nu\mu]}c^r) = \qquad (7.65)$$
$$(u_a^{i\mu}r - s_\lambda^A u_r^{\lambda\mu})c^r\mathcal{E}_A + d_\nu[(u_r^{A[\mu} - s_\lambda^A u_r^{\lambda[\mu})c^r\partial_A^{\nu]}\mathcal{L} + u_r^{[\nu\mu]}c^r\mathcal{L}].$$

7.5 Lagrangian BRST Theory

In contrast with the KT operator (7.26), the gauge operator **u** (7.38) need not be nilpotent. let us study its extension to a nilpotent graded derivation

$$\mathbf{b} = \mathbf{u} + \gamma = \mathbf{u} + \sum_{1\leq k\leq N+1}\gamma^{(k)} = \mathbf{u} + \sum_{1\leq k\leq N+1}\gamma^{r_{k-1}}\frac{\partial}{\partial c^{r_{k-1}}} \qquad (7.66)$$
$$= \left(u^A\frac{\partial}{\partial s^A} + \gamma^r\frac{\partial}{\partial c^r}\right) + \sum_{0\leq k\leq N-1}\left(u^{r_k}\frac{\partial}{\partial c^{r_k}} + \gamma^{r_{k+1}}\frac{\partial}{\partial c^{r_{k+1}}}\right)$$

of ghost number 1 by means of antifield-free terms $\gamma^{(k)}$ of higher polynomial degree in ghosts c^{r_i} and their jets $c_\Lambda^{r_i}$, $0 \leq i < k$. We call **b** (7.66) the *BRST operator*, where k-stage gauge SUSY are extended to k-stage BRST transformations acting both on $(k-1)$-stage and k-stage ghosts [12, 60, 134]. If the BRST operator exists, the cochain sequence (7.38) is brought into a *BRST complex*

$$0 \to \mathcal{P}_\infty^{0,n}\{N\}^1 \xrightarrow{\mathbf{b}} \mathcal{P}_\infty^{0,n}\{N\}^1 \xrightarrow{\mathbf{b}} \mathcal{P}_\infty^{0,n}\{N\}^2 \xrightarrow{\mathbf{b}} \cdots .$$

There is following necessary condition of the existence of such a BRST extension.

Theorem 7.12 *The gauge operator (7.38) admits the nilpotent extension (7.66) only if the higher-stage gauge SUSY conditions (7.46) and the higher-stage NI (7.29) are satisfied off-shell.*

Proof It is easily justified that, if the graded derivation **b** (7.66) is nilpotent, then the right hand sides of the equalities (7.46) equal zero, i.e.,

$$u^{(k+1)}(u^{(k)}) = 0, \qquad 0 \leq k \leq N-1, \qquad u^{(0)} = u. \qquad (7.67)$$

Using the relations (7.34)–(7.37), one can show that, in this case, the right hand sides of the higher-stage NI (7.29) also equal zero [9]. It follows that the summand G_{r_k} of each cocycle Δ_{r_k} (7.27) is δ_{k-1}-closed. Then its summand h_{r_k} also is δ_{k-1}-closed and, consequently, δ_{k-2}-closed. Hence it is δ_{k-1}-exact by virtue of the homology regularity condition (Definition 7.2). Therefore, Δ_{r_k} contains only a term G_{r_k} linear in antifields.

It follows at once from the equalities (7.67) that the *higher-stage gauge operator*

$$u_{\rm HS} = \mathbf{u} - u = u^{(1)} + \cdots + u^{(N)}$$

is nilpotent, and

$$\mathbf{u}(\mathbf{u}) = u(\mathbf{u}). \tag{7.68}$$

Therefore, the nilpotency condition for the BRST operator $\mathbf{b}$ (7.66) takes a form

$$\mathbf{b}(\mathbf{b}) = (u + \gamma)(\mathbf{u}) + (u + u_{\rm HS} + \gamma)(\gamma) = 0. \tag{7.69}$$

Let us denote

$$\begin{aligned}
&\gamma^{(0)} = 0,\\
&\gamma^{(k)} = \gamma^{(k)}_{(2)} + \cdots + \gamma^{(k)}_{(k+1)}, \quad k = 1, \ldots, N+1,\\
&\gamma^{r_{k-1}}_{(i)} = \sum_{k_1+\cdots+k_i=k+1-i} \left(\sum_{0\leq|\Lambda_{k_j}|} \gamma^{r_{k-1},\Lambda_{k_1},\ldots,\Lambda_{k_i}}_{(i)r_{k_1},\ldots,r_{k_i}} c^{r_{k_1}}_{\Lambda_{k_1}} \cdots c^{r_{k_i}}_{\Lambda_{k_i}} \right),\\
&\gamma^{(N+2)} = 0,
\end{aligned}$$

where $\gamma^{(k)}_{(i)}$ are terms of polynomial degree $2 \leq i \leq k+1$ in ghosts. Then the nilpotent property (7.69) of $\mathbf{b}$ falls into a set of equalities

$$u^{(k+1)}(u^{(k)}) = 0, \qquad 0 \leq k \leq N-1, \tag{7.70}$$

$$(u + \gamma^{(k+1)}_{(2)})(u^{(k)}) + u_{\rm HS}(\gamma^{(k)}_{(2)}) = 0, \qquad 0 \leq k \leq N+1, \tag{7.71}$$

$$\begin{aligned}
&\gamma^{(k+1)}_{(i)}(u^{(k)}) + u(\gamma^{(k)}_{(i-1)}) + u_{\rm HS}(\gamma^{k}_{(i)}) +\\
&\qquad \sum_{2\leq m\leq i-1} \gamma_{(m)}(\gamma^{(k)}_{(i-m+1)}) = 0, \qquad i-2 \leq k \leq N+1,
\end{aligned} \tag{7.72}$$

of ghost polynomial degree 1, 2 and $3 \leq i \leq N+3$, respectively.

The equalities (7.70) are exactly the conditions (7.67) in Theorem 7.12.

The equality (7.71) for $k = 0$ reads

$$(u + \gamma^{(1)})(u) = 0, \qquad \sum_{0\leq|\Lambda|} (d_\Lambda(u^A)\partial^\Lambda_A u^B + d_\Lambda(\gamma^r)u^{B,\Lambda}_r) = 0. \tag{7.73}$$

It takes a form of the Lie antibracket

$$[u, u] = -2\gamma^{(1)}(u) = -2 \sum_{0\leq|\Lambda|} d_\Lambda(\gamma^r)u^{B,\Lambda}_r \partial_B \tag{7.74}$$

of the odd gauge SUSY u. Its right-hand side is a nonlinear differential operator on a module $\mathscr{C}^{(0)}$ of ghosts taking values into a real space $\mathscr{SG}_L$ of SUSY. Following Remark 2.8, we treat it as a *generalized gauge* SUSY factorizing through gauge SUSY u. Thus, we come to the following.

Theorem 7.13 *The gauge operator (7.38) admits the nilpotent extension (7.66) only if the Lie antibracket of the odd gauge SUSY u (7.44) is generalized gauge SUSY factorizing through u.*

The equalities (7.71)–(7.72) for $k = 1$ take a form

$$(u + \gamma_{(2)}^{(2)})(u^{(1)}) + u^{(1)}(\gamma^{(1)}) = 0, \tag{7.75}$$

$$\gamma_{(3)}^{(2)}(u^{(1)}) + (u + \gamma^{(1)})(\gamma^{(1)}) = 0. \tag{7.76}$$

In particular, if a Lagrangian system is irreducible, i.e., $u^{(k)} = 0$ and $\mathbf{u} = u$, the BRST operator reads

$$\mathbf{b} = u + \gamma^{(1)} = u^A \partial_A + \gamma^r \partial_r = \sum_{0\leq|\Lambda|} u_r^{A,\Lambda} c_\Lambda^r \partial_A + \sum_{0\leq|\Lambda|,|\Xi|} \gamma_{pq}^{r,\Lambda,\Xi} c_\Lambda^p c_\Xi^q \partial_r.$$

In this case, the nilpotency conditions (7.75)–(7.76) are reduced to the equality

$$(u + \gamma^{(1)})(\gamma^{(1)}) = 0. \tag{7.77}$$

Furthermore, let gauge SUSY u be affine in fields s^A and their jets. It follows from the nilpotency condition (7.73) that the BRST term $\gamma^{(1)}$ is independent of original fields and their jets. Then the relation (7.77) takes a form of the Jacobi identity

$$\gamma^{(1)})(\gamma^{(1)}) = 0 \tag{7.78}$$

for coefficient functions $\gamma_{pq}^{r,\Lambda,\Xi}(x)$ in the Lie antibracket (7.74).

The relations (7.74) and (7.78) motivate us to think of the equalities (7.71)–(7.72) in a general case of reducible gauge symmetries as being *sui generis* generalized commutation relations and Jacobi identities of gauge SUSY, respectively [60]. Based on Theorem 7.13, we therefore say that nontrivial gauge symmetries are *algebraically closed* (in the terminology of [63]) if the gauge operator $\mathbf{u}$ (7.39) admits the nilpotent BRST extension $\mathbf{b}$ (7.66).

Remark 7.12 A Grassmann-graded Lagrangian system is called *Abelian* if its gauge SUSY u is *Abelian* and the higher-stage gauge SUSY are independent of original fields, i.e., if $u(\mathbf{u}) = 0$. It follows from the relation (7.68) that, in this case, the gauge operator itself is the BRST operator $\mathbf{u} = \mathbf{b}$. For instance, let a graded Lagrangian L be variationally trivial (Remark 7.10). Its variational derivatives $\mathscr{E}_i \equiv 0$ obey irreducible complete NI (7.55):

$$\overline{\delta}\Delta_i = 0, \qquad \Delta_i = \overline{s}_i.$$

The associated irreducible gauge SUSY is given by the gauge operator (7.56). Thus, a Lagrangian system with a variationally trivial Lagrangian is Abelian, and **u** (7.56) is the BRST operator. The topological BF theory exemplifies a reducible Abelian Lagrangian system (Chap. 12).

The DBGA $P^*_\infty\{N\}$ (7.33) is a particular *field-antifield theory* of the following type [6, 12, 63].

Let us consider a pull-back composite bundle

$$W = Z \underset{X}{\times} Z' \to Z \to X$$

where $Z' \to X$ is a vector bundle. Let us regard it as a graded vector bundle over Z possessing only odd part. The density-dual $\overline{VW}$ of the vertical tangent bundle VW of $W \to X$ is a graded vector bundle

$$\overline{VW} = ((\overline{Z}' \underset{Z}{\oplus} V^*Z) \underset{Z}{\otimes} \overset{n}{\wedge} T^*X) \underset{Y}{\oplus} Z'$$

over Z (cf. (7.5)). Let us consider the DBGA $\mathcal{P}^*_\infty[\overline{VW}; Z]$ (7.6) with the local generating basis $(z^a, \overline{z}_a)$, $[\overline{z}_a] = ([z^a] + 1)\mathrm{mod}\, 2$. Its elements z^a and $\overline{z}_a$ are called fields and antifields, respectively. Graded densities of this DBGA are endowed with the *antibracket*

$$\{\mathfrak{L}\omega, \mathfrak{L}'\omega\} = \left[\frac{\overleftarrow{\delta}\, \mathfrak{L}}{\delta \overline{z}_a}\frac{\delta \mathfrak{L}'}{\delta z^a} + (-1)^{[\mathfrak{L}']([\mathfrak{L}']+1)}\frac{\overleftarrow{\delta}\, \mathfrak{L}'}{\delta \overline{z}_a}\frac{\delta \mathfrak{L}}{\delta z^a}\right]\omega.$$

With this antibracket, one associates to any (even) Lagrangian $\mathfrak{L}\omega$ the odd vertical graded derivations

$$\upsilon_{\mathfrak{L}} = \overleftarrow{\mathcal{E}}^a \partial_a = \frac{\overleftarrow{\delta}\, \mathfrak{L}}{\delta \overline{z}_a}\frac{\partial}{\partial z^a}, \tag{7.79}$$

$$\overline{\upsilon}_{\mathfrak{L}} = \partial^a \mathcal{E}_a = \frac{\overleftarrow{\partial}}{\partial \overline{z}_a}\frac{\delta \mathfrak{L}}{\delta z^a}, \tag{7.80}$$

$$\vartheta_{\mathfrak{L}} = \upsilon_{\mathfrak{L}} + \overline{\upsilon}^l_{\mathfrak{L}} = (-1)^{[a]+1}\left(\frac{\delta \mathfrak{L}}{\delta \overline{z}^a}\frac{\partial}{\partial z_a} + \frac{\delta \mathfrak{L}}{\delta z^a}\frac{\partial}{\partial \overline{z}_a}\right), \tag{7.81}$$

such that

$$\vartheta_{\mathfrak{L}}(\mathfrak{L}'\omega) = \{\mathfrak{L}\omega, \mathfrak{L}'\omega\}.$$

Theorem 7.14 *The following conditions are equivalent.*
(i) The antibracket of a Lagrangian $\mathfrak{L}\omega$ *is* d_H*-exact, i.e.,*

$$\{\mathfrak{L}\omega, \mathfrak{L}\omega\} = 2\frac{\overleftarrow{\delta}\,\mathfrak{L}}{\delta \overline{z}_a}\frac{\delta\mathfrak{L}}{\delta z^a}\omega = d_H\sigma. \tag{7.82}$$

(ii) The graded derivation υ *(7.79) is a symmetry of a Lagrangian* $\mathfrak{L}\omega$.
(iii) The graded derivation $\overline{\upsilon}$ *(7.80) is a symmetry of* $\mathfrak{L}\omega$.
(iv) The graded derivation $\vartheta_{\mathfrak{L}}$ *(7.81) is nilpotent.*

Proof By virtue of the first variational formula (6.82), conditions (ii) and (iii) are equivalent to condition (i). The equality (7.82) is equivalent to that the odd density $\overleftarrow{\mathcal{E}}^a\mathcal{E}_a\omega$ is variationally trivial. Replacing right variational derivatives $\overleftarrow{\mathcal{E}}^a$ with $(-1)^{[a]+1}\mathcal{E}^a$, we obtain

$$2\sum_a(-1)^{[a]}\mathcal{E}^a\mathcal{E}_a\omega = d_H\sigma.$$

The variational operator acting on this relation results in the equalities

$$\sum_{0\leq|\Lambda|}(-1)^{[a]+|\Lambda|}d_\Lambda(\partial_b^\Lambda(\mathcal{E}^a\mathcal{E}_a)) = \sum_{0\leq|\Lambda|}(-1)^{[a]}[\eta(\partial_b\mathcal{E}^a)^\Lambda\mathcal{E}_{\Lambda a} + \eta(\partial_b\mathcal{E}_a)^\Lambda\mathcal{E}^a_\Lambda)] = 0,$$

$$\sum_{0\leq|\Lambda|}(-1)^{[a]+|\Lambda|}d_\Lambda(\partial^{\Lambda b}(\mathcal{E}^a\mathcal{E}_a)) = \sum_{0\leq|\Lambda|}(-1)^{[a]}[\eta(\partial^b\mathcal{E}^a)^\Lambda\mathcal{E}_{\Lambda a} + \eta(\partial^b\mathcal{E}_a)\mathcal{E}^a_\Lambda] = 0.$$

Due to the identity

$$(\delta\circ\delta)(L) = 0, \qquad \eta(\partial_B\mathcal{E}_A)^\Lambda = (-1)^{[A][B]}\partial_A^\Lambda\mathcal{E}_B,$$

we obtain

$$\sum_{0\leq|\Lambda|}(-1)^{[a]}[(-1)^{[b]([a]+1)}\partial^{\Lambda a}\mathcal{E}_b\mathcal{E}_{\Lambda a} + (-1)^{[b][a]}\partial_a^\Lambda\mathcal{E}_b\mathcal{E}^a_\Lambda] = 0,$$

$$\sum_{0\leq|\Lambda|}(-1)^{[a]+1}[(-1)^{([b]+1)([a]+1)}\partial^{\Lambda a}\mathcal{E}^b\mathcal{E}_{\Lambda a} + (-1)^{([b]+1)[a]}\partial_a^\Lambda\mathcal{E}^b\mathcal{E}^a_\Lambda] = 0$$

for all $\mathcal{E}_b$ and $\mathcal{E}^b$. This is exactly condition (iv).

The equality (7.82) is called the *classical master equation*. For instance, any variationally trivial Lagrangian satisfies the master equation. A solution of the master equation (7.82) is called nontrivial if both the derivations (7.79) and (7.80) do not vanish.

Being an element of the DBGA $P^*_\infty\{N\}$ (7.33), an original Lagrangian L obeys the master equation (7.82) and yields the graded derivations $\upsilon_L = 0$ (7.79) and $\overline{\upsilon}_L = \overline{\delta}$ (7.80), i.e., it is a trivial solution of the master equation.

The graded derivations (7.79)–(7.80) associated to the extended Lagrangian L_e (7.51) are extensions

$$\upsilon_e = \mathbf{u} + \frac{\overleftarrow{\delta} \mathcal{L}_1^*}{\delta \overline{s}_A}\frac{\partial}{\partial s^A} + \sum_{0\leq k\leq N} \frac{\overleftarrow{\delta} \mathcal{L}_1^*}{\delta \overline{c}_{r_k}}\frac{\partial}{\partial c^{r_k}},$$

$$\overline{\upsilon}_e = \delta_{\mathrm{KT}} + \frac{\partial}{\partial \overline{s}_A}\frac{\delta \mathcal{L}_1}{\delta s^A}$$

of the gauge and KT operators, respectively. However, the Lagrangian L_e need not satisfy the master equation. Therefore, let us consider its extension

$$L_E = L_e + L' = L + L_1 + L_2 + \cdots \tag{7.83}$$

by means of even densities $L_i, i \geq 2$, of zero antifield number and polynomial degree i in ghosts. The corresponding graded derivations (7.79)–(7.80) read

$$\upsilon_E = \upsilon_e + \frac{\overleftarrow{\delta} \mathcal{L}'}{\delta \overline{s}_A}\frac{\partial}{\partial s^A} + \sum_{0\leq k\leq N} \frac{\overleftarrow{\delta} \mathcal{L}'}{\delta \overline{c}_{r_k}}\frac{\partial}{\partial c^{r_k}}, \tag{7.84}$$

$$\overline{\upsilon}_E = \overline{\upsilon}_e + \frac{\partial}{\partial \overline{s}_A}\frac{\delta \mathcal{L}'}{\delta s^A} + \sum_{0\leq k\leq N} \frac{\partial}{\partial \overline{c}_{r_k}}\frac{\delta \mathcal{L}'}{\delta c^{r_k}}.$$

The Lagrangian L_E (7.83) where $L + L_1 = L_e$ is called a *proper extension* of an original Lagrangian L. The following is a corollary of Theorem 7.14.

Theorem 7.15 *A Lagrangian L is extended to a proper solution L_E (7.83) of the master equation only if the gauge operator* $\mathbf{u}$ *(7.38) admits a nilpotent extension.*

By virtue of condition (iv) of Theorem 7.14, this nilpotent extension is the derivation $\vartheta_E = \upsilon_E + \overline{\upsilon}'_E$ (7.81), called the *KT-BRST operator*. With this operator, the module of densities $P^{0,n}_\infty\{N\}$ is split into the *KT-BRST complex*

$$\cdots \longrightarrow P^{0,n}_\infty\{N\}_2 \longrightarrow P^{0,n}_\infty\{N\}_1 \longrightarrow P^{0,n}_\infty\{N\}_0 \longrightarrow \tag{7.85}$$
$$P^{0,n}_\infty\{N\}^1 \longrightarrow P^{0,n}_\infty\{N\}^2 \longrightarrow \cdots .$$

Putting all ghosts zero, we obtain a cochain morphism of this complex onto the KT complex, extended to $\overline{\mathscr{P}}^{0,n}_\infty\{N\}$ and reversed into the cochain one. Letting all antifields zero, we come to a cochain morphism of the KT-BRST complex (7.85) onto the cochain sequence (7.38), where the gauge operator is extended to the antifield-free part of the KT-BRST operator.

Theorem 7.16 *If the gauge operator* $\mathbf{u}$ *(7.38) can be extended to the BRST operator* $\mathbf{b}$ *(7.66), then the master equation has a nontrivial proper solution*

$$L_E = L_e + \sum_{1\leq k\leq N} \gamma^{r_{k-1}}\overline{c}_{r_{k-1}}\omega = \tag{7.86}$$
$$L + \mathbf{b}\left(\sum_{0\leq k\leq N} c^{r_{k-1}}\overline{c}_{r_{k-1}}\right)\omega + d_H\sigma.$$

Proof By virtue of Theorem 7.12, if the BRST operator $\mathbf{b}$ (7.66) exists, the densities Δ_{r_k} (7.27) contain only the terms G_{r_k} linear in antifields. It follows that the extended Lagrangian L_e (7.40) and, consequently, the Lagrangian L_E (7.86) are affine in antifields. In this case, we have

$$u^A = \overleftarrow{\delta}^A(\mathscr{L}_e), \qquad u^{r_k} = \overleftarrow{\delta}^{r_k}(\mathscr{L}_e)$$

for all indices A and r_k and, consequently,

$$\mathbf{b}^A = \overleftarrow{\delta}^A(\mathscr{L}_E), \qquad \mathbf{b}^{r_k} = \overleftarrow{\delta}^{r_k}(\mathscr{L}_E),$$

i.e., $\mathbf{b} = \upsilon_E$ is the graded derivation (7.84) defined by the Lagrangian L_E. Its nilpotency condition takes a form

$$\mathbf{b}(\overleftarrow{\delta}^A(\mathscr{L}_E)) = 0, \qquad \mathbf{b}(\overleftarrow{\delta}^{r_k}(\mathscr{L}_E)) = 0.$$

Hence, we obtain

$$\mathbf{b}(\mathscr{L}_E) = \mathbf{b}(\overleftarrow{\delta}^A(\mathscr{L}_E)\overline{s}_A + \overleftarrow{\delta}^{r_k}(\mathscr{L}_E)\overline{c}_{r_k}) = 0,$$

i.e., $\mathbf{b}$ is a variational symmetry of L_E. Consequently, L_E obeys the master equation.

For instance, let a gauge symmetry u be Abelian, and let the higher-stage gauge symmetries be independent of original fields, i.e., $u(\mathbf{u}) = 0$. Then $\mathbf{u} = \mathbf{b}$ and $L_E = L_e$.

The proper solution L_E (7.86) of the master equation is called the *BRST extension* of an original Lagrangian L. As was mentioned above, it is a necessary step towards BV quantization of classical Lagrangian field theory in terms of functional integrals [6, 13, 44, 50, 63].

Chapter 8
Yang–Mills Gauge Theory on Principal Bundles

In classical gauge field theory, gauge fields conventionally are described as principal connections on principal bundles [37, 61, 100, 102, 127]. We consider their first order Yang–Mills Lagrangian theory. This is irreducible degenerate Lagrangian theory (Sect. 8.6) where conserved Noether currents are reduced to a superpotential (Sect. 8.5). However, an energy-momentum current fails to be conserved because of a background world metric.

8.1 Geometry of Principal Bundles

We consider smooth principle bundles with a structure real Lie group of nonzero dimension [100, 134].

Principal Bundles

Let G, $\dim G > 0$, be a real Lie group which acts on a smooth manifold V on the left. A smooth fibre bundle $Y \to X$ with a typical fibre V is called a *bundle with a structure group* G or, in brief, the G*-bundle* if it possesses an atlas

$$\Psi = \{(U_\alpha, \psi_\alpha), \rho_{\alpha\beta}\}, \qquad \psi_\alpha = \rho_{\alpha\beta}\psi_\beta, \tag{8.1}$$

whose transition functions $\rho_{\alpha\beta}$ (B.3) factorize as

$$\rho_{\alpha\beta} : U_\alpha \cap U_\beta \times V \longrightarrow U_\alpha \cap U_\beta \times (G \times V) \overset{\mathrm{Id}\times\zeta}{\longrightarrow} U_\alpha \cap U_\beta \times V \tag{8.2}$$

through G-valued local smooth functions $\rho^G_{\alpha\beta} : U_\alpha \cap U_\beta \to G$ on X. This means that transition morphisms $\rho_{\alpha\beta}(x)$ (B.6) are elements of G acting on V. Transition functions (8.2) are called G*-valued.*

Let

$$\pi_P : P \to X \tag{8.3}$$

G. Sardanashvily, *Noether's Theorems*, Atlantis Studies in Variational Geometry 3, DOI 10.2991/978-94-6239-171-0_8

be a G-bundle whose typical fibre is the group space of G, which a group G acts on by left multiplications. It is called the *principal bundle* with a structure group G. Equivalently, a principal G-bundle is defined as a fibre bundle P (8.3) which admits an *action of G on P* on the right by a fibrewise morphism

$$R_{GP} : G \underset{X}{\times} P \underset{X}{\longrightarrow} P, \qquad R_g : p \to pg, \qquad \pi_P(p) = \pi_P(pg), \quad p \in P, \tag{8.4}$$

which is free and transitive on each fibre of P. As a consequence, the quotient of P with respect to the action (8.4) of G is diffeomorphic to a base X, i.e., $P/G = X$.

Remark 8.1 Given the principal bundle P (8.3), a fibre bundle $Y \to X$ with a structure group G and a typical fibre V is said to be *associated* to P (or, in brief, *P-associated*) if it is isomorphic to the quotient

$$Y = (P \times V)/G \tag{8.5}$$

of the product $P \times V$ by identification of elements (p, v) and $(pg, g^{-1}v)$ for all $g \in G$. Any G-bundle is associated to some principal G-bundle.

A principal G-bundle P is equipped with the bundle atlas (8.1):

$$\Psi_P = \{(U_\alpha, \psi_\alpha^P), \rho_{\alpha\beta}\} \tag{8.6}$$

whose trivialization morphisms ψ_α^P obey the condition

$$\psi_\alpha^P(pg) = g\psi_\alpha^P(p), \qquad g \in G.$$

Due to this property, every trivialization morphism ψ_α^P determines a unique local section $z_\alpha : U_\alpha \to P$ such that $(\psi_\alpha^P \circ z_\alpha)(x) = \mathbf{1}$. The transformation law for z_α reads

$$z_\beta(x) = z_\alpha(x)\rho_{\alpha\beta}(x), \qquad x \in U_\alpha \cap U_\beta. \tag{8.7}$$

Conversely, a family

$$\Psi_P = \{(U_\alpha, z_\alpha), \rho_{\alpha\beta}\} \tag{8.8}$$

of local sections of P which obey the transformation law (8.7) uniquely defines a bundle atlas Ψ_P of a principal bundle P. In particular, a principal bundle admits a global section if and only if it is trivial.

Remark 8.2 Every bundle atlas Ψ_P (8.8) of a principal G-bundle P defines a unique associated bundle atlas

$$\Psi = \{(U_\alpha, \psi_\alpha(x) = [z_\alpha(x)]^{-1})\}$$

of the P-associated bundle Y (8.5).

Let us consider the tangent morphism to the right action R_{GP} (8.4) of G on P. Since $TG = G \times \mathfrak{g}_l$ where $\mathfrak{g}_l$ is the left Lie algebra of G, this morphism takes a form

$$TR_{GP} : (G \times \mathfrak{g}_l) \underset{X}{\times} TP \underset{X}{\longrightarrow} TP, \tag{8.9}$$

and its restriction to $T_1 G \times_X TP$ provides a homomorphism

$$\mathfrak{g}_l \ni \varepsilon \to \xi_\varepsilon \in \mathcal{T}(P) \tag{8.10}$$

of the left Lie algebra $\mathfrak{g}_l$ of G to the Lie algebra $\mathcal{T}(P)$ of vector fields on P. Vector fields ξ_ε (8.10) are obviously vertical. They are called *fundamental vector fields* [82]. Given a basis $\{\varepsilon_r\}$ for $\mathfrak{g}_l$, the corresponding fundamental vector fields $\xi_r = \xi_{\varepsilon_r}$ form a family of $m = \dim \mathfrak{g}_l$ nowhere vanishing and linearly independent sections of the vertical tangent bundle VP of $P \to X$. Consequently, this bundle is trivial $VP = P \times \mathfrak{g}_l$ by virtue of Theorem B.12. Restricting the tangent morphism TR_{GP} (8.9) to

$$TR_{GP} : \widehat{0}(G) \underset{X}{\times} TP \underset{X}{\longrightarrow} TP, \tag{8.11}$$

we obtain the tangent prolongation of the structure group action R_{GP} (8.4). In brief, it is called the *action of* G *on* TP. Since the action of G (8.4) on P is fibrewise, its action (8.11) is restricted to the vertical tangent bundle VP of P.

A vector field ξ on P is called *equivariant* if $\xi(pg) = TR_g \circ \xi(p)$. Obviously, it is projectable. Taking the quotient of the tangent bundle $TP \to P$ and the vertical tangent bundle VP of P by G (8.11), we obtain the vector bundles

$$T_G P = TP/G, \qquad V_G P = VP/G \tag{8.12}$$

over X. Sections of $T_G P \to X$ are equivariant vector fields on P. Accordingly, sections of $V_G P \to X$ are equivariant vertical vector fields on P. Hence, a typical fibre of $V_G P \to X$ is the right Lie algebra $\mathfrak{g}_r$ of G subject to the adjoint representation of a structure group G. Therefore, $V_G P$ (8.12) is called the *Lie algebra bundle*.

Given the bundle atlas Ψ_P (8.6) of P, there is the corresponding atlas (8.1):

$$\Psi = \{(U_\alpha, \psi_\alpha), \mathrm{Ad}_{\rho_{\alpha\beta}}\} \tag{8.13}$$

of a Lie algebra bundle $V_G P$, which is provided with the corresponding bundle coordinates $(U_\alpha; x^\mu, \chi^m)$ with respect to the fibre frames $\{e_m = \psi_\alpha^{-1}(x)(\varepsilon_m)\}$, where $\{\varepsilon_m\}$ is a basis for the right Lie algebra $\mathfrak{g}_r$. These coordinates obey a transformation rule

$$\rho(\chi^m)\varepsilon_m = \chi^m \mathrm{Ad}_{\rho^{-1}}(\varepsilon_m).$$

A glance at this transformation rule shows that $V_G P$ is a G-bundle P.

Accordingly, the vector bundle T_GP (8.12) is endowed with bundle coordinates $(x^\mu, \dot{x}^\mu, \chi^m)$ with respect to the fibre frames $\{\partial_\mu, e_m\}$. Their transformation rule is

$$\rho(\chi^m)\varepsilon_m = \chi^m \mathrm{Ad}_{\rho^{-1}}(\varepsilon_m) + \dot{x}^\mu R^m_\mu \varepsilon_m. \tag{8.14}$$

For instance, if G is a matrix group, this transformation rule reads

$$\rho(\chi^m)\varepsilon_m = \chi^m \rho^{-1}\varepsilon_m \rho - \dot{x}^\mu \partial_\mu(\rho^{-1})\rho. \tag{8.15}$$

Since the second term in the right-hand sides of expressions (8.14) and (8.15) depend on derivatives of a G-valued function ρ on X, the vector bundle T_GP (8.12) fails to be a G-bundle.

The Lie bracket of equivariant vector fields on P is equivariant, and it induces the Lie bracket of sections of the vector bundle $T_GP \to X$. This bracket reads

$$\xi = \xi^\lambda \partial_\lambda + \xi^p e_p, \qquad \eta = \eta^\mu \partial_\mu + \eta^q e_q, \tag{8.16}$$

$$[\xi, \eta] = (\xi^\mu \partial_\mu \eta^\lambda - \eta^\mu \partial_\mu \xi^\lambda)\partial_\lambda + (\xi^\lambda \partial_\lambda \eta^r - \eta^\lambda \partial_\lambda \xi^r + c^r_{pq}\xi^p \eta^q)e_r. \tag{8.17}$$

Putting $\xi^\lambda = 0$ and $\eta^\mu = 0$ in the formulas (8.16) and (8.17), we obtain the Lie bracket

$$[\xi, \eta] = c^r_{pq}\xi^p \eta^q e_r \tag{8.18}$$

of sections of the Lie algebra bundle V_GP.

Principal Connections

In gauge field theory, gauge fields are conventionally described as principal connections on principal bundles. Principal connections on a principal bundle P (8.3) are connections on P which are equivariant with respect to the right action (8.4) of a structure group G on P.

Let J^1P be the first order jet manifold of a principal G-bundle $P \to X$ (8.3). Connections on a principal bundle $P \to X$ (Definition B.10) are global sections

$$A : P \to J^1P \tag{8.19}$$

of the affine jet bundle $J^1P \to P$ modelled over a vector bundle

$$T^*X \underset{P}{\otimes} VP = (T^*X \underset{P}{\otimes} \mathfrak{g}_l).$$

Let us consider the jet prolongation

$$J^1R_G : J^1(X \times G) \underset{X}{\times} J^1P \to J^1P$$

of the morphism R_{GP} (8.4). Restricting this morphism to

$$J^1R_G : \widehat{0}(G) \underset{X}{\times} J^1P \to J^1P,$$

we obtain the jet prolongation of the structure group action R_{GP} (8.4). In brief, it is called the *action of* G *on* J^1P. It reads

$$J^1R_g : j^1_x p \to (j^1_x p)g = j^1_x(pg). \tag{8.20}$$

The quotient of the affine jet bundle J^1P by G (8.20) is an affine bundle

$$C = J^1P/G \to X \tag{8.21}$$

modelled over a vector bundle $\overline{C} = T^*X \otimes_X V_G P \to X$. Hence, there is the vertical splitting

$$VC = C \otimes_X \overline{C} \tag{8.22}$$

of the vertical tangent bundle VC of $C \to X$.

Taking the quotient with respect to the action of a structure group G, one can reduce the canonical imbedding (B.48) (where $Y = P$) to a bundle monomorphism

$$\lambda_C : C \underset{X}{\longrightarrow} T^*X \underset{X}{\otimes} T_G P, \qquad \lambda_C : dx^\mu \otimes (\partial_\mu + \chi^m_\mu e_m). \tag{8.23}$$

It follows that, given atlases Ψ_P (8.6) of P and Ψ (8.13) of $T_G P$, the affine bundle C (8.21) is provided with bundle coordinates (x^λ, a^m_μ) possessing transition functions

$$\rho(a^m_\mu)\varepsilon_m = (a^m_\nu \mathrm{Ad}_{\rho^{-1}}(\varepsilon_m) + R^m_\nu \varepsilon_m)\frac{\partial x^\nu}{\partial x'^\mu}. \tag{8.24}$$

If G is a matrix group, these transition functions read

$$\rho(a^m_\mu)\varepsilon_m = (a^m_\nu \rho^{-1}(\varepsilon_m)\rho - \partial_\mu(\rho^{-1})\rho)\frac{\partial x^\nu}{\partial x'^\mu}.$$

A glance at this expression shows that C (8.21) as like as the vector bundle $T_G P$ (8.12) fails to be a bundle with a structure group G.

As was mentioned above, a connection A (8.19) on a principal bundle $P \to X$ is called a *principal connection* if it is *equivariant* under the action (8.20) of a structure group G, i.e.,

$$A(pg) = (J^1R_g \circ A)(p) \qquad g \in G.$$

There is obvious one-to-one correspondence between the principal connections on a principal G-bundle P and the global sections

$$A : X \to C \tag{8.25}$$

of the quotient bundle $C \to X$ (8.21), called the *bundle of principal connections*.

Theorem 8.1 *Since a bundle of principal connections $C \to X$ is affine, principal connections on a principal bundle always exist (Theorem B.4).*

Due to the bundle monomorphism (8.23), any principal connection A (8.25) is represented by a $T_G P$-valued *principal connection form*

$$A = dx^\lambda \otimes (\partial_\lambda + A^q_\lambda e_q) \tag{8.26}$$

(cf. (B.55)). Taking the quotient with respect to the action of a structure group G, one can reduce the exact sequence (B.21) (where $Y = P$) to the exact sequence

$$0 \to V_G P \underset{X}{\longrightarrow} T_G P \longrightarrow TX \to 0. \tag{8.27}$$

The principal connection A (8.26) yields a splitting of this exact sequence.

Remark 8.3 Any principal connection A (8.26) on a principal bundle $P \to X$ defines a unique connection on the P-associated fibre bundle Y (8.5) which reads

$$A = dx^\lambda \otimes (\partial_\lambda + A^p_\lambda I_p^i \partial_i),$$

where $\{I_p\}$ is a representation of the Lie algebra $\mathfrak{g}_r$ of G in V [83].

In gauge field theory, coefficients of the principal connection form (8.26) are treated as *gauge potentials*. We use this term in order to refer to sections A (8.25) of the bundle $C \to X$ of principal connections.

Let $P \to X$ be a principal G-bundle. The Frölicher–Nijenhuis bracket (B.37) on the space $\mathcal{O}^*(P) \otimes \mathcal{T}(P)$ of tangent-valued forms on P is compatible with the right action R_{GP} (8.4). Therefore, it induces the Frölicher–Nijenhuis bracket on a space $\mathcal{O}^*(X) \otimes T_G P(X)$ of $T_G P$-valued forms on X, where $T_G P(X)$ is a vector space of sections of a vector bundle $T_G P \to X$. Note that, as it follows from the exact sequence (8.27), there is an epimorphism $T_G P(X) \to \mathcal{T}(X)$. Let $A \in \mathcal{O}^1(X) \otimes T_G P(X)$ be the principal connection (8.26). The associated Nijenhuis differential (B.40) is

$$d_A : \mathcal{O}^r(X) \otimes T_G P(X) \to \mathcal{O}^{r+1}(X) \otimes V_G P(X),$$
$$d_A \phi = [A, \phi]_{\mathrm{FN}}, \quad \phi \in \mathcal{O}^r(X) \otimes T_G P(X).$$

The *strength* of a principal connection A is defined as a $V_G P$-valued two-form

$$F_A = \frac{1}{2} d_A A = \frac{1}{2}[A, A]_{\mathrm{FN}} \in \mathcal{O}^2(X) \otimes V_G P(X), \tag{8.28}$$

$$F_A = \frac{1}{2} F^r_{\lambda\mu} dx^\lambda \wedge dx^\mu \otimes e_r, \tag{8.29}$$

$$F^r_{\lambda\mu} = [\partial_\lambda + A^p_\lambda e_p, \partial_\mu + A^q_\mu e_q]^r = \partial_\lambda A^r_\mu - \partial_\mu A^r_\lambda + c^r_{pq} A^p_\lambda A^q_\mu.$$

Remark 8.4 The strength F_A (8.28) is not the standard curvature (B.63) of a principal connection because A (8.26) is not a tangent-valued form.

Canonical Principal Connection

Since gauge potentials are represented by sections of the bundle of principal connections $C \to X$ (8.21), gauge field theory is formulated as first order Lagrangian theory on a configuration space C.

A glance at the expression (8.20) shows that the fibre bundle $J^1P \to C$ is a trivial principal G-bundle. It is canonically isomorphic to the pull-back bundle

$$J^1P = P_C = C \underset{X}{\times} P \to C. \tag{8.30}$$

Given a principal G-bundle $P \to X$ and its jet manifold J^1P, let us consider the canonical morphism $\theta_{(1)}$ (B.48) where $Y = P$. By virtue of Remark B.3, this morphism defines a map $\theta : J^1P \times_P TP \to VP$. Taking its quotient with respect to G, we obtain a morphism

$$C \underset{X}{\times} T_G P \xrightarrow{\theta} V_G P, \qquad \theta(\partial_\lambda) = -a^p_\lambda e_p, \qquad \theta(e_p) = e_p. \tag{8.31}$$

This means that the exact sequence (8.27) admits the canonical splitting over C [51, 100]. In view of this fact, let us consider the pull-back bundle P_C (8.30). Since

$$V_G(C \underset{X}{\times} P) = C \underset{X}{\times} V_G P, \qquad T_G(C \underset{X}{\times} P) = TC \underset{X}{\times} T_G P, \tag{8.32}$$

the exact sequence (8.27) for the principal bundle P_C reads

$$0 \to C \underset{X}{\times} V_G P \underset{C}{\longrightarrow} TC \underset{X}{\times} T_G P \longrightarrow TC \to 0. \tag{8.33}$$

It is readily observed that the morphism (8.31) yields the horizontal splitting

$$TC \underset{X}{\times} T_G P \longrightarrow C \underset{X}{\times} T_G P \longrightarrow C \underset{X}{\times} V_G P,$$

of the exact sequence (8.33) and, consequently, it defines the principal connection

$$\mathscr{A} : TC \to TC \underset{X}{\times} T_G P,$$
$$\mathscr{A} = dx^\lambda \otimes (\partial_\lambda + a^p_\lambda e_p) + da^r_\lambda \otimes \partial^\lambda_r \in \mathcal{O}^1(C) \otimes T_G(C \underset{X}{\times} P)(X) \tag{8.34}$$

on the principal bundle $P_C \to C$ (8.30). It follows that the principal bundle P_C admits the *canonical principal connection* (8.34).

Following the expression (8.28), let us define the strength

$$F_{\mathscr{A}} = \frac{1}{2} d_{\mathscr{A}} \mathscr{A} = \frac{1}{2} [\mathscr{A}, \mathscr{A}] \in \mathcal{O}^2(C) \otimes V_G P(X),$$
$$F_{\mathscr{A}} = (da^r_\mu \wedge dx^\mu + \frac{1}{2} c^r_{pq} a^p_\lambda a^q_\mu dx^\lambda \wedge dx^\mu) \otimes e_r, \quad (8.35)$$

of the canonical principal connection $\mathscr{A}$ (8.34). It is called the *canonical strength* because, given the principal connection A (8.25) on $P \to X$, the pull-back $A^* F_{\mathscr{A}} = F_A$ is the strength (8.29) of A.

With the $V_G P$-valued two-form $F_{\mathscr{A}}$ (8.35) on C, let us define the $V_G P$-valued horizontal two-form $\mathscr{F} = h_0(F_{\mathscr{A}})$:

$$\mathscr{F} = \frac{1}{2} \mathscr{F}^r_{\lambda\mu} dx^\lambda \wedge dx^\mu \otimes \varepsilon_r, \qquad \mathscr{F}^r_{\lambda\mu} = a^r_{\lambda\mu} - a^r_{\mu\lambda} + c^r_{pq} a^p_\lambda a^q_\mu, \quad (8.36)$$

on $J^1 C$. It is called the *strength form*. For each principal connection A (8.25) on P, the pull-back $J^1 A^* \mathscr{F} = F_A$ is the strength (8.29) of A.

The strength form (8.36) yields an affine surjection

$$\mathscr{F}/2 : J^1 C \underset{C}{\longrightarrow} C \underset{X}{\times} (\overset{2}{\wedge} T^* X \otimes V_G P) \quad (8.37)$$

over C of the affine jet bundle $J^1 C \to C$ to the vector one

$$C \underset{X}{\times} (\overset{2}{\wedge} T^* X \otimes V_G P) \to C.$$

By virtue of Theorem B.10, its kernel $C_+ = \mathrm{Ker}\, \mathscr{F}/2$ is an affine subbundle of $J^1 C \to C$. Thus, we have the canonical splitting of the affine jet bundle

$$J^1 C = C_+ \underset{C}{\oplus} C_- = C_+ \underset{C}{\oplus} (C \underset{X}{\times} \overset{2}{\wedge} T^* X \otimes V_G P), \quad (8.38)$$
$$a^r_{\lambda\mu} = \frac{1}{2} (\mathscr{S}^r_{\lambda\mu} + \mathscr{F}^r_{\lambda\mu}) = \frac{1}{2} (a^r_{\lambda\mu} + a^r_{\mu\lambda} - c^r_{pq} a^p_\lambda a^q_\mu) \quad (8.39)$$
$$+ \frac{1}{2} (a^r_{\lambda\mu} - a^r_{\mu\lambda} + c^r_{pq} a^p_\lambda a^q_\mu).$$

The corresponding canonical projections are $\mathrm{pr}_2 = \mathscr{F}$ (8.37) and

$$\mathrm{pr}_1 = \mathscr{S} : J^1 C \to C_+. \quad (8.40)$$

The splitting (8.38) exemplifies that (3.100), but it is not related to a Lagrangian.

8.2 Principal Gauge Symmetries

In gauge field theory, *principal gauge transformations* are defined as principal automorphisms of a principal bundle P. An automorphism Φ_P of a principal G-bundle P is called *principal* if it is *equivariant* under the right action (8.4) of a structure group G on P, i.e.,

$$\Phi_P(pg) = \Phi_P(p)g, \qquad g \in G, \qquad p \in P. \tag{8.41}$$

In particular, every vertical principal automorphism of a principal bundle P looks like $\Phi_P(p) = pf(p)$, $p \in P$, where f is a G-valued *equivariant function* on P, i.e.,

$$f(pg) = g^{-1} f(p)g, \qquad g \in G. \tag{8.42}$$

There is one-to-one correspondence

$$s(\pi_P(p))p = pf(p), \qquad p \in P,$$

between the equivariant functions f (8.42) and the global sections s of the associated *group bundle*

$$\pi_{PG} : P^G \to X \tag{8.43}$$

whose typical fibre is a group G which acts on itself by the adjoint representation.

In order to describe gauge symmetries of gauge field theory on a principal bundle P, let us restrict our consideration to local one-parameter groups of local principal automorphisms of P. Their infinitesimal generators are equivariant vector fields ξ on P, and vice versa. We call ξ the *principal vector fields* or the *infinitesimal gauge transformation*. They are represented by sections ξ (8.16) of the vector bundle $T_G P$ (8.12). Principal vector fields constitute a real Lie algebra $T_G P(X)$ with respect to the Lie bracket (8.17). *Vertical principal vector fields* are sections

$$\xi = \xi^p e_p \tag{8.44}$$

of the Lie algebra bundle $V_G P \to X$ (8.12). They form a finite-dimensional Lie $C^\infty(X)$-algebra $\mathcal{G}(X) = V_G P(X)$ with respect to the bracket (8.18).

Remark 8.5 Any principal automorphism Φ_P (8.41) of P yields a unique *principal automorphism*

$$\Phi_Y : (p, v)/G \to (\Phi_P(p), v)/G, \qquad p \in P, \qquad v \in V,$$

of the P-associated bundle Y (8.5). As a consequence, any principal vector field ξ (8.44) yields an *associated principal vector field* on a P-associated bundle Y which is an infinitesimal generator of local one-parameter group of local principal automorphisms of Y.

As was mentioned above, a bundle of principal connections C fails to be a G-bundle, but principal automorphisms of P yield automorphisms of C as follows. Any local one-parameter group of local principal automorphism Φ_P (8.41) of a principal bundle P admits the jet prolongation $J^1\Phi_P$ (B.51) to a local one-parameter group of local equivariant automorphism of the jet manifold J^1P which, in turn, yields a one-parameter group of *principal automorphisms* Φ_C of the bundle of principal connections C (8.21) [80, 100]. Its infinitesimal generator is a vector field on C, called the *principal vector field* on C and regarded as an infinitesimal gauge transformation of C. Thus, any principal vector field ξ (8.16) on P yields a principal vector field u_ξ on C, which can be obtained as follows. Using the morphism (8.31), we get a morphism $\xi\rfloor\theta : C \to V_G P$, which is a section of of the Lie algebra bundle $V_G(C\times_X P)\to C$ in accordance with the first formula (8.32). Then the equation

$$u_\xi\rfloor F_{\mathscr{A}} = d_{\mathscr{A}}(\xi\rfloor\theta)$$

uniquely determines a principal vector field u_ξ on C. A direct computation leads to

$$u_\xi = \xi^\mu\partial_\mu + (\partial_\mu\xi^r + c^r_{pq}a^p_\mu\xi^q - a^r_\nu\partial_\mu\xi^\nu)\partial^\mu_r. \tag{8.45}$$

In particular, if ξ is the vertical principal vector field (8.44), we obtain the vertical principal vector field (8.45):

$$u_\xi = (\partial_\mu\xi^r + c^r_{pq}a^p_\mu\xi^q)\partial^\mu_r. \tag{8.46}$$

Remark 8.6 The jet prolongation (B.52) of the vector field u_ξ (8.45) onto J^1C reads

$$\begin{aligned} J^1u_\xi = u_\xi + (\partial_{\lambda\mu}\xi^r + c^r_{pq}a^p_\mu\partial_\lambda\xi^q + c^r_{pq}a^p_{\lambda\mu}\xi^q \\ - a^r_\nu\partial_{\lambda\mu}\xi^\nu - a^r_{\lambda\nu}\partial_\mu\xi^\nu - a^r_{\nu\mu}\partial_\lambda\xi^\nu)\partial^{\lambda\mu}_r. \end{aligned} \tag{8.47}$$

It is readily justified that the monomorphism

$$T_GP(X)\ni\xi\to u_\xi\in\mathcal{T}(C) \tag{8.48}$$

obeys the equality

$$u_{[\xi,\eta]} = [u_\xi, u_\eta], \tag{8.49}$$

i.e., it is a monomorphism of the real Lie algebra $T_GP(X)$ of sections of a vector bundle $T_GX\to$ to the real Lie algebra $\mathcal{T}(C)$ of vector fields on C. In particular, the image of the Lie algebra $\mathcal{G}(X)\subset T_GP(X)$ in $\mathcal{T}(C)$ also is a real Lie algebra, but not the $C^\infty(X)$-one because $u_{f\xi}\neq fu_\xi$, $f\in C^\infty(X)$.

Remark 8.7 A glance at the expression (8.45) shows that the monomorphism (8.48) is a linear first order differential operator which sends sections of the pull-back bundle $C\times_X T_GP\to C$ onto sections of the tangent bundle $TC\to C$. Refereing to Definition 2.3, we therefore can treat principal vector fields (8.45) as infinitesimal gauge transformations depending on gauge parameters $\xi \in T_G P(X)$.

8.3 Noether's Direct Second Theorem: Yang–Mills Lagrangian

Let us consider first order Lagrangian gauge theory on a principal bundle P. Its configuration space is the bundle of principal connections C (8.21), endowed with bundle coordinates (x^μ, a^m_μ) possessing transition functions (8.24). Given a first order Lagrangian

$$L = \mathscr{L}\omega : J^1C \to \overset{n}{\wedge} T^*X \tag{8.50}$$

on the jet manifold J^1C, the corresponding Euler–Lagrange operator (3.2) reads

$$\mathscr{E}_L = \mathscr{E}^\mu_r \theta^r_\mu \wedge \omega = (\partial^\mu_r - d_\lambda \partial^{\lambda\mu}_r)\mathscr{L}\theta^r_\mu \wedge \omega. \tag{8.51}$$

Its kernel defines the Euler–Lagrange equation

$$\mathscr{E}^\mu_r = (\partial^\mu_r - d_\lambda \partial^{\lambda\mu}_r)\mathscr{L} = 0.$$

Let us assume that the gauge theory Lagrangian L (8.50) on J^1C is invariant under vertical principal gauge transformations (or, in short, *gauge invariant*). This means that vertical principal vector fields u_ξ (8.46) are exact symmetries of L, i.e.,

$$\mathbf{L}_{J^1u_\xi}L = 0, \qquad J^1u_\xi = u_\xi + (\partial_{\lambda\mu}\xi^r + c^r_{pq}a^p_\mu\partial_\lambda\xi^q + c^r_{pq}a^p_{\lambda\mu}\xi^q)\partial^{\lambda\mu}_r, \tag{8.52}$$

(see the formula (8.47)). Then it follows from Remark 8.7 that vertical principal vector fields u_ξ (8.46) are gauge symmetries of L (Definition 2.3) whose gauge parameters are sections ξ (8.44) of the Lie algebra bundle $V_G P$. We call them *principal gauge symmetries*.

Let us apply Noether's direct second theorem to these gauge symmetries. In this case, the first variational formula (3.26) for the Lie derivative (8.52) takes a form

$$0 = (\partial_\mu\xi^r + c^r_{pq}a^p_\mu\xi^q)\mathscr{E}^\mu_r + d_\lambda[(\partial_\mu\xi^r + c^r_{pq}a^p_\mu\xi^q)\partial^{\lambda\mu}_r\mathscr{L})]. \tag{8.53}$$

It leads to the gauge invariance conditions

$$\partial^{\mu\lambda}_p\mathscr{L} + \partial^{\lambda\mu}_p\mathscr{L} = 0, \tag{8.54}$$

$$\mathscr{E}^\mu_r + d_\lambda\partial^{\lambda\mu}_r\mathscr{L} + c^q_{pr}a^p_\nu\partial^{\mu\nu}_q\mathscr{L} = 0, \tag{8.55}$$

$$c^r_{pq}(a^p_\mu\mathscr{E}^\mu_r + d_\lambda(a^p_\mu\partial^{\lambda\mu}_r\mathscr{L})) = 0. \tag{8.56}$$

One can regard the equalities (8.54) and (8.56) as the conditions of a Lagrangian L to be gauge invariant. They are brought into a form

$$\partial_p^{\mu\lambda}\mathcal{L} + \partial_p^{\lambda\mu}\mathcal{L} = 0, \tag{8.57}$$

$$\partial_q^{\mu}\mathcal{L} + c_{pq}^r a_\nu^p \partial_r^{\mu\nu}\mathcal{L} = 0, \tag{8.58}$$

$$c_{pq}^r (a_\mu^p \partial_r^\mu \mathcal{L} + a_{\lambda\mu}^p \partial_r^{\lambda\mu}\mathcal{L}) = 0. \tag{8.59}$$

Let us utilize the coordinates $(a_\mu^q, \mathcal{F}_{\lambda\mu}^r, \mathcal{S}_{\lambda\mu}^r)$ (8.39) which correspond to the canonical splitting (8.38) of the affine jet bundle $J^1C \to C$. With respect to these coordinates, the Eq. (8.57) reads

$$\frac{\partial \mathcal{L}}{\partial \mathcal{S}_{\mu\lambda}^p} = 0. \tag{8.60}$$

Then the Eq. (8.58) takes a form

$$\frac{\partial \mathcal{L}}{\partial a_\mu^q} = 0. \tag{8.61}$$

A glance at the equalities (8.60) and (8.61) shows that a gauge invariant Lagrangian factorizes through the strength $\mathcal{F}$ (8.36). Then the Eq. (8.59), written as

$$c_{pq}^r \mathcal{F}_{\lambda\mu}^p \frac{\partial \mathcal{L}}{\partial \mathcal{F}_{\lambda\mu}^r} = 0,$$

shows that a gauge symmetry u_ξ of L is exact. The following thus has been proved.

Theorem 8.2 *A gauge theory Lagrangian (8.50) possesses the exact gauge symmetry u_ξ (8.46) only if it factorizes through the strength $\mathcal{F}$ (8.36).*

A corollary of this result is the well-known *Utiyama theorem* [24].

Theorem 8.3 *There is a unique gauge invariant quadratic first order Lagrangian which is the Yang–Mills Lagrangian*

$$L_{\mathrm{YM}} = \frac{1}{4} a_{pq}^G g^{\lambda\mu} g^{\beta\nu} \mathcal{F}_{\lambda\beta}^p \mathcal{F}_{\mu\nu}^q \sqrt{|g|}\,\omega, \qquad g = \det(g_{\mu\nu}), \tag{8.62}$$

where a^G is a G-invariant bilinear form on the right Lie algebra $\mathfrak{g}_r$ and g is a world metric on X (Remark B.12).

The Euler–Lagrange operator (8.51) of the Yang–Mills Lagrangian L_{YM} reads

$$\mathcal{E}_{\mathrm{YM}} = \mathcal{E}_r^\mu \theta_r^\mu \wedge \omega = (\delta_r^n d_\lambda + c_{rp}^n a_\lambda^p)(a_{nq}^G g^{\mu\alpha} g^{\lambda\beta} \mathcal{F}_{\alpha\beta}^q \sqrt{|g|})\theta_\mu^r \wedge \omega. \tag{8.63}$$

Its kernel defines the *Yang–Mills equation*

$$\mathcal{E}_r^\mu = (\delta_r^n d_\lambda + c_{rp}^n a_\lambda^p)(a_{nq}^G g^{\mu\alpha} g^{\lambda\beta} \mathcal{F}_{\alpha\beta}^q \sqrt{|g|}) = 0. \tag{8.64}$$

We call a Lagrangian system $(\mathcal{S}^*_\infty[C], L_{\rm YM})$ the *Yang–Mills gauge theory.*

Remark 8.8 In gauge field theory, there are Lagrangians, e.g., the Chern–Simons ones (Chap. 11) which do not factorize through the strength of a gauge field, and whose gauge symmetries u_ξ (8.46) are not exact.

8.4 Noether's First Theorem: Conservation Laws

Since the gauge symmetry u_ξ (8.46) of the Yang–Mills Lagrangian (8.62) is exact, the first variational formula (8.53) leads to the gauge conservation law

$$0 \approx d_\lambda(-u_\xi{}^\mu_r \partial^{\lambda\mu}_r \mathcal{L}_{\rm YM})$$

of a Noether current

$$\mathcal{J}^\lambda_\xi = -(\partial_\mu \xi^r + c^r_{pq} a^p_\mu \xi^q)(a^G_{rq} g^{\mu\alpha} g^{\lambda\beta} \mathcal{F}^q_{\alpha\beta} \sqrt{|g|}). \tag{8.65}$$

In accordance with Theorem 2.8, the Noether current (8.65) is brought into the superpotential form (7.65) which reads

$$\begin{aligned} \mathcal{J}^\lambda_\xi &= \xi^r \mathcal{E}^\mu_r + d_\nu(\xi^r \partial^{[\nu\mu]}_r \mathcal{L}_{\rm YM}), \\ U^{\nu\mu} &= \xi^r a^G_{rq} g^{\nu\alpha} g^{\mu\beta} \mathcal{F}^q_{\alpha\beta} \sqrt{|g|}. \end{aligned} \tag{8.66}$$

Let us study energy-momentum conservation laws in Yang–Mills gauge theory. If a background world metric g is specified, one can find a particular vector field τ on X and its lift $\gamma\tau$ (B.27) onto C which is an exact symmetry of the Yang–Mills Lagrangian (8.62). Then the energy-momentum current (3.33) along such a symmetry is conserved. We here treat the energy-momentum conservation law as the gauge one in the case of an arbitrary world metric g and any vector field τ on X.

Let A (8.26) be a principal connection on P. For any vector field τ on X, this connection yields a section

$$\tau \rfloor A = \tau^\lambda \partial_\lambda + A^p_\lambda \tau^\lambda e_p$$

of the vector bundle $T_G P \to X$. It, in turn, defines the principal vector field (8.45) on the bundle of principal connection C which reads

$$A\tau = \tau^\lambda \partial_\lambda + (\partial_\mu(A^r_\nu \tau^\nu) + c^r_{pq} a^p_\mu A^q_\nu \tau^\nu - a^r_\nu \partial_\mu \tau^\nu)\partial^\mu_r. \tag{8.67}$$

Let us consider an energy-momentum current along this vector field [100, 119].

Since the Yang–Mills Lagrangian (8.62) depends on a background world metric g, the vector field $A\tau$ (8.67) is not its exact symmetry in general. Following the procedure in Sect. 2.4.6, let us consider the total Lagrangian

$$L = \frac{1}{4} a^G_{pq} \sigma^{\lambda\mu} \sigma^{\beta\nu} \mathcal{F}^p_{\lambda\beta} \mathcal{F}^q_{\mu\nu} \sqrt{|\sigma|} \omega, \qquad \sigma = \det(\sigma_{\mu\nu}), \tag{8.68}$$

on the jet manifold $J^1(C \times_X \overset{2}{\vee} TX)$ of the total configuration space $C \times_X \overset{2}{\vee} T^*X$, where a tensor bundle $\overset{2}{\vee} TX$ is provided with holonomic fibre coordinates $(\sigma^{\mu\nu})$. Given a vector field τ on X, there exists its canonical lift (B.29):

$$\widetilde{\tau} = \tau^\lambda \partial_\lambda + (\partial_\nu \tau^\alpha \sigma^{\nu\beta} + \partial_\nu \tau^\beta \sigma^{\nu\alpha}) \partial_{\alpha\beta},$$

onto a tensor bundle $\overset{2}{\vee} T^*X$. It is an infinitesimal general covariant transformations of $\overset{2}{\vee} T^*X$ (Sect. 10.1). Thus, we have the lift

$$\begin{aligned} \widetilde{\tau}_A = \widetilde{\tau} + A\tau = \tau^\lambda \partial_\lambda + (\partial_\mu (A^r_\nu \tau^\nu) + c^r_{pq} a^p_\mu A^q_\nu \tau^\nu - a^r_\nu \partial_\mu \tau^\nu) \partial^\mu_r \\ + (\partial_\nu \tau^\alpha \sigma^{\nu\beta} + \partial_\nu \tau^\beta \sigma^{\nu\alpha}) \partial_{\alpha\beta} \end{aligned} \tag{8.69}$$

of a vector field τ on X onto a product $C \times_X \overset{2}{\vee} T^*X$. Its is readily observed that the vector field $\widetilde{\tau}_A$ (8.69) is an exact symmetry of the total Lagrangian (8.68). One the shell (8.64), we then obtain the weak transformation law (3.43):

$$0 \approx (\partial_\nu \tau^\alpha g^{\nu\beta} + \partial_\nu \tau^\beta g^{\nu\alpha} - \partial_\lambda g^{\alpha\beta} \tau^\lambda) \partial_{\alpha\beta} \mathcal{L} - d_\lambda \mathcal{J}^\lambda_A \tag{8.70}$$

of the energy-momentum current

$$\mathcal{J}^\lambda_A = \partial^{\lambda\mu}_r \mathcal{L}_{\mathrm{YM}} [\tau^\nu a^r_{\nu\mu} - \partial_\mu (A^r_\nu \tau^\nu) - c^r_{pq} a^p_\mu A^q_\nu \tau^\nu + a^r_\nu \partial_\mu \tau^\nu] - \tau^\lambda \mathcal{L}_{\mathrm{YM}} \tag{8.71}$$

along the vector field $A\tau$ (8.67).

The weak identity (8.70) can be written in a form

$$0 \approx \partial_\lambda \tau^\mu t^\lambda_\mu \sqrt{|g|} - \tau^\mu \{_\mu{}^\beta{}_\lambda\} t^\lambda_\beta \sqrt{|g|} - d_\lambda \mathcal{J}^\lambda_A, \tag{8.72}$$

where $\{_\mu{}^\beta{}_\lambda\}$ are the Christoffel symbols (B.74) of a world metric g and

$$t^\mu_\beta \sqrt{|g|} = 2 g^{\mu\alpha} \partial_{\alpha\beta} \mathcal{L}_{\mathrm{YM}} = (\mathcal{F}^q_{\beta\nu} \partial^{\mu\nu}_q - \delta^\mu_\beta) \mathcal{L}_{\mathrm{YM}}$$

is the *metric energy-momentum tensor* of a gauge field. In particular, let a principal connection B be a classical solution of the Yang–Mills equation (8.64). Let us consider the lift (8.67) of a vector field τ on X onto C by means of a principal connection $A = B$. In this case, the energy-momentum current (8.71) reads

$$\mathcal{J}^\lambda_B \circ B = \tau^\mu (t^\lambda_\mu \circ B) \sqrt{|g|}. \tag{8.73}$$

Then the identity (8.72) on a solution B comes to a *covariant conservation law*

$$\nabla^\Gamma_\lambda((t^\lambda_\mu \circ B)\sqrt{|g|}) = 0, \tag{8.74}$$

where ∇ is the covariant differential (B.62) with respect to the Levi–Civita connection $\{_\mu{}^\beta{}_\lambda\}$ (B.74) of a background metric g.

Note that, considering a different lift $\gamma\tau$ of a vector field τ on X to a principal vector field on C, we obtain an energy-momentum current along $\gamma\tau$ which differs from $\mathcal{J}^\lambda_B$ (8.73) in a Noether current (8.66). Since such a current is reduced to a superpotential, one can always bring the energy-momentum transformation law (8.70) into the covariant conservation law (8.74).

8.5 Hamiltonian Gauge Theory

The Yang–Mills Lagrangian (8.62) exemplifies almost regular quadratic Lagrangians analyzed in Sect. 3.8. It admits a complete set of weakly associated Hamiltonian forms (8.78). A key point that neither these Hamiltonian forms nor the corresponding characteristic Lagrangians L_H (8.79) inherit principal gauge symmetries of a Yang–Mills Lagrangian so that the part (8.82) of the covariant Hamilton equation plays the role of gauge-type conditions. Restricted to the Lagrangian constraint space, they however lead to the gauge invariant constrained Lagrangian L_N (8.81).

The phase space of gauge field theory is the Legendre bundle (3.6) which due to the vertical splitting (8.22) of VC is

$$\pi_{\Pi C} : \Pi \to C, \qquad \Pi = \overset{n}{\wedge} T^*X \underset{C}{\otimes} TX \underset{C}{\otimes} [C \times \overline{C}]^*. \tag{8.75}$$

It is endowed with holonomic coordinates $(x^\lambda, a^p_\lambda, p^{\mu\lambda}_m)$. The Legendre bundle Π (8.75) admits the canonical decomposition (3.103):

$$\Pi = \Pi_+ \underset{C}{\oplus} \Pi_-,$$

$$p^{\mu\lambda}_m = \mathcal{R}^{(\mu\lambda)}_m + \mathcal{P}^{[\mu\lambda]}_m = p^{(\mu\lambda)}_m + p^{[\mu\lambda]}_m = \frac{1}{2}(p^{\mu\lambda}_m + p^{\lambda\mu}_m) + \frac{1}{2}(p^{\mu\lambda}_m - p^{\lambda\mu}_m).$$

The Legendre map (3.5) induced by the Yang–Mills Lagrangian (8.62) takes a form

$$p^{(\mu\lambda)}_m \circ \widehat{L}_{YM} = 0, \tag{8.76}$$

$$p^{[\mu\lambda]}_m \circ \widehat{L}_{YM} = a^G_{mn} g^{\mu\alpha} g^{\lambda\beta} \mathcal{F}^n_{\alpha\beta} \sqrt{|g|}. \tag{8.77}$$

It follows that $\mathrm{Ker}\, \widehat{L}_{YM} = C_+$, and the Lagrangian constraint space (3.9) is

$$N_L = \widehat{L}_{YM}(J^1C) = \Pi_-.$$

It is defined by the Eq. (8.76). Obviously, N_L is an imbedded submanifold of Π, and the Yang–Mills Lagrangian $L_{\rm YM}$ is almost regular.

Following the procedure in Sect. 3.8, let us consider connections Γ on a fibre bundle $C \to X$ which take their values into Ker $\widehat{L}$, i.e.,

$$\Gamma : C \to C_+, \qquad \Gamma^r_{\lambda\mu} - \Gamma^r_{\mu\lambda} + c^r_{pq} a^p_\lambda a^q_\mu = 0.$$

Given a symmetric world connection K (B.71) on T^*X, every principal connection B on a principal bundle $P \to X$ gives rise to the connection $\Gamma_B : C \to C_+$ such that $\Gamma_B \circ B = \mathcal{S} \circ J^1 B$ where $\mathcal{S}$ is the projection (8.40) [61]. It reads

$$\Gamma_B{}^r_{\lambda\mu} = \frac{1}{2}[\partial_\mu B^r_\lambda + \partial_\lambda B^r_\mu - c^r_{pq} a^p_\lambda a^q_\mu + c^r_{pq}(a^p_\lambda B^q_\mu + a^p_\mu B^q_\lambda)] - K_\lambda{}^\beta{}_\mu (a^r_\beta - B^r_\beta).$$

Given this connection, the corresponding Hamiltonian form (3.109):

$$H_B = p^{\lambda\mu}_r da^r_\mu \wedge \omega_\lambda - p^{\lambda\mu}_r \Gamma_B{}^r_{\lambda\mu}\omega - \widetilde{\mathcal{H}}_{YM}\omega, \tag{8.78}$$
$$\widetilde{\mathcal{H}}_{YM} = \frac{1}{4} a^{mn}_G g_{\mu\nu} g_{\lambda\beta} p^{[\mu\lambda]}_m p^{[\nu\beta]}_n \sqrt{|g|},$$

is weakly associated to the Yang–Mills Lagrangian $L_{\rm YM}$. It is the Poincaré–Cartan form of the Lagrangian

$$L_H = [p^{\lambda\mu}_r (a^r_{\lambda\mu} - \Gamma_B{}^r_{\lambda\mu}) - \widetilde{\mathcal{H}}_{YM}]\omega \tag{8.79}$$

on $\Pi \times_C J^1 C$. The pull-back of any Hamiltonian form H_B (8.78) onto the Lagrangian constraint space N_L is the constrained Hamiltonian form (3.76):

$$H_N = i^*_N H_B = p^{[\lambda\mu]}_r (da^r_\mu \wedge \omega_\lambda + \frac{1}{2} c^r_{pq} a^p_\lambda a^q_\mu \omega) - \widetilde{\mathcal{H}}_{YM}\omega. \tag{8.80}$$

The corresponding constrained Lagrangian L_N on $N_L \times_C J^1 C$ reads

$$L_N = (p^{[\lambda\mu]}_r \mathcal{F}^r_{\lambda\mu} - \widetilde{\mathcal{H}}_{YM})\omega. \tag{8.81}$$

Note that, in contrast with the Lagrangian (8.79), the constrained Lagrangian L_N (8.81) possesses gauge symmetries as follows. Gauge symmetries u_ξ (8.45) of the Yang–Mills Lagrangian give rise to vector fields

$$\widetilde{u}_\xi = (\partial_\mu \xi^r + c^r_{qp} a^q_\mu \xi^p)\partial^\mu_r - c^r_{qp}\xi^p p^{\lambda\mu}_r \partial^q_{\lambda\mu}$$

on Π. Their projection onto N_L provides gauge symmetries of the Lagrangian L_N in accordance with Theorem 3.21.

The Hamiltonian form H_B (8.78) yields the covariant Hamilton equation which consist of the equation (8.77) and the equations

$$a^m_{\lambda\mu} + a^m_{\mu\lambda} = 2\Gamma_B{}^m_{(\lambda\mu)}, \tag{8.82}$$

$$p^{\lambda\mu}_{\lambda r} = c^q_{pr} r^p_\lambda p^{[\lambda\mu]}_q - c^q_{rp} B^p_\lambda p^{(\lambda\mu)}_q + K_\lambda{}^\mu{}_\nu p^{(\lambda\nu)}_r. \tag{8.83}$$

The Hamilton equations (8.82)–(8.77) are similar to the equations (3.111) and (3.112), respectively. The Hamilton equations (8.77) and (8.83) restricted to the Lagrangian constraint space (8.76) are precisely the constrained Hamilton equations (3.77) for the constrained Hamiltonian form H_N (8.80), and they are equivalent to the Yang-Mills equation (8.64) for gauge potentials $A = \pi_{\Pi C} \circ r$.

Different Hamiltonian forms H_B lead to different equations (8.82). This equation is independent of momenta and, thus, it exemplifies the gauge-type condition $\Gamma_B \circ A = \mathscr{S} \circ J^1 A$ (3.111). A glance at this condition shows that, given a solution A of the Yang-Mills equations, there always exists a Hamiltonian form H_B (e.g., $H_{B=A}$) which obeys the condition (3.74), i.e.,

$$\widehat{H}_B \circ \widehat{L}_{YM} \circ J^1 A = J^1 A.$$

Consequently, the Hamiltonian forms H_B (8.78) parameterized by principal connections B constitute a complete set.

8.6 Noether's Inverse Second Theorem: BRST Extension

Gauge invariance conditions (8.54)–(8.56) lead to NI which the Euler–Lagrange operator $\mathscr{E}_{\rm YM}$ (8.63) of a Yang–Mills Lagrangian satisfies. These NI are associated to the gauge symmetry u_ξ (8.46). By virtue of the formula (2.19), they read

$$c^p_{rq} a^q_\mu \mathscr{E}^\mu_p + d_\mu \mathscr{E}^\mu_r = 0. \tag{8.84}$$

Theorem 8.4 *The Noether identities (8.84) are nontrivial.*

Proof Following the procedure in Sect. 7.1, let us consider the density dual

$$\overline{VC} = V^*C \underset{C}{\otimes} \overset{n}{\wedge} T^*X = (T^*X \underset{X}{\otimes} V_G P)^* \underset{C}{\otimes} \overset{n}{\wedge} T^*X \tag{8.85}$$

of the vertical tangent bundle VC of $C \to X$, and let us enlarge a DGA $\mathscr{S}^*_\infty[C]$ to the DBGA (7.6):

$$\mathscr{P}^*_\infty[\overline{VC}; C] = \mathscr{S}^*_\infty[\overline{VC}; C],$$

possessing the local generating basis $(a^r_\mu, \overline{a}^\mu_r)$ where $\overline{a}^\mu_r$ are odd antifields. Providing this DBGA with the nilpotent right graded derivation

$$\overline{\delta} = \frac{\partial}{\partial \overline{a}^\mu_r} \mathscr{E}^\mu_r,$$

let us consider the chain complex (7.7). Its one-chains

$$\Delta_r = c^p_{rq} a^q_\mu \overline{a}^\mu_p + d_\mu \overline{a}^\mu_r \tag{8.86}$$

are $\overline{\delta}$-cycles which define the NI (8.84). Clearly, they are not $\overline{\delta}$-boundaries. Therefore, the NI (8.84) are nontrivial.

Theorem 8.5 *The Noether identities (8.84) are complete.*

Proof The second order Euler–Lagrange operator $\mathcal{E}_{\rm YM}$ (8.63) takes its values into a space of sections of a vector bundle

$$(T^*X \underset{X}{\otimes} V_G P)^* \underset{X}{\otimes} \overset{n}{\wedge} T^*X \to X.$$

Let Φ be a first order differential operator on this vector bundle such that $\Phi \circ \mathcal{E}_{\rm YM} = 0$. This condition holds only if the highest derivative term of the composition $\Phi^1 \circ \mathcal{E}^2_{\rm YM}$ of the first order derivative term Φ^1 of Φ and the second order derivative term $\mathcal{E}^2_{\rm YM}$ of $\mathcal{E}_{\rm YM}$ vanishes. This is the case only of $\Phi^1 = \Delta^1_r = d_\mu \overline{a}^\mu_r$.

The graded densities $\Delta_r \omega$ (8.86) constitute a local basis for a $C^\infty(X)$-module $\mathcal{C}_{(0)}$ isomorphic to the module $\overline{V_G P}(X)$ of sections of the density dual $\overline{V_G P}$ of the Lie algebra bundle $V_G P \to X$. Let us enlarge a DBGA $\mathcal{P}^*_\infty[\overline{VC}; C]$ to a DBGA

$$\overline{\mathcal{P}}^*_\infty\{0\} = \mathcal{S}^*_\infty[\overline{VC}; C \underset{X}{\times} \overline{V_G P}]$$

with the local generating basis $(a^r_\mu, \overline{a}^\mu_r, \overline{c}_r)$ where $\overline{c}_r$ are even Noether antifields.

Theorem 8.6 *NI (8.84) are irreducible.*

Proof Providing a DBGA $\overline{\mathcal{P}}^*_\infty\{0\}$ with a nilpotent odd derivation

$$\delta_0 = \overline{\delta} + \frac{\partial}{\partial \overline{c}_r} \Delta_r,$$

let us consider the chain complex (7.15). Let us assume that Φ (7.16) is a two-cycle of this complex, i.e., the relation (7.17) holds. It is readily observed that Φ obeys this relation only if its first term G is $\overline{\delta}$-exact, i.e., the first-stage NI (7.17) are trivial.

By virtue of Theorems 8.4–8.6, Yang–Mills gauge theory is an irreducible degenerate Lagrangian theory characterized by complete NI (8.84).

Following Noether's inverse second Theorem 7.9, let us consider a DBGA

$$P^*_\infty\{0\} = \mathcal{S}^*_\infty[\overline{VC} \underset{C}{\oplus} V_G P; C \underset{X}{\times} \overline{V_G P}] \tag{8.87}$$

with the local generating basis $(a^r_\mu, \overline{a}^\mu_r, c^r, \overline{c}_r)$ where c_r are odd ghosts. The gauge operator $\mathbf{u}$ (7.39) associated to NI (8.84) reads

$$\mathbf{u} = u = (c^r_\mu + c^r_{pq} a^p_\mu c^q)\partial^\mu_r. \tag{8.88}$$

It is an odd gauge symmetry of the Yang–Mills Lagrangian L_{YM} which can be obtained from the gauge symmetry u_ξ (8.46) by replacement of gauge parameters ξ^r with odd ghosts c^r (Remark 7.7).

Since gauge symmetries u_ξ form the Lie algebra (8.49), the gauge operator $\mathbf{u}$ (8.88) admits the nilpotent BRST extension

$$\mathbf{b} = (c^r_\mu + c^r_{pq} a^p_\mu c^q)\frac{\partial}{\partial a^r_\mu} - \frac{1}{2} c^r_{pq} c^p c^q \frac{\partial}{\partial c^r},$$

which is the well-known BRST operator in Yang–Mills gauge theory [63]. Then, by virtue of Theorem 7.16, the Yang–Mills Lagrangian L_{YM} is extended to a proper solution of the master equation

$$L_E = L_{\mathrm{YM}} + (c^r_\mu + c^r_{pq} a^p_\mu c^q)\overline{a}^\mu_r \omega - \frac{1}{2} c^r_{pq} c^p c^q \overline{c}_r \omega.$$

Chapter 9
SUSY Gauge Theory on Principal Graded Bundles

By analogy to gauge field theory on principal bundles (Chap. 8), we develop SUSY gauge theory as theory of graded principal connections on principal graded bundles. A key point is that we consider simple principal graded bundles (Definition 9.2) subject to an action of an even Lie group, but not the whole graded Lie one. In this case, odd gauge potentials in comparison with the even ones are linear, but not affine objects (Remark 9.2). However, they admit affine gauge transformations (9.8) parameterized by ghosts. Our goal is Yang–Mills theory of graded gauge fields and its BRST extension.

Graded principal bundles and connections on these bundles [100, 140] can be studied similarly to principal superbundles and principal superconnections [7, 100]. Since the definition of a graded Lie group and its Lie superalgebra involves the notion of a Hopf algebra, we start with the following construction [58, 140].

Let $(Z,\mathfrak{A})$ be a graded manifold. One considers the *finite dual* $\mathfrak{A}(Z)^\circ$ of its structure ring $\mathfrak{A}(Z)$ which consists of elements a of the dual $\mathfrak{A}(Z)^*$ of $\mathfrak{A}(Z)$ vanishing on an ideal of $\mathfrak{A}(Z)$ of finite codimension. This is brought into a graded commutative coalgebra with a comultiplication

$$\Delta^\circ(a)(f\otimes f')=a(ff'),\qquad f,f'\in\mathfrak{A}(Z),\qquad a\in\mathfrak{A}(Z)^\circ,$$

and a counit $\epsilon^\circ(a)=a(\mathbf{1}_{\mathfrak{A}(Z)})$. In particular, $\mathfrak{A}(Z)^\circ$ includes the *evaluation elements* δ_z such that $\delta_z(f)=(\sigma(f))(z)$. Given an evaluation element δ_z, elements $u\in\mathfrak{A}(Z)^\circ$ are said to be *primitive* relative to δ_z if they obey a relation

$$\Delta^\circ(u)=u\otimes\delta_z+\delta_z\otimes u.$$

These elements are derivations of $\mathfrak{A}(Z)$ at z, i.e.,

$$u(ff')=u(f)\delta_z(f')+(-1)^{[u][f]}\delta_z(f)u(f').$$

G. Sardanashvily, *Noether's Theorems*, Atlantis Studies in Variational Geometry 3, DOI 10.2991/978-94-6239-171-0_9

Definition 9.1 A *graded Lie group* $(G, \mathcal{G})$ is defined as a graded manifold whose body G is a real Lie group, and the structure ring $\mathcal{G}(G)$ is a graded Hopf algebra (Δ, ϵ, S) such that a body epimorphism $\sigma : \mathcal{G}(G) \to C^\infty(G)$ is a Hopf algebra morphism where a ring $C^\infty(G)$ of smooth real functions on a Lie group G possesses a real Hopf algebra structure with the co-operations:

$$\Delta(f)(g, g') = f(gg'), \qquad \epsilon(f) = f(\mathbf{1}), \qquad S(f)(g) = f(g^{-1}), \qquad f \in C^\infty(G).$$

One can show that the finite dual $\mathcal{G}(G)^\circ$ of $\mathcal{G}(G)$ is equipped with the structure of a real Hopf algebra with the multiplication law

$$a \star b = (a \otimes b) \circ \Delta, \qquad a, b \in \mathcal{G}(G)^\circ. \tag{9.1}$$

With respect to this multiplication, the evaluation elements δ_g, $g \in G$, provided with the product $\delta_g \star \delta_{g'} = \delta_{gg'}$ (9.1), constitute a group isomorphic to G. They are called the *group-like elements*. It is readily observed that the set of primitive elements of $\mathcal{G}(G)^\circ$ relative to δ_1 is a real Lie superalgebra $\mathfrak{g}$ with respect to the bracket

$$[u, u'] = u \star u' - (-1)^{[u][u']} u' \star u.$$

It is called the *Lie superalgebra of a graded Lie group* $(G, \mathcal{G})$. Its even part $\mathfrak{g}_0$ is a Lie algebra of a Lie group G.

One says that a *graded Lie group* $(G, \mathcal{G})$ *acts on a graded manifold* $(Z, \mathfrak{A})$ on the right if there exists a morphism of graded manifolds

$$(\varphi, \Phi) : (Z, \mathfrak{A}) \times (G, \mathcal{G}) \to (Z, \mathfrak{A})$$

such that the corresponding ring morphism

$$\Phi : \mathfrak{A}(Z) \to \mathfrak{A}(Z) \otimes \mathcal{G}(G)$$

defines a structure of a right $\mathcal{G}(G)$-comodule on $\mathfrak{A}(Z)$, i.e.,

$$(\mathrm{Id} \otimes \Delta) \circ \Phi = (\Phi \otimes \mathrm{Id}) \circ \Phi, \qquad (\mathrm{Id} \otimes \epsilon) \circ \Phi = \mathrm{Id}.$$

For the right action (φ, Φ) and for each element $a \in \mathcal{G}(G)^\circ$, one can introduce a linear map

$$\Phi_a = (\mathrm{Id} \otimes a) \circ \Phi : \mathfrak{A}(Z) \to \mathfrak{A}(Z). \tag{9.2}$$

In particular, if a is a primitive element with respect to δ_e, then $\Phi_a \in \mathfrak{d}\mathfrak{A}(Z)$.

Let us consider the right action of a graded Lie group $(G, \mathcal{G})$ on itself. If $\Phi = \Delta$ and $a = \delta_g$ is a group-like element, then

$$r_g = (\mathrm{Id} \otimes \delta_g) \circ \Delta \tag{9.3}$$

is an even graded algebra isomorphism which corresponds to the right translation $G \to Gg$. Similarly, the left action

$$l_g = (\delta_g \otimes \mathrm{Id}) \circ \Delta \tag{9.4}$$

is defined. If $a \in \mathfrak{g}$, then l_a is a derivation of $\mathcal{G}(G)$. Given a basis $\{u_i\}$ for $\mathfrak{g}$, the derivations Φ_{u_i} constitute the global basis for $\mathfrak{d}\mathcal{G}(G)$, i.e., $\mathfrak{d}\mathcal{G}(G)$ is a free left $\mathcal{G}(G)$-module. In particular, there is a decomposition

$$\begin{aligned}
&\mathcal{G}(G) = \mathcal{G}'(G) \oplus_R \mathcal{G}''(G),\\
&\mathcal{G}'(G) = \{f \in \mathcal{G}(G) \,:\, \Phi_u(f) = 0, \;\; u \in \mathfrak{g}_0\},\\
&\mathcal{G}''(G) = \{f \in \mathcal{G}(G) \,:\, \Phi_u(f) = 0, \;\; u \in \mathfrak{g}_1\}.
\end{aligned}$$

Since $\mathcal{G}''(G) \cong C^\infty(G)$, one can show the following [1, 21].

Theorem 9.1 *A graded Lie group $(G, \mathcal{G})$ is a simple graded manifold modelled over a trivial bundle*

$$G \times \wedge \mathfrak{g}_1^* \to G, \tag{9.5}$$

endowed with the right action r_g, $g \in G$, (9.3) of a Lie group G.

Let us turn now to the notion of a principal graded bundle. Following Sect. 6.3, we however restrict our consideration to graded bundles over a smooth manifold X. The right action (φ, Φ) of a graded Lie group $(G, \mathcal{G})$ on a graded manifold $(Z, \mathfrak{A})$ is called *free* if, for each $z \in Z$, the morphism $\Phi_z : \mathfrak{A}(Z) \to \mathcal{G}(G)$ is such that the dual morphism $\Phi_{z*} : \mathcal{G}(G)^\circ \to \mathfrak{A}(Z)^\circ$ is injective. A right action (φ, Φ) of $(G, \mathcal{G})$ on $(Z, \mathfrak{A})$ is said to be *regular* if the morphism

$$(\varphi \times \mathrm{pr}_1) \circ \Delta : (Z, \mathfrak{A}) \times (G, \mathcal{G}) \to (Z, \mathfrak{A}) \times (Z, \mathfrak{A})$$

defines a closed graded submanifold of $(Z, \mathfrak{A}) \times (Z, \mathfrak{A})$.

Remark 9.1 Let us note that $(Z', \mathfrak{A}')$ is said to be a *graded submanifold* of $(Z, \mathfrak{A})$ if there exists a morphism $(Z', \mathfrak{A}') \to (Z, \mathfrak{A})$ such that the corresponding morphism $\mathfrak{A}'(Z')^\circ \to \mathfrak{A}(Z)^\circ$ is an inclusion. A graded submanifold is called closed if $\dim (Z', \mathfrak{A}') < \dim (Z, \mathfrak{A})$.

Then we come to the following variant of the well-known theorem on the quotient of a graded manifold [1, 140].

Theorem 9.2 *A right action (φ, Φ) of $(G, \mathcal{G})$ on $(Z, \mathfrak{A})$ is regular if and only if the quotient $(Z/G, \mathfrak{A}/\mathcal{G})$ is a graded manifold, i.e., there exists an epimorphism of graded manifolds $(Z, \mathfrak{A}) \to (Z/G, \mathfrak{A}/\mathcal{G})$ compatible with a surjection $Z \to Z/G$.*

In view of this theorem, a principal graded bundle $(P, \mathfrak{A})$ can be defined as a locally trivial submersion $(P, \mathfrak{A}) \to (P/G, \mathfrak{A}/\mathcal{G})$ with respect to the right regular free action of $(G, \mathcal{G})$ on $(P, \mathfrak{A})$. In an equivalent way, one can say that a principal graded bundle is a graded manifold $(P, \mathfrak{A})$ together with a free right action of a graded Lie group $(G, \mathcal{G})$ on $(P, \mathfrak{A})$ such that the quotient $(P/G, \mathfrak{A}/\mathcal{G})$ is a graded manifold and the natural surjection $(P, \mathfrak{A}) \to (P/G, \mathfrak{A}/\mathcal{G})$ is a submersion. Obviously, $P \to P/G$ is a familiar principal bundle with a structure group G.

As was mentioned above, we restrict our consideration to the case of a principal graded bundle whose base a trivial graded manifold $(X = P/G, \mathfrak{A}/\mathcal{G} = C^\infty_X)$. i.e., principal graded bundles over a smooth manifold X.

Definition 9.2 A *principal graded bundle* is defined to be as a simple graded manifold $(P, \mathfrak{A}_F)$ whose body is a principal bundle $P \to X$ with a structure group G, and which is modelled over a composite bundle

$$F = P \underset{X}{\times} W \to P \to X \tag{9.6}$$

where $W \to X$ is a P-associated bundle (Remark 8.1) with a typical fibre $\mathfrak{g}_1$ provided with the left action (9.4) of a group G. A principal graded bundle $(P, \mathfrak{A}_F)$ is subject to the right fibrewise action (9.2) of a group G.

Remark 9.2 Being kept as a simple graded manifold, a graded bundle $(P, \mathfrak{A}_F)$ fails to admit an action of a graded Lie group $(G, \mathcal{G})$. Therefore, it is not a principal graded bundle in a strict sense.

Turn now to Yang–Mills SUSY gauge theory.

Let $\mathfrak{g}$ be a finite-dimensional real Lie superalgebra with a basis $\{\varepsilon_r\}, r = 1, \dots, m$, and real structure constants c^r_{ij} such that

$$c^r_{ij} = -(-1)^{[i][j]} c^r_{ji}, \qquad [r] = [i] + [j],$$
$$(-1)^{[i][b]} c^r_{ij} c^j_{ab} + (-1)^{[a][i]} c^r_{aj} c^j_{bi} + (-1)^{[b][a]} c^r_{bj} c^j_{ia} = 0,$$

where $[r]$ denotes the Grassmann parity of ε_r. Its even part $\mathfrak{g}_0$ is a Lie algebra of some Lie group G. Let the adjoint representation of $\mathfrak{g}_0$ in $\mathfrak{g}$ is extended to the corresponding action of G on $\mathfrak{g}$.

Let P be a principal bundle with a structure group G. Given Yang–Mills gauge theory of principal connections on P (Sect. 8.3), we aim to extend it to Yang–Mills SUSY gauge theory associated to a Lie superalgebra $\mathfrak{g}$.

Let us consider a simple graded manifold $(G, \mathcal{G})$ modelled over the trivial vector bundle (9.5). In accordance with Theorem 9.1, this is a graded Lie group whose Lie superalgebra is $\mathfrak{g}$. Following Definition 9.2, let us define a principal graded bundle $(P, \mathfrak{A}_F)$ modelled over the composite bundle (9.6).

Let J^1F be a jet manifold of the fibre bundle $F \to X$ (9.6). It is a composite bundle

$$J^1F \to J^1P \to X, \tag{9.7}$$

where $J^1F \to J^1P$ is a vector bundle. Then a simple graded manifold $(J^1P, \mathfrak{A}_{J^1F})$ modelled over $J^1F \to J^1P$ is a graded jet manifold of a principal graded bundle $(P, \mathfrak{A}_F)$ in accordance with Definition 6.7. The composite bundle (9.7) is provided with the jet prolongation of the right action (9.2) of G onto $(P, \mathfrak{A}_F)$. By analogy with the bundle of principal connections C (8.21), let us consider the quotient

$$\mathscr{C} = J^1F/G \to C \to X.$$

Since $\mathscr{C} \to C$ is a vector bundle, we have a simple graded manifold $(C, \mathfrak{A}_{\mathscr{C}})$ modelled over a vector bundle $\mathscr{C} \to C$. It is a graded bundle over a manifold X.

We develop SUSY gauge theory as a Grassmann-graded Lagrangian theory on a graded bundle $(C, \mathfrak{A}_{\mathscr{C}})$ (see Sect. 6.5). Given a basis $\{\varepsilon_r\}$ for $\mathfrak{g}$, we have a local basis (x^λ, a^r_λ) of Grassmann parity $[a^r_\lambda] = [r]$ for a graded bundle $(C, \mathfrak{A}_{\mathscr{C}})$. Similarly to the splitting (8.39), jets of the elements a^r_λ admit the decomposition

$$a^r_{\lambda\mu} = \frac{1}{2}(\mathscr{F}^r_{\lambda\mu} + \mathscr{S}^r_{\lambda\mu}) =$$
$$\frac{1}{2}(a^r_{\lambda\mu} - a^r_{\mu\lambda} + c^r_{ij}a^i_\lambda a^j_\mu) + \frac{1}{2}(a^r_{\lambda\mu} + a^r_{\mu\lambda} - c^r_{ij}a^i_\lambda a^j_\mu).$$

Let us consider the DBGA $\mathscr{S}^*_\infty[\mathscr{C}; C]$ (6.39). Given the universal enveloping algebra $\overline{\mathfrak{g}}$ of $\mathfrak{g}$, we assume that there exists an even quadratic Casimir element $h^{ij}\varepsilon_i\varepsilon_j$ of $\overline{\mathfrak{g}}$ such that the matrix h^{ij} is nondegenerate. Let g be a world metric on X. Then a *graded Yang–Mills Lagrangian* is defined as

$$L_{\mathrm{YM}} = \frac{1}{4}h_{ij}g^{\lambda\mu}g^{\beta\nu}\mathscr{F}^i_{\lambda\beta}\mathscr{F}^j_{\mu\nu}\omega.$$

Its variational derivatives $\mathscr{E}^\lambda_r$ obey the irreducible complete NI

$$c^r_{ji}a^i_\lambda\mathscr{E}^\lambda_r + d_\lambda\mathscr{E}^\lambda_j = 0.$$

Therefore, let us enlarge the differential bigraded algebra $\mathscr{S}^*_\infty[\mathscr{C}; C]$ to the differential bigraded algebra

$$P^*_\infty\{0\} = \mathscr{S}^*_\infty[\mathscr{C} \underset{X}{\oplus} \overline{\mathscr{C}} \underset{X}{\oplus} VF/G \underset{X}{\oplus} \overline{VF/G}; P],$$

where $\overline{E}$ denotes the density dual (7.1) of a fibre bundle $E \to X$. Its local basis $(x^\lambda, a^r_\lambda, \overline{a}^\lambda_r, c^r, \overline{c}_r)$ contains gauge fields a^r_λ, their antifields $\overline{a}^\lambda_r$ of Grassmann parity $[\overline{a}^\lambda_r] = ([r]+1)\mathrm{mod}\,2$, the ghosts c^r of Grassmann parity $[c^r] = ([r]+1)\mathrm{mod}\,2$, and the Noether antifields $\overline{c}_r$ of Grassmann parity $[\overline{c}_r] = [r]$. Then the gauge operator (7.44) reads

$$u = (c^r_\lambda - c^r_{ji}c^j a^i_\lambda)\frac{\partial}{\partial a^r_\lambda} \tag{9.8}$$

(cf. (8.88)). It admits the nilpotent BRST extension

$$\mathbf{b} = (c^r_\lambda - c^r_{ji}c^j a^i_\lambda)\frac{\partial}{\partial a^r_\lambda} - \frac{1}{2}(-1)^{[i]}c^r_{ij}c^i c^j \frac{\partial}{\partial c^r}.$$

The corresponding proper solution (7.86) of the master equation takes a form

$$L_E = L_{\mathrm{YM}} + (c^r_\lambda - c^r_{ji}c^j a^i_\lambda)\overline{a}^\lambda_r\omega - \frac{1}{2}(-1)^{[i]}c^r_{ij}c^i c^j \overline{c}_r\omega.$$

Chapter 10
Gauge Gravitation Theory on Natural Bundles

Gravitation theory in the absence of matter fields can be formulated as gauge field theory on natural bundles over an oriented four-dimensional manifold X [61, 124, 130]. It is metric-affine gravitation theory whose dynamic variables are linear world connections and pseudo-Riemannian world metrics on X (Sect. 10.3). Its first order Lagrangians are invariant under general covariant transformations. Infinitesimal generators of local one-parameter groups of these transformations are the functorial lift (Definition B.6) of vector fields on X onto a natural bundle. They are exact gauge symmetries whose gauge parameters are vector fields on X. By virtue of Noether's first Theorem 2.7, the corresponding conserved symmetry current is the energy-momentum current (10.32) which is reduced to the generalized Komar superpotential (10.37) in accordance with Noether's third Theorem 2.8.

Throughout this chapter, by X is meant an oriented simply connected four-dimensional manifold, called a *world manifold*.

10.1 Relativity Principle: Natural Bundles

Gauge gravitation theory has been vigorously developed since 1960s in different variants which in common lead to above-mentioned metric-affine gravitation theory (Sect. 10.3). We refer the reader to [17, 71, 77, 124] for the history and the references on the subject.

It seems reasonable to believe that gauge gravitation theory must incorporates Einstein's General Relativity and, therefore, it should be based on Relativity and Equivalence Principles reformulated in fibre bundle terms [77, 116]. Relativity Principle states that gauge symmetries of classical gravitation theory are general covariant transformations. Fibre bundles possessing general covariant transformations constitute the category of so called natural bundles [83, 145].

Let $\pi : Y \to X$ be a smooth fibre bundle. Any automorphism (Φ, f) of Y, by definition, is projected as $\pi \circ \Phi = f \circ \pi$ onto a diffeomorphism f of its base X. The converse need not be true. A diffeomorphism of X need not give rise to an

G. Sardanashvily, *Noether's Theorems*, Atlantis Studies
in Variational Geometry 3, DOI 10.2991/978-94-6239-171-0_10

automorphism of Y, unless $Y \to X$ is a trivial bundle. Given a local one-parameter group (Φ_t, f_t) of local automorphisms of Y, its infinitesimal generator is a projectable vector field u on Y. This vector field is projected as $\tau \circ \pi = T\pi \circ u$ onto a vector field τ on X. Its flow is the one-parameter group (f_t) of diffeomorphisms of X which are projections of automorphisms (Φ_t, f_t) of Y. Conversely, let τ be a vector field on X. There is exists its lift $\gamma\tau$ onto Y (Definition B.5), e.g., by means of a connection on $Y \to X$, but this lift need not be functorial (Definition B.6). A fibre bundle $Y \to X$ is called the natural bundle if there exists the functorial lift of vector fields on X onto Y (Definition B.7). One treats this lift as an *infinitesimal general covariant transformation* of Y. Its flow is a local one-parameter group of local general covariant transformations.

Following Relativity Principle, one thus should develop gauge gravitation theory as a field theory on natural bundles [61, 130].

Remark 10.1 In general, there exist diffeomorphisms of X which do not belong to any one-parameter group of diffeomorphisms of X. In a general setting, one therefore considers a monomorphism $f \to \widetilde{f}$ of the group of diffeomorphisms of X to the group of bundle automorphisms of a natural bundle $T \to X$. Automorphisms $\widetilde{f}$ are called *general covariant transformations* of T. No vertical automorphism of T, unless it is the identity morphism, is a general covariant transformation. The group of automorphisms of a natural bundle is a semi-direct product of its subgroup of vertical automorphisms and the subgroup of general covariant transformations.

Tensor bundles, including the tangent and cotangent bundles of X, admit the functorial lifts $\widetilde{\tau}$ (B.29) and (B.30) of vector fields τ on X and, consequently, they exemplify natural bundles.

Remark 10.2 Any diffeomorphism f of X gives rise to the tangent automorphisms $\widetilde{f} = Tf$ of TX which is a general covariant transformation of TX as a natural bundle.

Tensor bundles over a world manifold X have the structure group

$$GL_4 = GL^+(4, \mathbb{R}). \tag{10.1}$$

The associated principal bundle is a fibre bundle

$$\pi_{LX} : LX \to X \tag{10.2}$$

of oriented linear frames in tangent spaces to a world manifold X. It is called the *frame bundle*. Its sections are termed the *frame fields*.

Given holonomic frames $\{\partial_\mu\}$ in the tangent bundle TX associated to the holonomic atlas Ψ_T (B.15), every element $\{H_a\}$ of a frame bundle LX takes a form $H_a = H_a^\mu \partial_\mu$, where H_a^μ is a matrix of the natural representation of a group GL_4 in $\mathbb{R}^4$. These matrices constitute bundle coordinates

$$(x^\lambda, H_a^\mu), \qquad H'^\mu_a = \frac{\partial x'^\mu}{\partial x^\lambda} H_a^\lambda,$$

on LX associated to its *holonomic atlas*

$$\Psi_T = \{(U_\iota, z_\iota = \{\partial_\mu\})\} \tag{10.3}$$

given by the local frame fields $z_\iota = \{\partial_\mu\}$. With respect to these coordinates, the right action (8.4) of GL_4 on LX reads

$$R_g : H_a^\mu \to H_b^\mu g^b{}_a, \qquad g \in GL_4.$$

A frame bundle LX is equipped with the canonical $\mathbb{R}^4$-valued one-form

$$\theta_{LX} = H_\mu^a dx^\mu \otimes t_a, \tag{10.4}$$

where $\{t_a\}$ is a fixed basis for $\mathbb{R}^4$ and H_μ^a is the inverse matrix of H_a^μ.

A frame bundle $LX \to X$ is a natural bundles. Any vector field $\tau = \tau^\lambda \partial_\lambda$ on X admits the functorial lift

$$\widetilde{\tau} = \tau^\mu \partial_\mu + \partial_\nu \tau^\alpha H_a^\nu \frac{\partial}{\partial H_a^\alpha}$$

onto LX defined by the condition $\mathbf{L}_{\widetilde{\tau}} \theta_{LX} = 0$. Moreover, all fibre bundles associated to the frame bundle LX (10.2) are natural bundles. However, there are natural bundles which are not associated to LX.

10.2 Equivalence Principle: Lorentz Reduced Structure

Gauge field theory deals with the three types of classical fields [61, 127]. These are gauge potentials, matter fields and Higgs fields. Higgs fields are responsible for spontaneous symmetry breaking. In classical gauge theory on a principal bundle $P \to X$, *spontaneous symmetry breaking* is characterized by the reduction of a structure Lie group G of this principal bundle to a closed subgroup H of exact symmetries [77, 80, 115, 146].

Let P (8.3) be a principal G-bundle, and let H, $\dim H > 0$, be a closed (and, consequently, Lie) subgroup of G. Then there is a composite bundle

$$P \to P/H \to X,$$

where

$$P_\Sigma = P \xrightarrow{\pi_{P\Sigma}} P/H \tag{10.5}$$

is a principal bundle with a structure group H and

$$\Sigma = P/H = (P \times G/H)/G \xrightarrow{\pi_{\Sigma X}} X \tag{10.6}$$

is a P-associated bundle with a structure group G and a typical fibre G/H on which G acts on the left.

One says that a structure Lie group G of the principal bundle $P \to X$ (8.3) is reduced to its closed subgroup H if the following equivalent conditions hold [61, 82, 124, 133].

- A principal bundle P admits the bundle atlas Ψ_P (8.6) with H-valued transition functions $\rho_{\alpha\beta}$.
- There exists a *reduced principal subbundle* P_H of P with a structure group H.

Remark 10.3 It is easily justified that these conditions are equivalent. If $P_H \subset P$ is a reduced subbundle, its atlas (8.8) given by local sections z_α of $P_H \to X$ is a desired atlas of P. Conversely, let $\Psi_P = \{(U_\alpha, z_\alpha), \rho_{\alpha\beta}\}$, be an atlas of P with H-valued transition functions $\rho_{\alpha\beta}$. For any $x \in U_\alpha \subset X$, let us define a submanifold $z_\alpha(x)H \subset P_x$. These submanifolds form a desired H-subbundle of P because $z_\alpha(x)H = z_\beta(x)H\rho_{\beta\alpha}(x)$ on the overlaps $U_\alpha \cap U_\beta$.

Theorem 10.1 *There is one-to-one correspondence*

$$P^h = \pi_{P\Sigma}^{-1}(h(X)) \tag{10.7}$$

between the reduced principal H-subbundles $i_h : P^h \to P$ of P and the global sections h of the quotient bundle $P/H \to X$ (10.6) [82].

A reduced principal H-subbundle of a principal G-bundle is called the *reduced H-structure*. A glance at the formula (10.7) shows that a reduced principal H-bundle P^h is the restriction $h^* P_\Sigma$ (B.12) of the principal H-bundle P_Σ (10.5) to $h(X) \subset \Sigma$.

In classical field theory, global sections of the quotient bundle $P/H \to X$ (10.6) are interpreted as *Higgs fields* [61, 115, 125].

In gravitation theory on natural bundles, Equivalence Principle reformulated in geometric terms requires that the structure group GL_4 (10.1) of the frame bundle LX (10.2) is reducible to a Lorentz subgroup $SO(1,3)$ [77, 116, 124]. It means that these fibre bundles admit atlases with $SO(1,3)$-valued transition functions or, equivalently, that there exist reduced principal subbundles of LX with a Lorentz structure group.

A reduced principle $SO(1,3)$-subbundle (or, shortly,a *Lorentz subbundle*) of a frame bundle LX is called the *reduced Lorentz structure*.

Remark 10.4 There is the well-known topological obstruction to the existence of a Lorentz reduced structure on a world manifold X. All noncompact manifolds and compact manifolds whose Euler characteristic equals zero admit a reduced Lorentz structure [33].

By virtue of Theorem 10.1, there is one-to-one correspondence between the reduced Lorentz subbundles $L^g X$ of a frame bundle LX and the global sections g of the quotient bundle

$$\Sigma_{\rm PR} = LX/SO(1,3). \tag{10.8}$$

Its typical fibre is $Gl_4/SO(1,3)$, and its global sections are *pseudo-Riemannian world metrics* g on X. Let us call $\Sigma_{\rm PR}$ (10.8) the *metric bundle*. For the sake of convenience, one usually identifies the metric bundle (10.8) with an open subbundle of the tensor bundle

$$\Sigma_{\rm PR} \subset \overset{2}{\vee} TX. \tag{10.9}$$

Therefore, the metric bundle $\Sigma_{\rm PR}$ (10.8) can be equipped with the bundle coordinates $(x^\lambda, \sigma^{\mu\nu})$.

In General Relativity, a pseudo-Riemannian world metric g describes a *gravitational field*. Hereafter, by a *metric* g is meant just a global section of the metric bundle (10.8).

Every metric g defines an associated *Lorentz bundle atlas*

$$\Psi^g = \{(U_\iota, z_\iota^g = \{h_a\})\} \tag{10.10}$$

of a frame bundle LX such that the corresponding local sections z_ι^g of LX take their values into the Lorentz subbundle L^gX and the transition functions of Ψ^g (10.10) between the frames $\{h_a\}$ are $SO(1,3)$-valued. The frames (10.10):

$$\{h_a = h_a^\mu(x)\partial_\mu\}, \qquad h_a^\mu = H_a^\mu \circ z_\iota^g, \qquad x \in U_\iota, \tag{10.11}$$

are called the *tetrad frames*. Certainly, a Lorentz bundle atlas Ψ^g is not unique.

Given a Lorentz bundle atlas Ψ^g, the pull-back

$$h = h^a \otimes t_a = z_\iota^{h*}\theta_{LX} = h_\lambda^a(x)dx^\lambda \otimes t_a \tag{10.12}$$

of the canonical form θ_{LX} (10.4) by a local section z_ι^g is called the (local) *tetrad form*. The tetrad form (10.12) determines the *tetrad coframes*

$$\{h^a = h_\mu^a(x)dx^\mu\}, \qquad x \in U_\iota, \tag{10.13}$$

in the cotangent bundle T^*X. They are the dual of the tetrad frames (10.11). The coefficients h_a^μ and h_μ^a of the tetrad frames (10.11) and coframes (10.13) are called the *tetrad functions*. They are transition functions between the holonomic atlas Ψ_T (10.3) and the Lorentz atlas Ψ^g (10.10) of a frame bundle LX.

With respect to the Lorentz atlas Ψ^h (10.10), a tetrad field h can be represented by the $\mathbb{R}^4$-valued tetrad form (10.12). Relative to this atlas, a pseudo-Riemannian world metric g takes the well-known form

$$g = \eta(h \otimes h) = \eta_{ab}h^a \otimes h^b, \qquad g_{\mu\nu} = h_\mu^a h_\nu^b \eta_{ab},$$

where η is the *Minkowski metric* of signature $(+,---)$ in $\mathbb{R}^4$.

10.3 Metric-Affine Gauge Gravitation Theory

Based on Relativity and Equivalence Principles formulated in geometric terms (Sects. 10.1 and 10.2), gravitation theory can be developed as gauge field theory on natural bundles provided with reduced Lorentz structures [61, 77, 116, 130]. It is metric-affine gravitation theory whose Lagrangian $L_{\rm MA}$ is invariant under general covariant transformations. In the absence of matter fields, its dynamic variables are world connections Γ and pseudo-Riemannian metrics g on a world manifold X.

Since the tangent bundle TX is associated to a frame bundle LX, every world connection Γ (B.70):

$$\Gamma = dx^\lambda \otimes (\partial_\lambda + \Gamma_\lambda{}^\mu{}_\nu \dot x^\nu \dot\partial_\mu), \tag{10.14}$$

on TX is associated to a principal connection on LX. It follows that world connections are represented by sections of the quotient bundle

$$C_{\rm W} = J^1LX/GL_4, \tag{10.15}$$

called the *bundle of world connections*. With respect to the holonomic atlas Ψ_T (10.3), the bundle of world connections $C_{\rm W}$ (10.15) is provided with coordinates

$$(x^\lambda, k_\lambda{}^\nu{}_\alpha), \qquad k'_\lambda{}^\nu{}_\alpha = \left[\frac{\partial x'^\nu}{\partial x^\gamma}\frac{\partial x^\beta}{\partial x'^\alpha}k_\mu{}^\gamma{}_\beta + \frac{\partial x^\beta}{\partial x'^\alpha}\frac{\partial^2 x'^\nu}{\partial x^\mu \partial x^\beta}\right]\frac{\partial x^\mu}{\partial x'^\lambda},$$

so that, for any section Γ of $C_{\rm W} \to X$, its coordinates $k_\lambda{}^\nu{}_\alpha \circ \Gamma = \Gamma_\lambda{}^\nu{}_\alpha$ are components of the linear connection Γ (10.14).

Though the bundle of world connections $C_{\rm W} \to X$ (10.15) is not LX-associated, it is a natural bundle. It admits the functorial lift

$$\widetilde\tau_C = \tau^\mu\partial_\mu + [\partial_\nu\tau^\alpha k_\mu{}^\nu{}_\beta - \partial_\beta\tau^\nu k_\mu{}^\alpha{}_\nu - \partial_\mu\tau^\nu k_\nu{}^\alpha{}_\beta + \partial_{\mu\beta}\tau^\alpha]\frac{\partial}{\partial k_\mu{}^\alpha{}_\beta} \tag{10.16}$$

of vector fields τ on X [100].

The first order jet manifold $J^1C_{\rm W}$ of the bundle of world connections admits the canonical splitting (8.38). In order to obtain its coordinate expression, let us consider the strength (8.29) of the linear world connection Γ (10.14). It reads

$$F_\Gamma = \frac{1}{2}F_{\lambda\mu}{}^b{}_a I_b{}^a dx^\lambda \wedge dx^\mu = \frac{1}{2}R_{\mu\nu}{}^\alpha{}_\beta dx^\lambda \wedge dx^\mu,$$

where $(I_b{}^a)^\alpha{}_\beta = H^\alpha_b H^a_\beta$ are generators of the group GL_4 (10.1) in fibres of TX with respect to the holonomic frames, and

$$R_{\lambda\mu}{}^\alpha{}_\beta = \partial_\lambda\Gamma_\mu{}^\alpha{}_\beta - \partial_\mu\Gamma_\lambda{}^\alpha{}_\beta + \Gamma_\lambda{}^\gamma{}_\beta\Gamma_\mu{}^\alpha{}_\gamma - \Gamma_\mu{}^\gamma{}_\beta\Gamma_\lambda{}^\alpha{}_\gamma$$

are components of the curvature (B.72) of a world connection Γ. Accordingly, the above mentioned canonical splitting (8.38) of J^1C_W can be written in a form

$$\begin{aligned} k_{\lambda\mu}{}^\alpha{}_\beta = &\frac{1}{2}(\mathscr{R}_{\lambda\mu}{}^\alpha{}_\beta + \mathscr{S}_{\lambda\mu}{}^\alpha{}_\beta) = \\ &\frac{1}{2}(k_{\lambda\mu}{}^\alpha{}_\beta - k_{\mu\lambda}{}^\alpha{}_\beta + k_\lambda{}^\gamma{}_\beta k_\mu{}^\alpha{}_\gamma - k_\mu{}^\gamma{}_\beta k_\lambda{}^\alpha{}_\gamma) + \\ &\frac{1}{2}(k_{\lambda\mu}{}^\alpha{}_\beta + k_{\mu\lambda}{}^\alpha{}_\beta - k_\lambda{}^\gamma{}_\beta k_\mu{}^\alpha{}_\gamma + k_\mu{}^\gamma{}_\beta k_\lambda{}^\alpha{}_\gamma). \end{aligned} \tag{10.17}$$

It is readily observed that, if Γ is a section of $C_W \to X$, then

$$\mathscr{R}_{\lambda\mu}{}^\alpha{}_\beta \circ J^1\Gamma = R_{\lambda\mu}{}^\alpha{}_\beta.$$

Due the canonical vertical splitting (B.41) of the vertical tangent bundle *VTX* of *TX*, the curvature (B.72) of a world connection Γ can be represented by a tangent-valued two-form

$$R = \frac{1}{2}R_{\lambda\mu}{}^\alpha{}_\beta \dot{x}^\beta dx^\lambda \wedge dx^\mu \otimes \partial_\alpha$$

on *TX*. Then the Ricci tensor

$$R_c = \frac{1}{2}R_{\lambda\mu}{}^\lambda{}_\beta dx^\mu \otimes dx^\beta$$

of a world connection Γ is defined.

Owing to the above mentioned vertical splitting (B.41) of *VTX*, the torsion form T (B.73) of Γ defines the *tangent-valued torsion two-form*

$$T_T = \frac{1}{2}T_\mu{}^\nu{}_\lambda dx^\lambda \wedge dx^\mu \otimes \partial_\nu, \qquad T_\mu{}^\nu{}_\lambda = \Gamma_\mu{}^\nu{}_\lambda - \Gamma_\lambda{}^\nu{}_\mu. \tag{10.18}$$

Remark 10.5 Given a pseudo-Riemannian metric g, every world connection Γ (10.14) admits a decomposition

$$\Gamma_{\mu\nu\alpha} = \{_{\mu\nu\alpha}\} + S_{\mu\nu\alpha} + \frac{1}{2}C_{\mu\nu\alpha}$$

in the Christoffel symbols $\{_{\mu\nu\alpha}\}$ (B.74), the *nonmetricity tensor*

$$C_{\mu\nu\alpha} = C_{\mu\alpha\nu} = \nabla^\Gamma_\mu g_{\nu\alpha} = \partial_\mu g_{\nu\alpha} + \Gamma_{\mu\nu\alpha} + \Gamma_{\mu\alpha\nu}$$

and the *contorsion*

$$S_{\mu\nu\alpha} = -S_{\mu\alpha\nu} = \frac{1}{2}(T_{\nu\mu\alpha} + T_{\nu\alpha\mu} + T_{\mu\nu\alpha} + C_{\alpha\nu\mu} - C_{\nu\alpha\mu}),$$

where $T_{\mu\nu\alpha} = -T_{\alpha\nu\mu}$ are coefficients of the torsion form (10.18) of Γ. A world connection Γ is called the *metric connection* for a pseudo-Riemannian world metric g if the covariant differential $\nabla^\Gamma g$ (B.62) of a metric g with respect to a connection Γ vanishes, i.e., *metricity condition* $\nabla^\Gamma_\mu g_{\nu\alpha} = 0$ holds. A metric connection reads

$$\Gamma_{\mu\nu\alpha} = \{_{\mu\nu\alpha}\} + \frac{1}{2}(T_{\nu\mu\alpha} + T_{\nu\alpha\mu} + T_{\mu\nu\alpha}).$$

For instance, the Levi–Civita connection (B.74) is a symmetric metric connection.

As was mentioned above, world connections are represented by sections of the bundle of world connections $C_{\rm W}$ (10.15), whereas pseudo-Riemannian world metrics are described by sections of the open subbundle (10.9). Therefore, a configuration space of metric-affine gauge theory is a bundle product

$$Y = \Sigma_{\rm PR} \underset{X}{\times} C_{\rm W} \tag{10.19}$$

coordinated by $(x^\lambda, \sigma^{\mu\nu}, k_\mu{}^\alpha{}_\beta)$. Its first order jet manifold is

$$J^1Y = J^1\Sigma_{\rm PR} \underset{X}{\times} J^1C_{\rm W}, \tag{10.20}$$

where $J^1C_{\rm W}$ admits the canonical splitting (8.38) given by the coordinate expression (10.17). Let us consider the DGA (1.7):

$$\mathcal{S}^*_\infty[Y] = \mathcal{O}^*_\infty Y, \tag{10.21}$$

possessing the local generating basis $(\sigma^{\alpha\beta}, k_\mu{}^\alpha{}_\beta)$.

A Lagrangian of metric-affine gauge gravitation theory is a first order Lagrangian

$$L_{\rm MA} = \mathcal{L}_{\rm MA}(\sigma^{\alpha\beta}, k_\mu{}^\alpha{}_\beta, \sigma^{\alpha\beta}_\lambda, k_{\lambda\mu}{}^\alpha{}_\beta)\omega \tag{10.22}$$

on the jet manifold (10.20). Its Euler–Lagrange operator reads

$$\delta L_{\rm MA} = (\mathcal{E}_{\alpha\beta} d\sigma^{\alpha\beta} + \mathcal{E}^\mu{}_\alpha{}^\beta dk_\mu{}^\alpha{}_\beta) \wedge \omega. \tag{10.23}$$

The fibre bundle (10.19) is a natural bundle admitting the functorial lift

$$\begin{aligned}\widetilde{\tau}_{\Sigma C} = \tau^\mu\partial_\mu + (\sigma^{\nu\beta}\partial_\nu\tau^\alpha + \sigma^{\alpha\nu}\partial_\nu\tau^\beta)\frac{\partial}{\partial\sigma^{\alpha\beta}} + \\ (\partial_\nu\tau^\alpha k_\mu{}^\nu{}_\beta - \partial_\beta\tau^\nu k_\mu{}^\alpha{}_\nu - \partial_\mu\tau^\nu k_\nu{}^\alpha{}_\beta + \partial_{\mu\beta}\tau^\alpha)\frac{\partial}{\partial k_\mu{}^\alpha{}_\beta}\end{aligned} \tag{10.24}$$

of vector fields τ on X (cf. (10.16)). Following Definition 2.3, one can treat vector fields $\widetilde{\tau}_{\Sigma C}$ (10.24) as infinitesimal gauge transformations whose gauge parameters are vector fields τ on X.

Lagrangians $L_{\rm MA}$ of metric-affine gauge gravitation theory are assumed to be invariant under general covariant transformations. This means that infinitesimal gauge transformations $\widetilde{\tau}_{\Sigma C}$ (10.24) are exact symmetries of a Lagrangian $L_{\rm MA}$. By analogy with Theorem 8.2, one can show that, if the first order Lagrangian $L_{\rm MA}$ (10.22) does not depend on jet coordinates $\sigma_\lambda^{\alpha\beta}$ and if it possesses exact gauge symmetries (10.24), it factorizes as

$$L_{\rm MA} = \mathscr{L}_{\rm MA}(\sigma^{\alpha\beta}, t_\nu{}^\alpha{}_\sigma, \mathscr{R}_{\lambda\mu}{}^\alpha{}_\beta)\omega \tag{10.25}$$

through the terms $\mathscr{R}_{\lambda\mu}{}^\alpha{}_\beta$ (10.17) and the torsion terms

$$t_\nu{}^\alpha{}_\sigma = k_\nu{}^\alpha{}_\sigma - k_\sigma{}^\alpha{}_\nu. \tag{10.26}$$

In particular, a *Hilbert–Einstein Lagrangian* in metric-affine gauge gravitation theory (see Remark 10.7) reads

$$L_{\rm HE} = \frac{1}{2\kappa}\mathscr{R}\sqrt{-\sigma}\omega, \qquad \mathscr{R} = \sigma^{\lambda\nu}\mathscr{R}_{\alpha\lambda}{}^\alpha{}_\nu, \qquad \sigma = \det(\sigma_{\alpha\beta}). \tag{10.27}$$

10.4 Energy-Momentum Gauge Conservation Law

Since infinitesimal general covariant transformations $\widetilde{\tau}_{\Sigma C}$ (10.24) are assumed to be exact gauge symmetries of a metric-affine gravitation Lagrangian, let us study the corresponding conservation law. This is an energy-momentum conservation law (Definition 3.3) because the vector fields $\widetilde{\tau}_{\Sigma C}$ are the lift (B.27) of vector fields on X [52, 120, 130]. There are several approaches to discover an energy-momentum conservation law in gravitation theory. Here we treat this conservation law as a particular gauge conservation law . Accordingly, the energy-momentum of gravity is seen as a particular symmetry current (see, e.g., [5, 19, 71, 74]). Since infinitesimal general covariant transformations $\widetilde{\tau}_{\Sigma C}$ (10.24) are gauge symmetries, the corresponding energy-momentum current reduces to a superpotential (Theorem 2.8).

Let us assume that the metric-affine gravitation Lagrangian $L_{\rm MA}$ (10.25) is independent of torsion terms (10.26). Then the following relations take place:

$$\pi^{\lambda\nu}{}_\alpha{}^\beta = -\pi^{\nu\lambda}{}_\alpha{}^\beta, \qquad \pi^{\lambda\nu}{}_\alpha{}^\beta = \frac{\partial\mathscr{L}_{\rm MA}}{\partial k_{\lambda\nu}{}^\alpha{}_\beta}, \tag{10.28}$$

$$\frac{\partial\mathscr{L}_{\rm MA}}{\partial k_\nu{}^\alpha{}_\beta} = \pi^{\lambda\nu}{}_\alpha{}^\sigma k_\lambda{}^\beta{}_\sigma - \pi^{\lambda\nu}{}_\sigma{}^\beta k_\lambda{}^\sigma{}_\alpha. \tag{10.29}$$

Let us introduce the compact notation

$$y^A = k_\mu{}^\alpha{}_\beta, \quad u_\mu{}^\alpha{}_{\beta\gamma}{}^{\varepsilon\sigma} = \delta^\varepsilon_\mu \delta^\sigma_\beta \delta^\alpha_\gamma, \quad u_\mu{}^\alpha{}_{\beta\gamma}{}^\varepsilon = k_\mu{}^\varepsilon{}_\beta \delta^\alpha_\gamma - k_\mu{}^\alpha{}_\gamma \delta^\varepsilon_\beta - k_\gamma{}^\alpha{}_\beta \delta^\varepsilon_\mu.$$

Then the vector fields (10.24) take a form

$$\widetilde{\tau}_{\Sigma C} = \tau^\lambda \partial_\lambda + (\sigma^{\nu\beta}\partial_\nu \tau^\alpha + \sigma^{\alpha\nu}\partial_\nu \tau^\beta)\partial_{\alpha\beta} + (u^{A\beta}{}_\alpha \partial_\beta \tau^\alpha + u^{A\beta\mu}{}_\alpha \partial_{\beta\mu}\tau^\alpha)\partial_A.$$

We also have the equalities

$$\pi^\lambda_A u^{A\beta\mu}{}_\alpha = \pi^{\lambda\mu}{}_\alpha{}^\beta, \qquad \pi^\varepsilon_A u^{A\beta}{}_\alpha = -\partial^\varepsilon{}_\alpha{}^\beta \mathscr{L}_{\rm MA} - \pi^{\varepsilon\beta}{}_\sigma{}^\gamma k_\alpha{}^\sigma{}_\gamma.$$

Since $\mathbf{L}_{J^1\widetilde{\tau}_{\Sigma C}} L_{\rm MA} = 0$, the first variational formula (3.26) takes a form

$$\begin{aligned} 0 = {} & (\sigma^{\nu\beta}\partial_\nu\tau^\alpha + \sigma^{\alpha\nu}\partial_\nu\tau^\beta - \tau^\lambda\sigma^{\alpha\beta}_\lambda)\delta_{\alpha\beta}\mathscr{L}_{\rm MA} + \\ & (u^{A\beta}{}_\alpha\partial_\beta\tau^\alpha + u^{A\beta\mu}{}_\alpha\partial_{\beta\mu}\tau^\alpha - \tau^\lambda y^A_\lambda)\delta_A\mathscr{L}_{\rm MA} - \\ & d_\lambda[\pi^\lambda_A(y^A_\alpha\tau^\alpha - u^{A\beta}{}_\alpha\partial_\beta\tau^\alpha - u^{A\varepsilon\beta}{}_\alpha\partial_{\varepsilon\beta}\tau^\alpha) - \tau^\lambda\mathscr{L}_{\rm MA}]. \end{aligned} \tag{10.30}$$

The first variational formula (10.30) on-shell leads to a weak conservation law

$$0 \approx -d_\lambda[\pi^\lambda_A(y^A_\alpha\tau^\alpha - u^{A\beta}{}_\alpha\partial_\beta\tau^\alpha - u^{A\varepsilon\beta}{}_\alpha\partial_{\varepsilon\beta}\tau^\alpha) - \tau^\lambda\mathscr{L}_{\rm MA}] \tag{10.31}$$

of the *energy-momentum current* of metric-affine gravity

$$\mathscr{J}_{\rm MA}{}^\lambda = \pi^\lambda_A(y^A_\alpha\tau^\alpha - u^{A\beta}{}_\alpha\partial_\beta\tau^\alpha - u^{A\varepsilon\beta}{}_\alpha\partial_{\varepsilon\beta}\tau^\alpha) - \tau^\lambda\mathscr{L}_{\rm MA}. \tag{10.32}$$

Due to the arbitrariness of gauge parameters τ^λ, the first variational formula (10.30) falls into the set of equalities (4.52)–(4.55) which read

$$\pi^{(\lambda\varepsilon}{}_\gamma{}^{\sigma)} = 0, \tag{10.33}$$

$$(u^{A\varepsilon\sigma}{}_\gamma\partial_A + u^{A\varepsilon}{}_\gamma\partial^\sigma_A)\mathscr{L}_{\rm MA} = 0, \tag{10.34}$$

$$\delta^\beta_\alpha\mathscr{L}_{\rm MA} + (2\sigma^{\beta\mu}\delta_{\alpha\mu} + u^{A\beta}{}_\alpha\delta_A)\mathscr{L}_{\rm MA} + d_\mu(\pi^\mu_A u^{A\beta}{}_\alpha) - y^A_\alpha\pi^\beta_A = 0, \tag{10.35}$$

$$\partial_\lambda\mathscr{L}_{\rm MA} = 0.$$

It is readily observed that the equalities (10.33) and (10.34) hold due to the relations (10.28) and (10.29), respectively. Substituting the term $y^A_\alpha\pi^\beta_A$ from the expression (10.35) in the energy-momentum conservation law (10.31), one brings this conservation law into a form

$$\begin{aligned} 0 \approx -d_\lambda[& 2\sigma^{\lambda\mu}\tau^\alpha\delta_{\alpha\mu}\mathscr{L}_{\rm MA} + u^{A\lambda}{}_\alpha\tau^\alpha\delta_A\mathscr{L}_{\rm MA} - \pi^\lambda_A u^{A\beta}{}_\alpha\partial_\beta\tau^\alpha + \\ & d_\mu(\pi^{\lambda\mu}{}_\alpha{}^\beta)\partial_\beta\tau^\alpha + d_\mu(\pi^\mu_A u^{A\lambda}{}_\alpha)\tau^\alpha - d_\mu(\pi^{\lambda\mu}{}_\alpha{}^\beta\partial_\beta\tau^\alpha)]. \end{aligned} \tag{10.36}$$

After separating a variational derivatives, the energy-momentum conservation law (10.36) of metric-affine gravity takes the following superpotential form [52, 120]:

$$
\begin{aligned}
0 \approx -d_\lambda[&2\sigma^{\lambda\mu}\tau^\alpha\delta_{\alpha\mu}\mathscr{L}_{\mathrm{MA}} + \\
&(k_\mu{}^\lambda{}_\gamma\delta^\mu{}_\alpha{}^\gamma\mathscr{L}_{\mathrm{MA}} - k_\mu{}^\sigma{}_\alpha\delta^\mu{}_\sigma{}^\lambda\mathscr{L}_{\mathrm{MA}} - k_\alpha{}^\sigma{}_\gamma\delta^\lambda{}_\sigma{}^\gamma\mathscr{L}_{\mathrm{MA}})\tau^\alpha + \\
&\delta^\lambda{}_\alpha{}^\mu\mathscr{L}_{\mathrm{MA}}\partial_\mu\tau^\alpha - d_\mu(\delta^\mu{}_\alpha{}^\lambda\mathscr{L}_{\mathrm{MA}})\tau^\alpha + d_\mu(\pi^{\mu\lambda}{}_\alpha{}^\nu(\partial_\nu\tau^\alpha - k_\sigma{}^\alpha{}_\nu\tau^\sigma))],
\end{aligned}
$$

where the energy-momentum current on-shell reduces to the *generalized Komar superpotential*

$$
U_{\mathrm{MA}}{}^{\mu\lambda} = \pi^{\mu\lambda}{}_\alpha{}^\nu(\partial_\nu\tau^\alpha - k_\sigma{}^\alpha{}_\nu\tau^\sigma). \tag{10.37}
$$

We can rewrite this superpotential as

$$
U_{\mathrm{MA}}{}^{\mu\lambda} = 2\frac{\partial\mathscr{L}_{\mathrm{MA}}}{\partial\mathscr{R}_{\mu\lambda}{}^\alpha{}_\nu}(D_\nu\tau^\alpha + t_\nu{}^\alpha{}_\sigma\tau^\sigma),
$$

where D_ν is the covariant derivative relative to the connection $k_\nu{}^\alpha{}_\sigma$ and $t_\nu{}^\alpha{}_\sigma$ is its torsion (10.26).

Remark 10.6 Let us consider the Hilbert–Einstein Lagrangian (10.27) of metric-affine gauge gravitation theory. Then the generalized Komar superpotential (10.37) comes to the well known Komar superpotential if we substitute the Levi–Civita connection $k_\nu{}^\alpha{}_\sigma = \{_\nu{}^\alpha{}_\sigma\}$ (10.40).

10.5 BRST Gravitation Theory

Since infinitesimal general covariant transformations $\widetilde{\tau}_{\Sigma C}$ (10.24) are assumed to be exact gauge symmetries of a metric-affine gravitation Lagrangian, let us study the corresponding NI.

Following the procedure in Sect. 7.1, let us consider a pull-back bundle

$$
TX \underset{X}{\times} Y = TX \underset{X}{\times} \Sigma_{\mathrm{PR}} \underset{X}{\times} C_{\mathrm{W}},
$$

and let us enlarge the DGA $\mathscr{S}^*_\infty[Y]$ (10.21) to a DBGA $\mathscr{P}^*_\infty[TX; Y]$ possessing the local generating basis $(\sigma^{\alpha\beta}, k_\mu{}^\alpha{}_\beta, c^\mu)$ of even fields $(\sigma^{\alpha\beta}, k_\mu{}^\alpha{}_\beta)$ and odd ghosts (c^μ).

Taking the vertical part of vector fields $\widetilde{\tau}_{\Sigma C}$ (10.24) and replacing gauge parameters τ^λ with ghosts c^λ, we obtain the odd vertical graded derivation

$$
\begin{aligned}
u &= u^{\alpha\beta}\frac{\partial}{\partial\sigma^{\alpha\beta}} + u_\mu{}^\alpha{}_\beta\frac{\partial}{\partial k_\mu{}^\alpha{}_\beta} = \\
&(\sigma^{\nu\beta}c^\alpha_\nu + \sigma^{\alpha\nu}c^\beta_\nu - c^\lambda\sigma^{\alpha\beta}_\lambda)\frac{\partial}{\partial\sigma^{\alpha\beta}} +
\end{aligned} \tag{10.38}
$$

$$(c^\alpha_\nu k_\mu{}^\nu{}_\beta - c^\nu_\beta k_\mu{}^\alpha{}_\nu - c^\nu_\mu k_\nu{}^\alpha{}_\beta + c^\alpha_{\mu\beta} - c^\lambda k_{\lambda\mu}{}^\alpha{}_\beta)\frac{\partial}{\partial k_\mu{}^\alpha{}_\beta}.$$

Since it is SUSY of a metric-affine Lagrangian $L_{\rm MA}$, its Euler–Lagrange operator $\delta L_{\rm MA}$ (10.23) obeys complete NI

$$-\sigma^{\alpha\beta}_\lambda \mathcal{E}_{\alpha\beta} - 2d_\mu(\sigma^{\mu\beta}\mathcal{E}_{\lambda\beta} - k_{\lambda\mu}{}^\alpha{}_\beta\mathcal{E}^\mu{}_\alpha{}^\beta - \qquad (10.39)$$
$$d_\mu[(k_\nu{}^\mu{}_\beta\delta^\alpha_\lambda - k_\nu{}^\alpha{}_\lambda\delta^\mu_\beta - k_\lambda{}^\alpha{}_\beta\delta^\mu_\nu)\mathcal{E}^\nu{}_\alpha{}^\beta] + d_{\mu\beta}\mathcal{E}^\mu{}_\lambda{}^\beta = 0.$$

Remark 10.7 Let us note that the Hilbert–Einstein Lagrangian $L_{\rm HE}$ of General Relativity depends only on metric variables $\sigma^{\alpha\beta}$. It is a reduced second order Lagrangian which differs from the first order one $L'_{\rm HE}$ in a variationally trivial term. The infinitesimal gauge covariant transformations $\widetilde{\tau}_{\Sigma C}$ (10.24) are variational (but not exact) symmetries of the first order Lagrangian $L'_{\rm HE}$, and the graded derivation u (10.38) is so. It reads

$$u = (\sigma^{\nu\beta}c^\alpha_\nu + \sigma^{\alpha\nu}c^\beta_\nu - c^\lambda\sigma^{\alpha\beta}_\lambda)\frac{\partial}{\partial\sigma^{\alpha\beta}}.$$

Then the corresponding NI (10.39) take a familiar form

$$\nabla_\mu\mathcal{E}^\mu_\lambda = (d_\mu + \{_\mu{}^\beta{}_\lambda\})\mathcal{E}^\mu_\beta = 0,$$

where $\mathcal{E}^\mu_\lambda = \sigma^{\mu\alpha}\mathcal{E}_{\alpha\lambda}$ and

$$\{_\mu{}^\beta{}_\lambda\} = -\frac{1}{2}\sigma^{\beta\nu}(d_\mu\sigma_{\nu\lambda} + d_\lambda\sigma_{\mu\nu} - d_\nu\sigma_{\mu\lambda}) \qquad (10.40)$$

are the Christoffel symbols expressed into function $\sigma_{\alpha\beta}$ of $\sigma^{\mu\nu}$ given by the relations $\sigma^{\mu\alpha}\sigma_{\alpha\beta} = \delta^\mu_\beta$.

The NI (10.39) are irreducible. Therefore, the graded derivation (10.38) is a gauge operator of gauge gravitation theory. It possesses a nilpotent BRST extension

$$\mathbf{b} = u + c^\lambda_\mu c^\mu\frac{\partial}{\partial c^\lambda}.$$

Accordingly, an original gravitation Lagrangian $L_{\rm MA}$ admits a BRST extension to a proper solution of the master equation which reads

$$L_E = L_{\rm MA} + u^{\alpha\beta}\overline{\sigma}_{\alpha\beta}\omega + u_\mu{}^\alpha{}_\beta\overline{k}^\mu{}_\alpha{}^\beta\omega + c^\lambda_\mu c^\mu\overline{c}_\lambda\omega,$$

where $\overline{\sigma}_{\alpha\beta}$, $\overline{k}^\mu{}_\alpha{}^\beta$ and $\overline{c}_\lambda$ are the corresponding antifields [8, 61].

Chapter 11
Chern–Simons Topological Field Theory

In classical field theory, there are Lagrangian field models whose Lagrangians are independent of a world metric on a base X. These are topological field theories of Schwartz type [16]. Here we address Chern–Simons topological field theory on a principal bundle. In comparison with Yang–Mills gauge theory in Chap. 8, it possesses principal gauge symmetries which are neither vertical nor exact (Remark 8.8). The corresponding conservation laws (11.11) and (11.12) involve Noether currents and the energy-momentum ones.

Note that one usually considers the local Chern–Simons Lagrangian which is the local Chern–Simons form derived from the local transgression formula for the Chern characteristic form. The global Chern–Simons Lagrangian is well defined, but depends on a background gauge potential [20, 41, 57, 61].

Given a principal bundle $P \to X$ with a structure Lie group G, let $C \to X$ be the bundle of principal connections (8.21), $\mathscr{A}$ the canonical principal connection (8.34) on the principal G-bundle P_C (8.30), and $F_{\mathscr{A}}$ (8.35) its strength. Let

$$I_k(\chi) = b_{r_1 \dots r_k} \chi^{r_1} \cdots \chi^{r_k} \tag{11.1}$$

be a G-invariant polynomial of degree $k > 1$ on the right Lie algebra $\mathfrak{g}_r$ of G. With $F_{\mathscr{A}}$ (8.35), one can associate to this polynomial I_k the closed $2k$-form

$$P_{2k}(F_{\mathscr{A}}) = b_{r_1 \dots r_k} F_{\mathscr{A}}^{r_1} \wedge \cdots \wedge F_{\mathscr{A}}^{r_k}, \qquad 2k \leq n, \tag{11.2}$$

on C which is invariant under automorphisms of C induced by vertical principal automorphisms of P. Given a section A of $C \to X$, the pull-back

$$P_{2k}(F_A) = A^* P_{2k}(F_{\mathscr{A}}) \tag{11.3}$$

G. Sardanashvily, *Noether's Theorems*, Atlantis Studies
in Variational Geometry 3, DOI 10.2991/978-94-6239-171-0_11

of the form $P_{2k}(F_{\mathscr{A}})$ (11.2) is a closed $2k$-form on X where F_A is the strength (8.29) of a principal connection A. It is called the *characteristic form*. The characteristic forms (11.3) possess the following important properties [37, 82, 102].

- Every characteristic form $P_{2k}(F_A)$ (11.3) is a closed form, i.e., $dP_{2k}(F_A) = 0$;
- The difference $P_{2k}(F_A) - P_{2k}(F_{A'})$ of characteristic forms is an exact form, whenever A and A' are different principal connections on a principal bundle P.

It follows that characteristic forms $P_{2k}(F_A)$ possess the same de Rham cohomology class $[P_{2k}(F_A)]$ for all principal connections A on P. The association

$$I_k(\chi) \to [P_{2k}(F_A)] \in H^*_{\mathrm{DR}}(X)$$

is the well-known *Weil homomorphism*. The de Rham cohomology class $[P_{2k}(F_A)]$ is a topological invariant. Choosing a certain family of characteristic forms (11.3), one therefore can obtain characteristic classes of a principal bundle P.

Let I_k (11.1) be a G-invariant polynomial of degree $k > 1$ on the right Lie algebra $\mathfrak{g}_r$ of G. Let $P_{2k}(F_{\mathscr{A}})$ (11.2) be the corresponding closed $2k$-form on C and $P_{2k}(F_A)$ (11.3) its pullback onto X by means of a section A of $C \to X$. Let the same symbol $P_{2k}(F_A)$ stand for its pull-back onto C. Since $C \to X$ is an affine bundle and, consequently, the de Rham cohomology of C equals that of X, the exterior forms $P_{2k}(F_{\mathscr{A}})$ and $P_{2k}(F_A)$ possess the same de Rham cohomology class $[P_{2k}(F_{\mathscr{A}})] = [P_{2k}(F_A)]$ for any principal connection A. Consequently, the exterior forms $P_{2k}(F_{\mathscr{A}})$ and $P_{2k}(F_A)$ on C differ from each other in an exact form

$$P_{2k}(F_{\mathscr{A}}) - P_{2k}(F_A) = d\mathfrak{S}_{2k-1}(a, A). \tag{11.4}$$

This relation is called the *transgression formula* on C [57, 61]. Its pull-back by means of a section B of $C \to X$ gives the *transgression formula on a base* X:

$$P_{2k}(F_B) - P_{2k}(F_A) = d\mathfrak{S}_{2k-1}(B, A).$$

For instance, let

$$c(F_{\mathscr{A}}) = \det\left(\mathbf{1} + \frac{i}{2\pi}F_{\mathscr{A}}\right) = 1 + c_1(F_{\mathscr{A}}) + c_2(F_{\mathscr{A}}) + \cdots$$

be the *total Chern form* on a bundle of principal connections C. Its components $c_k(F_{\mathscr{A}})$ are *Chern characteristic forms* on C. If $P_{2k}(F_{\mathscr{A}}) = c_k(F_{\mathscr{A}})$ is the characteristic Chern $2k$-form, then $\mathfrak{S}_{2k-1}(a, A)$ (11.4) is a *Chern–Simons* $(2k-1)$*-form*.

In particular, one can choose a local section $A = 0$. In this case, $\mathfrak{S}_{2k-1}(a, 0)$ is called the *local Chern–Simons form*. Let $\mathfrak{S}_{2k-1}(A, 0)$ be its pull-back onto X by means of a section A of $C \to X$. Then the Chern–Simons form $\mathfrak{S}_{2k-1}(a, A)$ (11.4) admits the decomposition

$$\mathfrak{S}_{2k-1}(a, A) = \mathfrak{S}_{2k-1}(a, 0) - \mathfrak{S}_{2k-1}(A, 0) + dK_{2k-1}. \tag{11.5}$$

The transgression formula (11.4) also yields the transgression formula

$$h_0(P_{2k}(F_{\mathscr{A}}) - P_{2k}(F_A)) = d_H(h_0\mathfrak{S}_{2k-1}(a, A)),$$
$$h_0\mathfrak{S}_{2k-1}(a, A) = k\int_0^1 \mathscr{P}_{2k}(t, A)dt, \tag{11.6}$$
$$\mathscr{P}_{2k}(t, A) = b_{r_1\ldots r_k}(a^{r_1}_{\mu_1} - A^{r_1}_{\mu_1})dx^{\mu_1} \wedge \mathscr{F}^{r_2}(t, A) \wedge \cdots \wedge \mathscr{F}^{r_k}(t, A),$$
$$\mathscr{F}^{r_j}(t, A) = \frac{1}{2}[ta^{r_j}_{\lambda_j\mu_j} + (1-t)\partial_{\lambda_j}A^{r_j}_{\mu_j} - ta^{r_j}_{\mu_j\lambda_j} - (1-t)\partial_{\mu_j}A^{r_j}_{\lambda_j} + \frac{1}{2}c^{r_j}_{pq}(ta^p_{\lambda_j} + (1-t)A^p_{\lambda_j})(ta^q_{\mu_j} + (1-t)A^q_{\mu_j}]dx^{\lambda_j} \wedge dx^{\mu_j} \otimes e_r,$$

on J^1C (where $b_{r_1\ldots r_k}$ are coefficients of the invariant polynomial (11.1)).

If $2k-1 = \dim X$, the density (11.6) is the global *Chern–Simons Lagrangian*

$$L_{CS}(A) = h_0\mathfrak{S}_{2k-1}(a, A) \tag{11.7}$$

of Chern–Simons topological field theory. It depends on a background gauge field A. The decomposition (11.5) induces the decomposition

$$L_{CS}(A) = h_0\mathfrak{S}_{2k-1}(a, 0) - h_0\mathfrak{S}_{2k-1}(A, 0) + d_H h_0 K_{2k-1},$$

where

$$L_{CS} = h_0\mathfrak{S}_{2k-1}(a, 0)$$

is the local *Chern–Simons Lagrangian.*

For instance, if $\dim X = 3$, the global Chern–Simons Lagrangian (11.7) reads

$$L_{CS}(A) = \left[\frac{1}{2}h_{mn}\varepsilon^{\alpha\beta\gamma}a^m_\alpha(\mathscr{F}^n_{\beta\gamma} - \frac{1}{3}c^n_{pq}a^p_\beta a^q_\gamma)\right]\omega \tag{11.8}$$
$$-\left[\frac{1}{2}h_{mn}\varepsilon^{\alpha\beta\gamma}A^m_\alpha(F_A{}^n_{\beta\gamma} - \frac{1}{3}c^n_{pq}A^p_\beta A^q_\gamma)\right]\omega - d_\alpha(h_{mn}\varepsilon^{\alpha\beta\gamma}a^m_\beta A^n_\gamma)\omega,$$

where $\varepsilon^{\alpha\beta\gamma}$ is the skew-symmetric Levi–Civita tensor.

Since the density

$$-\mathfrak{S}_{2k-1}(A, 0) + d_H h_0 K_{2k-1}$$

is variationally trivial, the global Chern–Simons Lagrangian (11.7) possesses the same NI and gauge symmetries as the local one (11.7). They are the following.

Infinitesimal generators of local one-parameter groups of principal automorphisms of a principal bundle P are equivariant vector fields on P. They are identified with sections ξ (8.16) of the vector bundle $T_G P \to X$ (8.12), and yield the principal vector fields u_ξ (8.45) on the bundle of principal connections C. Sections ξ (8.16) play a role of gauge parameters.

Lemma 11.1 *Principal vector fields (8.45) locally are symmetries of the global Chern–Simons Lagrangian $L_{CS}(B)$ (11.7).*

Proof Since $\dim X = 2k - 1$, the transgression formula (11.4) takes a form

$$P_{2k}(F_{\mathscr{A}}) = d\mathfrak{S}_{2k-1}(a, A).$$

The Lie derivative $\mathbf{L}_{J^1 \upsilon}$ acting on its sides results in an equality

$$0 = d(u_\xi \rfloor d\mathfrak{S}_{2k-1}(a, A)) = d(\mathbf{L}_{J^1 u_\xi} \mathfrak{S}_{2k-1}(a, A)),$$

i.e., the Lie derivative $\mathbf{L}_{J^1 u_\xi} \mathfrak{S}_{2k-1}(a, A)$ locally is d-exact. Consequently, the horizontal form $h_0 \mathbf{L}_{J^1 u_\xi} \mathfrak{S}_{2k-1}(a, A)$ locally is d_H-exact. A direct computation shows that

$$h_0 \mathbf{L}_{J^1 u_\xi} \mathfrak{S}_{2k-1}(a, A) = \mathbf{L}_{J^1 u_\xi}(h_0 \mathfrak{S}_{2k-1}(a, A)) + d_H S.$$

It follows that the Lie derivative $\mathbf{L}_{J^1 \upsilon} L_{CS}(A)$ of the global Chern–Simons Lagrangian along any vector field u_ξ (8.45) locally is d_H-exact, i.e., this vector field is locally a symmetry of $L_{CS}(A)$.

By virtue of item (iii) of Theorem 2.4, a vertical part

$$u_V = (-c^r_{pq} \xi^p a^q_\lambda + \partial_\lambda \xi^r - a^r_\mu \partial_\lambda \xi^\mu - \xi^\mu a^r_{\mu\lambda}) \partial^\lambda_r \tag{11.9}$$

of the vector field u_ξ (8.45) locally is a symmetry of $L_{CS}(A)$, too.

Given the fibre bundle $T_G P \to X$ (8.12), let the same symbol also stand for the pull-back of $T_G P$ onto C. Let us consider the DBGA (7.6):

$$\mathscr{P}^*_\infty[T_G P; C] = \mathscr{S}^*_\infty[T_G P; C],$$

possessing the local generating basis $(a^r_\lambda, c^\lambda, c^r)$ of even fields a^r_λ and odd ghosts c^λ, c^r. Substituting these ghosts for gauge parameters in the vector field u_V (11.9) (Remark 7.7), we obtain the odd vertical graded derivation

$$u = (-c^r_{pq} c^p a^q_\lambda + c^r_\lambda - c^\mu_\lambda a^r_\mu - c^\mu a^r_{\mu\lambda}) \partial^\lambda_r \tag{11.10}$$

of a DBGA $\mathscr{P}^*_\infty[T_G P; C]$. This graded derivation as like as the vector fields υ_V (11.9) locally is a symmetry of the global Chern–Simons Lagrangian $L_{CS}(A)$ (11.7), i.e., an odd density $\mathbf{L}_{J^1 u}(L_{CS}(A))$ locally is d_H-exact. Hence, it is δ-closed and, consequently, d_H-exact in accordance with Theorem 6.12. Thus, we have proved the following.

Lemma 11.2 *The graded derivation u (11.10) is a gauge symmetry of the global Chern–Simons Lagrangian $L_{CS}(A)$ (11.7), and the vector fields u_V (11.9) and u_ξ (8.45) are well.*

Therefore, let us turn to Noether's first theorem.

Let us consider the vertical principal vector field ξ (8.44). In accordance with Lemma 11.2, the corresponding vertical principal vector field u_ξ (8.46) is a gauge symmetry of the Chern–Simons Lagrangian (11.7), i.e.,

$$\mathbf{L}_{J^1 u_\xi} L_{CS}(A) = d_H \sigma.$$

As a consequence, Chern–Simons topological field theory with the Lagrangian (11.7) admits a weak conservation law

$$0 \approx -d_H(\mathscr{J}_{u_\xi} + \sigma) \tag{11.11}$$

of a symmetry current $\mathscr{J}_{u_\xi} + \sigma$ where

$$\mathscr{J}_{u_\xi} = -\partial_r^{\lambda\mu}[\mathscr{L}_{CS}(A)](\partial_\mu \xi^r + c^r_{pq} a^p_\mu \xi^q)$$

is the Noether current (3.32) along a vertical principal vector field u_ξ. By virtue of Noether's third Theorem 2.8, a conserved current $\mathscr{J}_{u_\xi} + \sigma$ is reduced to a superpotential.

Let us study energy-momentum conservation laws in Chern–Simons topological field theory. Given a gauge field A, lets us consider the lift $A\tau$ (8.67) of vector field τ on X onto the bundle of principal connection C. It is a principal vector field on C and, thus, is a symmetry of the Chern–Simons Lagrangian (11.7) by virtue of Lemma 11.2. Then we come to a weak conservation law

$$0 \approx -d_H(\mathscr{J}_A + \sigma') \tag{11.12}$$

of a symmetry current $\mathscr{J}_A + \sigma'$ where

$$\mathscr{J}_A^\lambda = \partial_r^{\lambda\mu} \mathscr{L}_{CS}[\tau^\nu a^r_{\nu\mu} - \partial_\mu(A^r_\nu \tau^\nu) - c^r_{pq} a^p_\mu A^q_\nu \tau^\nu + a^r_\nu \partial_\mu \tau^\nu] - \tau^\lambda \mathscr{L}_{CS}$$

is an energy-momentum current along the vector field $A\tau$ (8.67).

Turn now to Noether's second theorem where the graded vector field u (11.10) is a gauge symmetry. By virtue of the formulas (2.18)–(2.19), the corresponding NI read

$$\overline{\delta}\Delta_j = -c^r_{ji} a^i_\lambda \mathscr{E}^\lambda_r - d_\lambda \mathscr{E}^\lambda_j = 0, \tag{11.13}$$

$$\overline{\delta}\Delta_\mu = -a^r_{\mu\lambda} \mathscr{E}^\lambda_r + d_\lambda(a^r_\mu \mathscr{E}^\lambda_r) = 0. \tag{11.14}$$

They are irreducible and nontrivial, unless dim $X = 3$. Therefore, the gauge operator (7.39) is $\mathbf{u} = u$. It admits a nilpotent BRST extension

$$\mathbf{b} = (-c^r_{ji}c^j a^i_\lambda + c^r_\lambda - c^\mu_\lambda a^r_\mu - c^\mu a^r_{\mu\lambda})\frac{\partial}{\partial a^r_\lambda} - \frac{1}{2}c^r_{ij}c^i c^j \frac{\partial}{\partial c^r} + c^\lambda_\mu c^\mu \frac{\partial}{\partial c^\lambda}. \quad (11.15)$$

In order to include antifields $(\overline{a}^\lambda_r, \overline{c}_r, \overline{c}_\mu)$, let us enlarge a DBGA $\mathscr{P}^*_\infty[T_G P; C]$ to a DBGA

$$\mathscr{P}^*_\infty\{0\} = \mathscr{S}^*_\infty[\overline{VC} \underset{C}{\oplus} T_G P; C \underset{X}{\times} \overline{T_G P}]$$

where $\overline{VC}$ is the density dual (8.85) of the vertical tangent bundle VC of $C \to X$ and $\overline{T_G P}$ is the density dual of $T_G P \to X$ (cf. (8.87)). By virtue of Theorem 7.16, given the BRST operator $\mathbf{b}$ (11.15), the global Chern–Simons Lagrangian $L_{CS}(A)$ (11.7) is extended to the proper solution (7.86) of the master equation which reads

$$L_E = L_{CS}(A) + [(-c^r_{pq}c^p a^q_\lambda + c^r_\lambda - c^\mu_\lambda a^r_\mu - c^\mu a^r_{\mu\lambda})\overline{a}^\lambda_r - \frac{1}{2}c^r_{ij}c^i c^j \overline{c}_r + c^\lambda_\mu c^\mu \overline{c}_\lambda]\omega.$$

If dim $X = 3$, the global Chern–Simons Lagrangian takes the form (11.8). Its Euler–Lagrange operator is

$$\delta L_{CS}(B) = \mathscr{E}^\lambda_r \theta^r_\lambda \wedge \omega, \qquad \mathscr{E}^\lambda_r = h_{rp}\varepsilon^{\lambda\beta\gamma}\mathscr{F}^p_{\beta\gamma}.$$

A glance at the NI (11.13)–(11.14) shows that they are equivalent to NI

$$\begin{aligned} \overline{\delta}\Delta_j &= -c^r_{ji}a^i_\lambda \mathscr{E}^\lambda_r - d_\lambda \mathscr{E}^\lambda_j = 0, \\ \overline{\delta}\Delta'_\mu &= \overline{\delta}\Delta_\mu + a^r_\mu \overline{\delta}\Delta_r = c^\mu \mathscr{F}^r_{\lambda\mu}\mathscr{E}^\lambda_r = 0. \end{aligned} \quad (11.16)$$

These NI define the gauge symmetry u (11.10) written in a form

$$u = (-c^r_{pq}c'^p a^q_\lambda + c'^r_\lambda + c^\mu \mathscr{F}^r_{\lambda\mu})\partial^\lambda_r \quad (11.17)$$

where $c'^r = c^r - a^r_\mu c^\mu$. It is readily observed that, if dim $X = 3$, the NI $\overline{\delta}\Delta'_\mu$ (11.16) are trivial. Then the corresponding part $c^\mu \mathscr{F}^r_{\lambda\mu}\partial^\lambda_r$ of the gauge symmetry u (11.17) also is trivial. Consequently, the nontrivial gauge symmetry of the Chern–Simons Lagrangian (11.8) is

$$u = (-c^r_{pq}c'^p a^q_\lambda + c'^r_\lambda)\partial^\lambda_r.$$

Chapter 12
Topological BF Theory

We consider topological BF theory of two exterior forms A and B of form degree $|A| + |B| = \dim X - 1$ on a smooth manifold X [16]. It is reducible degenerate Lagrangian theory which satisfies the homology regularity condition (Definition 7.2) [11, 137]. Its dynamic variables A and B are sections of a fibre bundle

$$Y = \overset{p}{\wedge} T^*X \oplus \overset{q}{\wedge} T^*X, \qquad p + q = n - 1 > 1,$$

coordinated by $(x^\lambda, A_{\mu_1 \dots \mu_p}, B_{\nu_1 \dots \nu_q})$. Without a loss of generality, let q be even and $q \geq p$. The corresponding DGA is $\mathcal{O}^*_\infty Y$ (1.7).

There are the canonical p- and q-forms

$$A = A_{\mu_1 \dots \mu_p} dx^{\mu_1} \wedge \dots \wedge dx^{\mu_p}, \qquad B = B_{\nu_1 \dots \nu_q} dx^{\nu_1} \wedge \dots \wedge dx^{\nu_q}$$

on Y. A Lagrangian of topological BF theory is

$$L_{\mathrm{BF}} = A \wedge d_H B = \varepsilon^{\mu_1 \dots \mu_n} A_{\mu_1 \dots \mu_p} d_{\mu_{p+1}} B_{\mu_{p+2} \dots \mu_n} \omega, \tag{12.1}$$

where ε is the Levi–Civita symbol. It is a reduced first order Lagrangian whose Euler–Lagrange operator (3.2) is of first order. It reads

$$\delta L = \mathcal{E}_A^{\mu_1 \dots \mu_p} dA_{\mu_1 \dots \mu_p} \wedge \omega + \mathcal{E}_B^{\nu_{p+2} \dots \nu_n} dB_{\nu_{p+2} \dots \nu_n} \wedge \omega, \tag{12.2}$$

$$\mathcal{E}_A^{\mu_1 \dots \mu_p} = \varepsilon^{\mu_1 \dots \mu_n} d_{\mu_{p+1}} B_{\mu_{p+2} \dots \mu_n}, \tag{12.3}$$

$$\mathcal{E}_B^{\mu_{p+2} \dots \mu_n} = -\varepsilon^{\mu_1 \dots \mu_n} d_{\mu_{p+1}} A_{\mu_1 \dots \mu_p}. \tag{12.4}$$

The corresponding Euler–Lagrange equation can be written in a form

$$d_H B = 0, \qquad d_H A = 0. \tag{12.5}$$

G. Sardanashvily, *Noether's Theorems*, Atlantis Studies
in Variational Geometry 3, DOI 10.2991/978-94-6239-171-0_12

It obeys the NI

$$d_H d_H B = 0, \qquad d_H d_H A = 0. \tag{12.6}$$

One can regard the components $\mathscr{E}_A^{\mu_1\ldots\mu_p}$ (12.3) and $\mathscr{E}_B^{\mu_{p+2}\ldots\mu_n}$ (12.4) of the Euler–Lagrange operator (12.2) as a $(\overset{p}{\wedge} TX)\otimes_X(\overset{n}{\wedge} T^*X)$-valued differential operator on a fibre bundle $\overset{q}{\wedge} T^*X$ and a $(\overset{q}{\wedge} TX)\otimes_X(\overset{n}{\wedge} T^*X)$-valued differential operator on a fibre bundle $\overset{p}{\wedge} T^*X$, respectively. They are of the same type as the $\overset{n-1}{\wedge} TX$-valued differential operator (D.19) in Example in Appendix D (cf. the equations (12.5)) and (D.18). Therefore, the analysis of NI of the differential operators (12.3) and (12.4) is a repetition of that of NI of the operator (D.19) (cf. NI (12.6) and (D.23)).

Following Example in Appendix D, let us consider the family of vector bundles

$$\begin{aligned}
&E_k = \overset{p-k-1}{\wedge} T^*X \underset{X}{\times} \overset{q-k-1}{\wedge} T^*X, \qquad 0\leq k < p-1,\\
&E_k = \mathbb{R} \underset{X}{\times} \overset{q-p}{\wedge} T^*X, \qquad k = p-1,\\
&E_k = \overset{q-k-1}{\wedge} T^*X, \quad p-1 < k < q-1,\\
&E_{q-1} = X\times\mathbb{R}.
\end{aligned}$$

Let us enlarge the DGA $\mathscr{O}^*_\infty Y$ to the DBGA $P^*_\infty\{q-1\}$ (7.33) which is

$$P^*_\infty\{q-1\} = \mathscr{P}^*_\infty[\overline{VY}\underset{Y}{\oplus} E_0\oplus\cdots\underset{Y}{\oplus} E_{q-1}\underset{Y}{\oplus}\overline{E}_0\underset{Y}{\oplus}\cdots\underset{Y}{\oplus}\overline{E}_{q-1};Y].$$

It possesses the local generating basis

$$\begin{aligned}
&\{A_{\mu_1\ldots\mu_p}, B_{\nu_1\ldots\nu_q}, \varepsilon_{\mu_2\ldots\mu_p},\ldots,\varepsilon_{\mu_p},\varepsilon,\xi_{\nu_2\ldots\nu_q},\ldots,\xi_{\nu_q},\xi,\\
&\quad \overline{A}^{\mu_1\ldots\mu_p}, \overline{B}^{\nu_1\ldots\nu_q}, \overline{\varepsilon}^{\mu_2\ldots\mu_p},\ldots,\overline{\varepsilon}^{\mu_p},\overline{\varepsilon},\overline{\xi}^{\nu_2\ldots\nu_q},\ldots,\overline{\xi}^{\nu_q},\overline{\xi}\}
\end{aligned}$$

of Grassmann parity

$$\begin{aligned}
&[\varepsilon_{\mu_k\ldots\mu_p}] = [\xi_{\nu_k\ldots\nu_q}] = (k+1)\,\mathrm{mod}\,2, \qquad [\varepsilon] = p\,\mathrm{mod}\,2, \qquad [\xi]=0,\\
&[\overline{\varepsilon}^{\mu_k\ldots\mu_p}] = [\overline{\xi}^{\nu_k\ldots\nu_q}] = k\,\mathrm{mod}\,2, \qquad [\overline{\varepsilon}] = (p+1)\,\mathrm{mod}\,2, \qquad [\overline{\xi}] = 1,
\end{aligned}$$

of ghost number

$$\mathrm{gh}[\varepsilon_{\mu_k\ldots\mu_p}] = \mathrm{gh}[\xi_{\nu_k\ldots\nu_q}] = k, \qquad \mathrm{gh}[\varepsilon] = p+1, \qquad \mathrm{gh}[\xi] = q+1,$$

and of antifield number

$$\begin{aligned}
&\mathrm{Ant}[\overline{A}^{\mu_1\ldots\mu_p}] = \mathrm{Ant}[\overline{B}^{\nu_{p+1}\ldots\nu_q}] = 1,\\
&\mathrm{Ant}[\overline{\varepsilon}^{\mu_k\ldots\mu_p}] = \mathrm{Ant}[\overline{\xi}^{\nu_k\ldots\nu_q}] = k+1,
\end{aligned}$$

$$\mathrm{Ant}[\overline{\varepsilon}] = p, \qquad \mathrm{Ant}[\overline{\varepsilon}] = q.$$

One can show that the homology regularity condition holds (see Lemma D.5) and that the DBGA $P^*_\infty\{q-1\}$ is endowed with the KT operator

$$\delta_{\mathrm{KT}} = \frac{\partial}{\partial \overline{A}^{\mu_1\ldots\mu_p}}\mathscr{E}_A^{\mu_1\ldots\mu_p} + \frac{\partial}{\partial \overline{B}^{\nu_1\ldots\nu_q}}\mathscr{E}_B^{\nu_1\ldots\nu_q} +$$
$$\sum_{2\leq k\leq p}\frac{\partial}{\partial\overline{\varepsilon}^{\mu_k\ldots\mu_p}}\Delta_A^{\mu_k\ldots\mu_p} + \frac{\partial}{\partial\overline{\varepsilon}}d_{\mu_p}\overline{\varepsilon}^{\mu_p} + \sum_{2\leq k\leq q}\frac{\partial}{\partial\overline{\xi}^{\nu_k\ldots\nu_q}}\Delta_B^{\nu_k\ldots\nu_q} + \frac{\partial}{\partial\overline{\xi}}d_{\nu_q}\overline{\xi}^{\nu_q},$$
$$\Delta_A^{\mu_2\ldots\mu_p} = d_{\mu_1}\overline{A}^{\mu_1\ldots\mu_p}, \qquad \Delta_A^{\mu_{k+1}\ldots\mu_p} = d_{\mu_k}\overline{\varepsilon}^{\mu_k\mu_{k+1}\ldots\mu_p}, \qquad 2\leq k<p,$$
$$\Delta_B^{\nu_2\ldots\nu_q} = d_{\nu_1}\overline{B}^{\nu_1\ldots\nu_q}, \quad \Delta_B^{\nu_{k+1}\ldots\nu_q} = d_{\nu_k}\overline{\xi}^{\nu_k\nu_{k+1}\ldots\nu_q}, \quad 2\leq k<q.$$

Its nilpotentness provides the complete NI (12.5):

$$d_{\mu_1}\mathscr{E}_A^{\mu_1\ldots\mu_p} = 0, \qquad d_{\nu_1}\mathscr{E}_B^{\nu_1\ldots\nu_q} = 0,$$

and the $(k-1)$-stage ones

$$d_{\mu_k}\Delta_A^{\mu_k\ldots\mu_p} = 0, \qquad k = 2,\ldots,p,$$
$$d_{\nu_k}\Delta_B^{\nu_k\ldots\nu_q} = 0, \qquad k = 2,\ldots,q,$$

(cf. (D.25)). It follows that the topological BF theory is $(q-1)$-reducible.

Applying Noether's inverse second Theorem 7.9, one obtains the gauge operator (7.39) which reads

$$\mathbf{u} = d_{\mu_1}\varepsilon_{\mu_2\ldots\mu_p}\frac{\partial}{\partial A_{\mu_1\mu_2\ldots\mu_p}} + d_{\nu_1}\xi_{\nu_2\ldots\nu_q}\frac{\partial}{\partial B_{\nu_1\nu_2\ldots\nu_q}} + \tag{12.7}$$
$$\left[d_{\mu_2}\varepsilon_{\mu_3\ldots\mu_p}\frac{\partial}{\partial\varepsilon_{\mu_2\mu_3\ldots\mu_p}} + \cdots + d_{\mu_p}\varepsilon\frac{\partial}{\partial\varepsilon_{\mu_p}}\right] +$$
$$\left[d_{\nu_2}\xi_{\nu_3\ldots\nu_q}\frac{\partial}{\partial\xi_{\nu_2\nu_3\ldots\nu_q}} + \cdots + d_{\nu_q}\xi\frac{\partial}{\partial\xi_{\nu_q}}\right].$$

In particular, a gauge symmetry of the Lagrangian L_{BF} (12.1) is

$$u = d_{\mu_1}\varepsilon_{\mu_2\ldots\mu_p}\frac{\partial}{\partial A_{\mu_1\mu_2\ldots\mu_p}} + d_{\nu_1}\xi_{\nu_2\ldots\nu_q}\frac{\partial}{\partial B_{\nu_1\nu_2\ldots\nu_q}}.$$

This gauge symmetry is Abelian. It also is readily observed that higher-stage gauge symmetries are independent of original fields. Consequently, topological BF theory is Abelian, and its gauge operator **u** (12.7) is nilpotent. Thus, it is the BRST operator

$\mathbf{b} = \mathbf{u}$. As a result, the Lagrangian L_{BF} is extended to the proper solution of the master equation $L_E = L_e$ (7.40) which reads

$$L_e = L_{\mathrm{BF}} + \varepsilon_{\mu_2 \ldots \mu_p} d_{\mu_1} \overline{A}^{\mu_1 \ldots \mu_p} + \sum_{1<k<p} \varepsilon_{\mu_{k+1} \ldots \mu_p} d_{\mu_k} \overline{\varepsilon}^{\mu_k \ldots \mu_p} + \varepsilon d_{\mu_p} \overline{\varepsilon}^{\mu_p} +$$
$$\xi_{\nu_2 \ldots \nu_q} d_{\nu_1} \overline{B}^{\nu_1 \ldots \nu_q} + \sum_{1<k<q} \xi_{\nu_{k+1} \ldots \nu_q} d_{\nu_k} \overline{\xi}^{\nu_k \ldots \mu_q} + \xi d_{\nu_q} \overline{\xi}^{\nu_q}.$$

Glossary

Acronyms

BRST	Becchi–Rouet–Stora–Tyitin
CIS	Completely integrable system
DBGA	Differential bigraded algebra
DGA	Differential graded algebra
KT	Koszul–Tate
NI	Noether identities
SUSY	Supersymmetry

Notation

$C^\infty(Z)$	ring of smooth real functions on a smooth manifold Z
∂_A^B	partial derivative with respect to coordinates with indices ${}^B_A$
d_λ, d_Λ	total derivative and multiderivative
d	exterior differential
d_H	total (horizontal) differential
d_V	vertical differential
D_Γ	covariant differential with respect to a connection Γ
$\mathfrak{d}\mathcal{A}$	derivation module of a ring $\mathcal{A}$
$\{\partial_\lambda\}$, $\{dx^\lambda\}$	holonomic frame and coframe
E^*	the dual of a vector bundle E
$\mathcal{E}_L$	Euler–Lagrange operator of a Lagrangian L
f^*Y	pull-back bundle
$f^*\phi$	pull-back exterior form
$\mathfrak{g}_l$, $\mathfrak{g}_r$	left and right Lie algebras
h_0	horizontal projection
$J^1\Phi$	jet prolongation of a bundle morphism Φ

G. Sardanashvily, *Noether's Theorems*, Atlantis Studies
in Variational Geometry 3, DOI 10.2991/978-94-6239-171-0

$J^1 s$	jet prolongation of a section s
$J^1 u$	jet prolongation of a vector filed u
$J^k Y$	k-order jet manifold
$J^\infty Y$	infinite order jet manifold
$\mathcal{J}_\upsilon$	symmetry current along a generalized vector field υ
LX	frame bundle
$\mathcal{O}^* \mathcal{A}$	de Rham complex of a commutative ring $\mathcal{A}$
$\mathcal{O}^*(Z)$	DGA of exterior forms on a smooth manifold Z
$\mathcal{O}^*_\infty Y$	DGA of exterior forms on all jet manifolds of a fibre bundle Y
TZ, T^*Z	tangent and cotangent bundles of a manifold Z
$\mathcal{T}(Z)$	real Lie algebra of vector fields on a manifold Z
Tf	tangent prolongation of a manifold morphism f
VY, V^*Y	vertical tangent and cotangent bundles of a fibre bundle Y
$(Z, \mathfrak{U}_E)$	simple graded manifold modelled over a vector bundle $E \to Z$
$\mathcal{S}^*(E, Z)$	DBGA of exterior forms on a simple graded manifold $(Z, \mathfrak{U}_E)$
$Y(X)$	vector space of global sections of a vector bundle $Y \to X$
$(Y, F, \mathfrak{U}_F)$	graded bundle modelled over a composite bundle $F \to Y \to X$
$\mathcal{S}^*_\infty(F, Y)$	DBGA of exterior forms on all jet manifolds of a graded bundle
δ	variational operator
ω	local volume form
θ^i_Λ	local contact one-form
$\widetilde{\tau}$	functorial lift of a vector field τ
$\Gamma\tau$	horizontal lift of a vector field τ
$\mathbf{1}$	unit element of algebras and groups
$\widehat{0}$	canonical zero-valued section of a vector bundle
$\otimes P, \wedge P$	tensor and exterior algebras of a module P
$\wedge E$	exterior bundle
$\lvert . \rvert$	form degree
$[.]$	Grassmann parity

Convention

Throughout the book, all algebras, except the Lie and graded Lie ones, are associative, and modules over commutative algebras are commutative bimodules.

Unless otherwise stated, all morphisms are smooth (i.e., of class C^∞), and manifolds are smooth real and finite-dimensional. They are Hausdorff second-countable and, consequently, locally compact and paracompact topological spaces (Remark B.1).

Only proper covers throughout the book are considered (Remark C.4).

By the gradation is meant the $\mathbb{N}$-gradation, unless it is the Grassmann $\mathbb{Z}_2$-gradation.

The standard symbols $\otimes$, $\vee$, and $\wedge$ stand for the tensor, symmetric, and exterior products, respectively. The interior product (contraction) is denoted by $\rfloor$.

Appendix A
Differential Calculus over Commutative Rings

In this Appendix, the relevant basics on differential operators and the Chevalley–Eilenberg differential calculus over commutative rings are summarized [58, 85, 131].

The differential calculus over a ring $C^\infty(X)$ of smooth real functions on a manifold X especially is considered (Sect. A.4). A key point is the well-known Serre–Swan theorem (Theorem A.10) which states the categorial equivalence between the vector bundles over a manifold X and the projective $C^\infty(X)$-modules of finite rank. In particular, the Chevalley–Eilenberg differential calculus over a real ring $C^\infty(X)$ is just the DGA of exterior forms on a smooth manifold X.

A.1 Commutative Algebra

This section addresses the algebraic calculus over commutative algebras [93, 97].

Algebras and Modules

An *algebra* $\mathcal{A}$ is an additive group which is additionally provided with distributive multiplication. All algebras throughout the book are associative, unless they are the Lie and graded Lie ones. By a *ring* is meant a *unital* algebra, i.e., it contains the unit element $\mathbf{1} \neq 0$. A *field* is a commutative ring whose nonzero elements arc invertible.

A subset $\mathcal{I}$ of an algebra $\mathcal{A}$ is called the left (resp. right) *ideal* if it is a subgroup of an additive group $\mathcal{A}$ and $ab \in \mathcal{I}$ (resp. $ba \in \mathcal{I}$) for all $a \in \mathcal{A}$, $b \in \mathcal{I}$. If $\mathcal{I}$ is both a left and right ideal, it is termed the two-sided ideal. A two-sided ideal is a subalgebra, but a *proper* ideal (i.e., $\mathcal{I} \neq \mathcal{A}$) of a ring is not a subring.

Let $\mathcal{A}$ be a commutative ring. Of course, its ideals are two-sided. Its proper ideal is said to be *maximal* if it does not belong to another proper ideal. A commutative ring $\mathcal{A}$ is called *local* if it has a unique maximal ideal. This ideal consists of all non-invertible elements of $\mathcal{A}$. A proper two-sided ideal $\mathcal{I}$ of a commutative ring is

G. Sardanashvily, *Noether's Theorems*, Atlantis Studies
in Variational Geometry 3, DOI 10.2991/978-94-6239-171-0

termed *prime* if $ab \in \mathcal{I}$ implies either $a \in \mathcal{I}$ or $b \in \mathcal{I}$. Any maximal two-sided ideal is prime. Given a two-sided ideal $\mathcal{I} \subset \mathcal{A}$, an additive quotient group $\mathcal{A}/\mathcal{I}$ is an algebra, called the *quotient algebra*. If $\mathcal{A}$ is a ring, then $\mathcal{A}/\mathcal{I}$ is so.

Given an algebra $\mathcal{A}$, an additive group P is said to be the left (resp. right) $\mathcal{A}$-*module* if it is provided with distributive multiplication $\mathcal{A} \times P \to P$ by elements of $\mathcal{A}$ such that $(ab)p = a(bp)$ (resp. $(ab)p = b(ap)$) for all $a, b \in \mathcal{A}$ and $p \in P$. If $\mathcal{A}$ is a ring, one additionally assumes that $\mathbf{1}p = p = p\mathbf{1}$ for all $p \in P$. Left and right module structures are usually written by means of left and right multiplications $(a, p) \to ap$ and $(a, p) \to pa$, respectively. If P is both a left module over an algebra $\mathcal{A}$ and a right module over an algebra $\mathcal{A}'$, it is called the $(\mathcal{A} - \mathcal{A}')$-bimodule (an $\mathcal{A}$-bimodule if $\mathcal{A} = \mathcal{A}'$). If $\mathcal{A}$ is a commutative algebra, an $(\mathcal{A} - \mathcal{A})$-bimodule P is said to be *commutative* if $ap = pa$ for all $a \in \mathcal{A}$ and $p \in P$. Any left or right module over a commutative algebra $\mathcal{A}$ can be brought into a commutative bimodule. Therefore, unless otherwise stated, any module over a commutative algebra $\mathcal{A}$ is termed the $\mathcal{A}$-module.

A module over a field is called the *vector space*. If an algebra $\mathcal{A}$ is a module over a ring $\mathcal{K}$, it is said to be the $\mathcal{K}$-*algebra*. Any algebra can be seen as a $\mathbb{Z}$-algebra.

Remark A.1 Any $\mathcal{K}$-algebra $\mathcal{A}$ can be extended to a ring $\widetilde{\mathcal{A}}$ by the adjunction of the unit element $\mathbf{1}$ to $\mathcal{A}$. It is defined as a direct sum of $\mathcal{K}$-modules $\mathcal{K} \oplus \mathcal{A}$ provided with the multiplication

$$(\lambda_1, a_1)(\lambda_2, a_2) = (\lambda_1\lambda_2, \lambda_1 a_2 + \lambda_2 a_1 + a_1 a_2), \quad \lambda_1, \lambda_2 \in \mathcal{K}, \quad a_1, a_2 \in \mathcal{A}.$$

Elements of $\widetilde{\mathcal{A}}$ can be written as $(\lambda, a) = \lambda\mathbf{1} + a$, $\lambda \in \mathcal{K}$, $a \in \mathcal{A}$. Let us note that, if $\mathcal{A}$ is a ring, the unit element $\mathbf{1}_{\mathcal{A}}$ in $\mathcal{A}$ fails to be that in $\widetilde{\mathcal{A}}$. In this case, a ring $\widetilde{\mathcal{A}}$ is isomorphic to a product of $\mathcal{A}$ and an algebra $\mathcal{K}(\mathbf{1} - \mathbf{1}_{\mathcal{A}})$.

Hereafter, all associative algebras in this Appendix are assumed to be commutative, unless they are graded.

The following are standard constructions of new modules from the old ones.

- A *direct sum* $P_1 \oplus P_2$ of $\mathcal{A}$-modules P_1 and P_2 is an additive group $P_1 \times P_2$ provided with an $\mathcal{A}$-module structure

  $$a(p_1, p_2) = (ap_1, ap_2), \qquad p_{1,2} \in P_{1,2}, \qquad a \in \mathcal{A}.$$

  Let $\{P_i\}_{i\in I}$ be a set of modules. Their direct sum $\oplus P_i$ contains elements $(\ldots, p_i, \ldots)$ of a Cartesian product $\prod P_i$ so that $p_i \neq 0$ at most for a finite number of indices $i \in I$.
- A *tensor product* $P \otimes Q$ of $\mathcal{A}$-modules P and Q is an additive group which is generated by elements $p \otimes q$, $p \in P$, $q \in Q$, obeying relations

  $$(p + p') \otimes q = p \otimes q + p' \otimes q, \quad p \otimes (q + q') = p \otimes q + p \otimes q',$$
  $$pa \otimes q = p \otimes aq, \qquad p \in P, \qquad q \in Q, \qquad a \in \mathcal{A}.$$

It is provided with an $\mathscr{A}$-module structure

$$a(p \otimes q) = (ap) \otimes q = p \otimes (qa) = (p \otimes q)a.$$

If a ring $\mathscr{A}$ is treated as an $\mathscr{A}$-module, a tensor product $\mathscr{A} \otimes_{\mathscr{A}} Q$ is canonically isomorphic to Q via the assignment

$$\mathscr{A} \underset{\mathscr{A}}{\otimes} Q \ni a \otimes q \leftrightarrow aq \in Q.$$

- Given a submodule Q of an $\mathscr{A}$-module P, the quotient P/Q of an additive group P with respect to its subgroup Q also is provided with an $\mathscr{A}$-module structure. It is called the *quotient module*.
- A set $\mathrm{Hom}_{\mathscr{A}}(P, Q)$ of $\mathscr{A}$-linear morphisms of an $\mathscr{A}$-module P to an $\mathscr{A}$-module Q is naturally an $\mathscr{A}$-module. An $\mathscr{A}$-module $P^* = \mathrm{Hom}_{\mathscr{A}}(P, \mathscr{A})$ is termed the *dual* of an $\mathscr{A}$-module P. There is a natural monomorphism $P \to P^{**}$.

An $\mathscr{A}$-module P is called *free* if it admits a *basis*, i.e., a linearly independent subset $I \subset P$ spanning P such that each element of P has a unique expression as a linear combination of elements of I with a finite number of nonzero coefficients from an algebra $\mathscr{A}$. Any vector space is free. Any module is isomorphic to a quotient of a free module. A module is said to be *finitely generated* (or of *finite rank*) if it is a quotient of a free module with a finite basis.

One says that a module P is *projective* if it is a direct summand of a free module, i.e., there exists a module Q such that $P \oplus Q$ is a free module. A module P is projective if and only if $P = \mathbf{p}S$ where S is a free module and $\mathbf{p}$ is a projector of S, i.e., $\mathbf{p} \circ \mathbf{p} = \mathbf{p}$. If P is a projective module of finite rank over a ring, then its dual P^* is so, and P^{**} is isomorphic to P.

Sequences and Limits of Modules

Now we focus exposition on sequences, direct and inverse limits of modules [97, 104].

A composition of module morphisms

$$P \xrightarrow{i} Q \xrightarrow{j} T$$

is said to be *exact* at Q if $\mathrm{Ker}\, j = \mathrm{Im}\, i$. A composition of module morphisms

$$0 \to P \xrightarrow{i} Q \xrightarrow{j} T \to 0 \tag{A.1}$$

is called the *short exact sequence* if it is exact at all the terms P, Q, and T. This condition implies that: (i) i is a monomorphism, (ii) $\mathrm{Ker}\, j = \mathrm{Im}\, i$, and (iii) j is an epimorphism onto a quotient $T = Q/P$.

Theorem A.1 *Given an exact sequence of modules (A.1) and another $\mathcal{A}$-module R, the sequence of modules*

$$0 \to \mathrm{Hom}_{\mathcal{A}}(T, R) \xrightarrow{j^*} \mathrm{Hom}_{\mathcal{A}}(Q, R) \xrightarrow{i^*} \mathrm{Hom}\,(P, R)$$

is exact at the first and second terms, i.e., j^* is a monomorphism, but i^* need not be an epimorphism.

One says that the exact sequence (A.1) is *split* if there exists a monomorphism $s : T \to Q$ such that $j \circ s = \mathrm{Id}\, T$ or, equivalently,

$$Q = i(P) \oplus s(T) \cong P \oplus T.$$

Theorem A.2 *The exact sequence (A.1) is always split if T is a projective module.*

A *directed set* I is a set with an order relation $<$ which satisfies the following three conditions: (i) $i < i$, for all $i \in I$; (ii) if $i < j$ and $j < k$, then $i < k$; (iii) for any $i, j \in I$, there exists $k \in I$ such that $i < k$ and $j < k$. It may happen that $i \neq j$, but $i < j$ and $j < i$ simultaneously.

A family of $\mathcal{A}$-modules $\{P_i\}_{i \in I}$, indexed by a directed set I, is termed the *direct system* if, for any pair $i < j$, there exists a morphism $r^i_j : P_i \to P_j$ such that

$$r^i_i = \mathrm{Id}\, P_i, \qquad r^i_j \circ r^j_k = r^i_k, \qquad i < j < k.$$

A direct system of modules admits a *direct limit*.

Definition A.1 This is a module P_∞ together with morphisms $r^i_\infty : P_i \to P_\infty$ such that $r^i_\infty = r^j_\infty \circ r^i_j$ for all $i < j$. A module P_∞ consists of elements of a direct sum $\oplus_I P_i$ modulo the identification of elements of P_i with their images in P_j for all $i < j$.

An example of a direct system is a *direct sequence*

$$P_0 \longrightarrow P_1 \longrightarrow \cdots P_i \xrightarrow{r^i_{i+1}} \cdots, \qquad I = \mathbb{N}.$$

It should be noted that direct limits also exist in the categories of commutative algebras and rings, but not in a category of noncommutative groups.

Theorem A.3 *Direct limits commute with direct sums and tensor products of modules. Namely, let $\{P_i\}$ and $\{Q_i\}$ be two direct systems of modules over the same algebra which are indexed by the same directed set I, and let P_∞ and Q_∞ be their direct limits. Then direct limits of direct systems $\{P_i \oplus Q_i\}$ and $\{P_i \otimes Q_i\}$ are $P_\infty \oplus Q_\infty$ and $P_\infty \otimes Q_\infty$, respectively.*

Theorem A.4 *A morphism of a direct system $\{P_i, r^i_j\}_I$ to a direct system $\{Q_{i'}, \rho^{i'}_{j'}\}_{I'}$ consists of an order preserving map $f : I \to I'$ and morphisms $F_i : P_i \to Q_{f(i)}$ which obey compatibility conditions $\rho^{f(i)}_{f(j)} \circ F_i = F_j \circ r^i_j$. If P_∞ and Q_∞ are limits of these direct systems, there exists a unique morphism $F_\infty : P_\infty \to Q_\infty$ such that $\rho^{f(i)}_\infty \circ F_i = F_\infty \circ r^i_\infty$.*

Theorem A.5 *Moreover, direct limits preserve monomorphisms and epimorphisms. Namely, if all $F_i : P_i \to Q_{f(i)}$ are monomorphisms or epimorphisms, so is $\Phi_\infty : P_\infty \to Q_\infty$. Let short exact sequences*

$$0 \to P_i \xrightarrow{F_i} Q_i \xrightarrow{\Phi_i} T_i \to 0 \tag{A.2}$$

for all $i \in I$ define a short exact sequence of direct systems of modules $\{P_i\}_I$, $\{Q_i\}_I$, and $\{T_i\}_I$ which are indexed by the same directed set I. Then there exists a short exact sequence of their direct limits

$$0 \to P_\infty \xrightarrow{F_\infty} Q_\infty \xrightarrow{\Phi_\infty} T_\infty \to 0. \tag{A.3}$$

In particular, a direct limit of quotient modules Q_i/P_i is a quotient module Q_∞/P_∞. By virtue of Theorem A.3, if all the exact sequences (A.2) are split, the exact sequence (A.3) is well.

Remark A.2 Let P be an $\mathscr{A}$-module. We denote $P^{\otimes k} = \overset{k}{\otimes} P$. Let us consider a direct system of $\mathscr{A}$-modules with respect to monomorphisms

$$\mathscr{A} \longrightarrow (\mathscr{A} \oplus P) \longrightarrow \cdots (\mathscr{A} \oplus P \oplus \cdots \oplus P^{\otimes k}) \longrightarrow \cdots .$$

Its direct limit

$$\otimes P = \mathscr{A} \oplus P \oplus \cdots \oplus P^{\otimes k} \oplus \cdots \tag{A.4}$$

is an $\mathbb{N}$-graded $\mathscr{A}$-algebra with respect to a tensor product $\otimes$. It is called the *tensor algebra* of a module P. Its quotient

$$\wedge P = \mathscr{A} \oplus P \oplus \cdots \oplus \overset{k}{\wedge} P \oplus \cdots$$

by an ideal generated by elements $p \otimes p' + p' \otimes p$, $p, p' \in P$, is an $\mathbb{N}$-graded commutative algebra with respect to the *exterior product*

$$p \wedge p' = \frac{1}{2}(p \otimes p' - p' \otimes p).$$

It is called the *exterior algebra* of a module P.

We restrict our consideration of inverse systems of modules to *inverse sequences*

$$P^0 \longleftarrow P^1 \longleftarrow \cdots P^i \overset{\pi_i^{i+1}}{\longleftarrow} \cdots . \tag{A.5}$$

It admits an inverse limit.

Definition A.2 The *inverse limit* of the inverse sequence of modules (A.5) is a module P^∞ together with morphisms $\pi_i^\infty : P^\infty \to P^i$ so that $\pi_i^\infty = \pi_i^j \circ \pi_j^\infty$ for all $i < j$. It consists of elements $(\ldots, p^i, \ldots)$, $p^i \in P^i$, of the Cartesian product $\prod P^i$ such that $p^i = \pi_i^j(p^j)$ for all $i < j$.

Theorem A.6 *Inverse limits preserve monomorphisms, but not epimorphisms. If a sequence*

$$0 \to P^i \xrightarrow{F^i} Q^i \xrightarrow{\Phi^i} T^i, \qquad i \in \mathbb{N},$$

of inverse systems of modules $\{P^i\}$, $\{Q^i\}$ *and* $\{T^i\}$ *is exact, so is a sequence of inverse limits*

$$0 \to P^\infty \xrightarrow{F^\infty} Q^\infty \xrightarrow{\Phi^\infty} T^\infty.$$

In contrast with direct limits (Definition A.1), the inverse ones (Definition A.2) exist in the category of groups which are not necessarily commutative.

Remark A.3 Let $\{P_i\}$ be a direct sequence of modules. Given another module Q, modules $\mathrm{Hom}\,(P_i, Q)$ constitute an inverse system such that its inverse limit (Definition A.2) is isomorphic to $\mathrm{Hom}\,(P_\infty, Q)$.

Complexes

Turn now to complexes of modules over a commutative ring [97, 104].

Let $\mathcal{K}$ be a commutative ring. An inverse sequence

$$0 \leftarrow B_0 \overset{\partial_1}{\longleftarrow} B_1 \overset{\partial_2}{\longleftarrow} \cdots B_p \overset{\partial_{p+1}}{\longleftarrow} \cdots \tag{A.6}$$

of $\mathcal{K}$-modules B_p and homomorphisms ∂_p is said to be the *chain complex* if

$$\partial_p \circ \partial_{p+1} = 0, \qquad p \in \mathbb{N},$$

i.e., $\mathrm{Im}\,\partial_{p+1} \subset \mathrm{Ker}\,\partial_p$. Homomorphisms ∂_p are called the *boundary operators*. Elements of a module B_p, its submodules $\mathrm{Ker}\,\partial_p \subset B_p$ and $\mathrm{Im}\,\partial_{p+1} \subset \mathrm{Ker}\,\partial_p$ are termed *p-chains*, *p-cycles* and *p-boundaries*, respectively. The *p-homology group* of the chain complex B_* (A.6) is defined as the $\mathcal{K}$-quotient module

$$H_p(B_*) = \operatorname{Ker} \partial_p / \operatorname{Im} \partial_{p+1}.$$

The chain complex (A.6) is exact at a term B_p if and only if $H_p(B_*) = 0$. This complex is said to be *k-exact* if its homology groups $H_{p\leq k}(B_*)$ are trivial. It is called *exact* if all its homology groups are trivial, i.e., it is an exact sequence.

A direct sequence

$$0 \to B^0 \xrightarrow{\delta^0} B^1 \xrightarrow{\delta^1} \cdots B^p \xrightarrow{\delta^p} \cdots \tag{A.7}$$

of modules B^p and their homomorphisms δ^p is said to be the *cochain complex* (or, shortly, the *complex*) if $\delta^{p+1} \circ \delta^p = 0$, $p \in \mathbb{N}$, i.e., $\operatorname{Im} \delta^p \subset \operatorname{Ker} \delta^{p+1}$. Homomorphisms δ^p are termed the *coboundary operators*. For the sake of convenience, let us denote $\delta^{p=-1} : 0 \to B^0$. Elements of a module B^p, its submodules $\operatorname{Ker} \delta^p \subset B^p$ and $\operatorname{Im} \delta^{p-1}$ are called *p-cochains*, *p-cocycles* and *p-coboundaries*, respectively. A *p-cohomology group* of the complex B^* (A.7) is the quotient $\mathcal{K}$-module

$$H^p(B^*) = \operatorname{Ker} \delta^p / \operatorname{Im} \delta^{p-1}.$$

The complex (A.7) is exact at a term B^p if and only if $H^p(B^*) = 0$. This complex is an exact sequence if all its cohomology groups are trivial.

A complex (B^*, δ^*) is called *acyclic* if its cohomology groups $H^{0<p}(B^*)$ are trivial. It is acyclic if there exists a *homotopy operator* $\mathbf{h}$, defined as a set of module morphisms

$$\mathbf{h}^{p+1} : B^{p+1} \to B^p, \qquad p \in \mathbb{N},$$
$$\mathbf{h}^{p+1} \circ \delta^p + \delta^{p-1} \circ \mathbf{h}^p = \operatorname{Id} B^p, \qquad p \in \mathbb{N}_+.$$

Indeed, if $\delta^p b^p = 0$, then $b^p = \delta^{p-1}(\mathbf{h}^p b^p)$, and $H^{p>0}(B^*) = 0$. A complex (B^*, δ^*) is said to be a *resolution* of a module B if it is acyclic and $H^0(B^*) = \operatorname{Ker} \delta^0 = B$.

The following are the standard constructions of new complexes from the old ones.

- Given complexes (B_1^*, δ_1^*) and (B_2^*, δ_2^*), their *direct sum* $B_1^* \oplus B_2^*$ is a complex of modules

$$(B_1^* \oplus B_2^*)^p = B_1^p \oplus B_2^p$$

with respect to the coboundary operators

$$\delta_\oplus^p (b_1^p + b_2^p) = \delta_1^p b_1^p + \delta_2^p b_2^p.$$

- Given a subcomplex (C^*, δ^*) of a complex (B^*, δ^*), the *quotient complex* B^*/C^* is defined as a complex of quotient modules B^p/C^p provided with the coboundary operators $\delta^p[b^p] = [\delta^p b^p]$, where $[b^p] \in B^p/C^p$ denotes the coset of an element b^p.

- Given complexes (B_1^*, δ_1^*) and (B_2^*, δ_2^*), their *tensor product* $B_1^* \otimes B_2^*$ is a complex of modules

$$(B_1^* \otimes B_2^*)^p = \underset{k+r=p}{\oplus} B_1^k \otimes B_2^r$$

with respect to the coboundary operators

$$\delta_\otimes^p(b_1^k \otimes b_2^r) = (\delta_1^k b_1^k) \otimes b_2^r + (-1)^k b_1^k \otimes (\delta_2^r b_2^r).$$

A *cochain morphism* of complexes $\gamma : B_1^* \to B_2^*$ is defined as a family of degree-preserving homomorphisms

$$\gamma^p : B_1^p \to B_2^p, \qquad \delta_2^p \circ \gamma^p = \gamma^{p+1} \circ \delta_1^p, \qquad p \in \mathbb{N}.$$

It follows that if $b^p \in B_1^p$ is a cocycle or a coboundary, then $\gamma^p(b^p) \in B_2^p$ is so. Therefore, a cochain morphism of complexes yields an induced homomorphism of their cohomology groups

$$[\gamma]^* : H^*(B_1^*) \to H^*(B_2^*).$$

Let short exact sequences

$$0 \to C^p \xrightarrow{\gamma_p} B^p \xrightarrow{\zeta_p} F^p \to 0$$

for all $p \in \mathbb{N}$ define a short exact sequence of complexes

$$0 \to C^* \xrightarrow{\gamma} B^* \xrightarrow{\zeta} F^* \to 0, \tag{A.8}$$

where γ is a cochain monomorphism and ζ is a cochain epimorphism onto the quotient $F^* = B^*/C^*$.

Theorem A.7 *The short exact sequence of complexes (A.8) yields the long exact sequence of their cohomology groups*

$$0 \to H^0(C^*) \xrightarrow{[\gamma]^0} H^0(B^*) \xrightarrow{[\zeta]^0} H^0(F^*) \xrightarrow{\tau^0} H^1(C^*) \longrightarrow \cdots$$
$$\longrightarrow H^p(C^*) \xrightarrow{[\gamma]^p} H^p(B^*) \xrightarrow{[\zeta]^p} H^p(F^*) \xrightarrow{\tau^p} H^{p+1}(C^*) \longrightarrow \cdots .$$

Theorem A.8 *A direct sequence of complexes*

$$B_0^* \longrightarrow B_1^* \longrightarrow \cdots B_k^* \xrightarrow{\gamma_{k+1}^k} B_{k+1}^* \longrightarrow \cdots$$

admits a direct limit B^*_∞ *which is a complex whose cohomology* $H^*(B^*_\infty)$ *is a direct limit of the direct sequence of cohomology groups*

$$H^*(B^*_0) \longrightarrow H^*(B^*_1) \longrightarrow \cdots H^*(B^*_k) \stackrel{[\gamma^k_{k+1}]}{\longrightarrow} H^*(B^*_{k+1}) \longrightarrow \cdots .$$

A.2 Differential Operators on Modules and Rings

This section addresses the notion of a (linear) differential operator on a module over a commutative ring [53, 85, 131].

Let $\mathcal{K}$ be a commutative ring and $\mathcal{A}$ a commutative $\mathcal{K}$-ring. Let P and Q be $\mathcal{A}$-modules. A $\mathcal{K}$-module $\mathrm{Hom}_{\mathcal{K}}(P, Q)$ of $\mathcal{K}$-module homomorphisms $\Phi : P \to Q$ can be endowed with two different $\mathcal{A}$-module structures

$$(a\Phi)(p) = a\Phi(p), \qquad (\Phi \bullet a)(p) = \Phi(ap), \qquad a \in \mathcal{A}, \quad p \in P. \tag{A.9}$$

For the sake of convenience, we will refer to the second one as an $\mathcal{A}^\bullet$-module structure. Let us put

$$\delta_a \Phi = a\Phi - \Phi \bullet a, \qquad a \in \mathcal{A}.$$

Definition A.3 An element $\Delta \in \mathrm{Hom}_{\mathcal{K}}(P, Q)$ is called the (linear) Q-valued *differential operator* of order s on P if $\delta_{a_0} \circ \cdots \circ \delta_{a_s} \Delta = 0$ for any tuple of $s+1$ elements $a_0, \ldots, a_s$ of $\mathcal{A}$.

A set $\mathrm{Diff}_s(P, Q)$ of these operators inherits the $\mathcal{A}$- and $\mathcal{A}^\bullet$-module structures (A.9). Of course, an s-order differential operator also is of $(s+1)$-order.

In particular, zero order differential operators obey a condition

$$\delta_a \Delta(p) = a\Delta(p) - \Delta(ap) = 0, \qquad a \in \mathcal{A}, \qquad p \in P,$$

and, consequently, they coincide with $\mathcal{A}$-module morphisms $P \to Q$. A first order differential operator Δ satisfies a condition

$$\delta_b \circ \delta_a \, \Delta(p) = ba\Delta(p) - b\Delta(ap) - a\Delta(bp) + \Delta(abp) = 0, \quad a, b \in \mathcal{A}.$$

Let $P = \mathcal{A}$. Any zero order Q-valued differential operator Δ on $\mathcal{A}$ is defined by its value $\Delta(\mathbf{1})$. Then there is an isomorphism $\mathrm{Diff}_0(\mathcal{A}, Q) = Q$ via the association

$$Q \ni q \to \Delta_q \in \mathrm{Diff}_0(\mathcal{A}, Q),$$

where Δ_q is given by an equality $\Delta_q(\mathbf{1}) = q$. A first order Q-valued differential operator Δ on $\mathcal{A}$ fulfils a condition

$$\Delta(ab) = b\Delta(a) + a\Delta(b) - ba\Delta(\mathbf{1}), \qquad a, b \in \mathcal{A}.$$

Definition A.4 A first order Q-valued differential operator Δ on $\mathscr{A}$ is termed the Q-valued *derivation* of $\mathscr{A}$ if $\Delta(\mathbf{1}) = 0$, i.e., it obeys the *Leibniz rule*

$$\Delta(ab) = \Delta(a)b + a\Delta(b), \qquad a, b \in \mathscr{A}.$$

One obtains at once that any first order differential operator on $\mathscr{A}$ falls into a sum

$$\Delta(a) = a\Delta(\mathbf{1}) + [\Delta(a) - a\Delta(\mathbf{1})]$$

of a zero order differential operator $a\Delta(\mathbf{1})$ and a derivation $\Delta(a) - a\Delta(\mathbf{1})$. If ∂ is a derivation of $\mathscr{A}$, then $a\partial$ is well for any $a \in \mathscr{A}$. Hence, derivations of $\mathscr{A}$ constitute an $\mathscr{A}$-module $\mathfrak{d}(\mathscr{A}, Q)$, called the *derivation module*. There is an $\mathscr{A}$-module decomposition

$$\mathrm{Diff}_1(\mathscr{A}, Q) = Q \oplus \mathfrak{d}(\mathscr{A}, Q). \qquad \text{(A.10)}$$

If $Q = \mathscr{A}$, the derivation module $\mathfrak{d}\mathscr{A}$ of $\mathscr{A}$ also is a Lie algebra over a ring $\mathscr{K}$ with respect to a Lie bracket

$$[u, u'] = u \circ u' - u' \circ u, \qquad u, u' \in \mathscr{A}.$$

Accordingly, the decomposition (A.10) takes a form

$$\mathrm{Diff}_1(\mathscr{A}) = \mathscr{A} \oplus \mathfrak{d}\mathscr{A}.$$

A derivation u of A is termed *inner* if there exists an element $q \in \mathscr{A}$ such that $u(a) = qa - aq$.

Remark A.4 Let us note that, given a (noncommutative) $\mathscr{K}$-algebra $\mathscr{A}$, by a derivation of $\mathscr{A}$ usually is meant a $\mathscr{K}$-module morphism $u : \mathscr{A} \to \mathscr{A}$ which obeys the Leibniz rule

$$u(ab) = u(a)b + au(b), \qquad a, b \in \mathscr{A}.$$

However, this derivation rule differs from that (6.2) of derivations of a graded commutative ring and that in [93].

A.3 Chevalley–Eilenberg Differential Calculus

Any commutative $\mathscr{K}$-ring $\mathscr{A}$ admits the differential calculus as follows.

By a gradation throughout this section is meant the $\mathbb{N}$-gradation.

A *graded algebra* Ω^* over a commutative ring $\mathscr{K}$ is defined as a direct sum

$$\Omega^* = \bigoplus_k \Omega^k, \qquad k \in \mathbb{N},$$

of $\mathcal{K}$-modules Ω^k, provided with an associative multiplication law $\alpha \cdot \beta, \alpha, \beta \in \Omega^*$, such that $\alpha \cdot \beta \in \Omega^{|\alpha|+|\beta|}$, where $|\alpha|$ denotes the degree of an element $\alpha \in \Omega^{|\alpha|}$. In particular, it follows that Ω^0 is a (noncommutative) $\mathcal{K}$-algebra $\mathcal{A}$, while $\Omega^{k>0}$ are $\mathcal{A}$-bimodules and Ω^* is an $(\mathcal{A} - \mathcal{A})$-algebra. A graded algebra is said to be *graded commutative* if

$$\alpha \cdot \beta = (-1)^{|\alpha||\beta|}\beta \cdot \alpha, \qquad \alpha, \beta \in \Omega^*.$$

Definition A.5 A graded algebra Ω^* is called the *differential graded algebra* (DGA) if it is a cochain complex of $\mathcal{K}$-modules

$$0 \to \mathcal{K} \longrightarrow \mathcal{A} \xrightarrow{\delta} \Omega^1 \xrightarrow{\delta} \cdots \Omega^k \xrightarrow{\delta} \cdots \tag{A.11}$$

with respect to a coboundary operator δ which obeys the *graded Leibniz rule*

$$\delta(\alpha \cdot \beta) = \delta\alpha \cdot \beta + (-1)^{|\alpha|}\alpha \cdot \delta\beta. \tag{A.12}$$

In particular, $\delta : \mathcal{A} \to \Omega^1$ is a Ω^1-valued derivation of a $\mathcal{K}$-algebra $\mathcal{A}$.

The cochain complex (A.11) is termed the *de Rham complex* of a DGA (Ω^*, δ). This algebra also is said to be the *differential calculus* over $\mathcal{A}$. Cohomology $H^*(\Omega^*)$ of the complex (A.11) is called the *de Rham cohomology* of a DGA (Ω^*, δ).

A morphism γ between two DGAs (Ω^*, δ) and (Ω'^*, δ') is defined as a cochain morphism, i.e., $\gamma \circ \delta = \gamma \circ \delta'$. It yields the corresponding morphism of the de Rham cohomology groups of these algebras.

One considers a minimal differential graded subalgebra $\Omega^*\mathcal{A}$ of a DGA Ω^* which contains $\mathcal{A}$. Seen as an $(\mathcal{A} - \mathcal{A})$-algebra, it is generated by elements δa, $a \in \mathcal{A}$, and consists of monomials $\alpha = a_0\delta a_1 \cdots \delta a_k$, $a_i \in \mathcal{A}$, whose product obeys the *juxtaposition rule*

$$(a_0\delta a_1) \cdot (b_0\delta b_1) = a_0\delta(a_1 b_0) \cdot \delta b_1 - a_0 a_1 \delta b_0 \cdot \delta b_1$$

in accordance with the equality (A.12). A DGA $(\Omega^*\mathcal{A}, \delta)$ is termed the *minimal differential calculus* over $\mathcal{A}$.

Now, let us show that any commutative $\mathcal{K}$-ring $\mathcal{A}$ defines the differential calculus.

As was mentioned above, the derivation module $\mathfrak{d}\mathcal{A}$ of $\mathcal{A}$ also is a Lie $\mathcal{K}$-algebra. Given its Chevalley–Eilenberg complex $C^*[\mathfrak{d}\mathcal{A}; \mathcal{A}]$ [47, 131], let us consider the extended Chevalley–Eilenberg complex

$$0 \to \mathcal{K} \xrightarrow{\text{in}} C^*[\mathfrak{d}\mathcal{A}; \mathcal{A}]$$

of the Lie algebra $\mathfrak{d}\mathcal{A}$ with coefficients in a ring $\mathcal{A}$, regarded as a $\mathfrak{d}\mathcal{A}$-module [58, 131]. This complex contains a subcomplex $\mathcal{O}^*[\mathfrak{d}\mathcal{A}]$ of $\mathcal{A}$-multilinear skew-symmetric maps

$$\phi^k : \overset{k}{\times}\mathfrak{d}\mathcal{A} \to \mathcal{A}$$

with respect to the *Chevalley–Eilenberg coboundary operator*

$$d\phi(u_0, \ldots, u_k) = \sum_{i=0}^{k}(-1)^i u_i(\phi(u_0, \ldots, \widehat{u_i}, \ldots, u_k)) + \tag{A.13}$$
$$\sum_{i<j}(-1)^{i+j}\phi([u_i, u_j], u_0, \ldots, \widehat{u}_i, \ldots, \widehat{u}_j, \ldots, u_k).$$

Indeed, a direct verification shows that if ϕ is an $\mathcal{A}$-multilinear map, so is $d\phi$. In particular,

$$(da)(u) = u(a), \qquad a \in \mathcal{A}, \qquad u \in \mathfrak{d}\mathcal{A},$$
$$(d\phi)(u_0, u_1) = u_0(\phi(u_1)) - u_1(\phi(u_0)) - \phi([u_0, u_1]), \quad \phi \in \mathcal{O}^1[\mathfrak{d}\mathcal{A}],$$
$$\mathcal{O}^0[\mathfrak{d}\mathcal{A}] = \mathcal{A}, \qquad \mathcal{O}^1[\mathfrak{d}\mathcal{A}] = \mathrm{Hom}_{\mathcal{A}}(\mathfrak{d}\mathcal{A}, \mathcal{A}).$$

It follows that $d(\mathbf{1}) = 0$ and d is an $\mathcal{O}^1[\mathfrak{d}\mathcal{A}]$-valued derivation of $\mathcal{A}$.

A graded module $\mathcal{O}^*[\mathfrak{d}\mathcal{A}]$ is provided with the structure of a graded $\mathcal{A}$-algebra with respect to the product

$$\phi \wedge \phi'(u_1, ..., u_{r+s}) =$$
$$\sum_{i_1<\cdots<i_r;\, j_1<\cdots<j_s} \mathrm{sgn}^{i_1\cdots i_r j_1\cdots j_s}_{1\cdots r+s}\, \phi(u_{i_1}, \ldots, u_{i_r})\phi'(u_{j_1}, \ldots, u_{j_s}),$$
$$\phi \in \mathcal{O}^r[\mathfrak{d}\mathcal{A}], \qquad \phi' \in \mathcal{O}^s[\mathfrak{d}\mathcal{A}], \qquad u_k \in \mathfrak{d}\mathcal{A},$$

where $\mathrm{sgn}^{\cdots}_{\cdots}$ denotes the sign of a permutation. This product obeys the relations

$$d(\phi \wedge \phi') = d(\phi) \wedge \phi' + (-1)^{|\phi|}\phi \wedge d(\phi'), \quad \phi, \phi' \in \mathcal{O}^*[\mathfrak{d}\mathcal{A}],$$
$$\phi \wedge \phi' = (-1)^{|\phi||\phi'|}\phi' \wedge \phi. \tag{A.14}$$

By virtue of the first one, $(\mathcal{O}^*[\mathfrak{d}\mathcal{A}], d)$ is a $\mathcal{K}$-DGA, whereas the relation (A.14) shows that $\mathcal{O}^*[\mathfrak{d}\mathcal{A}]$ is a graded commutative algebra.

Definition A.6 A DGA $(\mathcal{O}^*[\mathfrak{d}\mathcal{A}], d)$ is called the *Chevalley–Eilenberg differential calculus* over a $\mathcal{K}$-ring $\mathcal{A}$ [58, 131].

Definition A.7 The *minimal Chevalley–Eilenberg differential calculus* $\mathcal{O}^*\mathcal{A}$ over a ring $\mathcal{A}$ consists of the monomials $a_0 da_1 \wedge \cdots \wedge da_k$, $a_i \in \mathcal{A}$.

Its complex

$$0 \to \mathcal{K} \longrightarrow \mathcal{A} \xrightarrow{d} \mathcal{O}^1\mathcal{A} \xrightarrow{d} \cdots \mathcal{O}^k\mathcal{A} \xrightarrow{d} \cdots \tag{A.15}$$

is termed the *de Rham complex* of a $\mathcal{K}$-ring $\mathcal{A}$, and its cohomology $H^*(\mathcal{A})$ is said to be the *de Rham cohomology* of $\mathcal{A}$.

A.4 Differential Calculus over $C^\infty(X)$. Serre–Swan Theorem

Let X be a smooth manifold (Remark B.1) and $C^\infty(X)$ a real *ring of smooth real functions* on X.

Let C^∞_X be a sheaf of germs of local smooth real functions on a manifold X (Remark C.1). Its structure module of global sections is a ring $C^\infty(X)$. Similarly to a sheaf C^0_X of germs of continuous real functions, a stalk C^∞_x of a sheaf C^∞_X at a point $x \in X$ has a unique maximal ideal of germs of smooth functions vanishing at x. Consequently, C^∞_X is a local-ringed space (Remark C.7). Though a sheaf C^∞_X is defined on a topological space X, it fixes a unique smooth manifold structure on X as follows.

Theorem A.9 *Let X be a paracompact topological space and $(X, \mathfrak{A})$ a local-ringed space. Let X admit an open cover $\{U_i\}$ such that a sheaf $\mathfrak{A}$ restricted to each U_i is isomorphic to a local-ringed space $(\mathbb{R}^n, C^\infty_{R^n})$. Then X is an n-dimensional smooth manifold together with a natural isomorphism of local-ringed spaces $(X, \mathfrak{A})$ and (X, C^∞_X).*

One can think of this result as being an alternative definition of smooth real manifolds in terms of local-ringed spaces. A smooth manifold X also is algebraically reproduced as a certain subspace of the spectrum of a real ring $C^\infty(X)$ of smooth real functions on X [4, 58]. In particular, there is one-to-one correspondence between the manifold morphisms $X \to X'$ and the $\mathbb{R}$-ring morphisms $C^\infty(X') \to C^\infty(X)$.

Since a manifold admits the partition of unity by smooth functions (Remark B.1), it follows from Theorem C.8 that any sheaf of C^∞_X-modules on X, including a sheaf C^∞_X itself, is fine and, consequently, acyclic (Remark C.6).

For instance, let $Y \to X$ be a vector bundle. The germs of its sections form a sheaf of $C^\infty(X)$-modules, called the *structure sheaf* S_Y of a vector bundle $Y \to X$. The sheaf S_Y is fine. Its structure module of global sections coincides with the structure module $Y(X)$ of global sections of a vector bundle $Y \to X$. The forthcoming *Serre–Swan theorem* shows that these modules exhaust all projective modules of finite rank over $C^\infty(X)$. This theorem originally has been proved in the case of a compact manifold X, but it is generalized to an arbitrary manifold [58, 111].

Theorem A.10 *Let X be a smooth manifold. A $C^\infty(X)$-module P is isomorphic to the structure module of a smooth vector bundle over X if and only if it is a projective module of finite rank.*

Theorem A.10 states the categorial equivalence between the vector bundles over a manifold X and the projective modules of finite rank over a ring $C^\infty(X)$ of smooth real functions on X. The following are corollaries of this equivalence

- The structure module $Y^*(X)$ of the dual $Y^* \to X$ of a vector bundle $Y \to X$ is the $C^\infty(X)$-dual $Y(X)^*$ of the structure module $Y(X)$ of $Y \to X$.
- Any exact sequence of vector bundles

$$0 \to Y \longrightarrow Y' \longrightarrow Y'' \to 0 \tag{A.16}$$

over the same base X yields an exact sequence

$$0 \to Y(X) \longrightarrow Y'(X) \longrightarrow Y''(X) \to 0 \tag{A.17}$$

of their structure modules, and vice versa. In accordance with Theorem A.2, the exact sequence (A.17) always is split. Every its splitting defines that of the exact sequence (A.16), and vice versa.
- The derivation module of a real ring $C^\infty(X)$ coincides with a $C^\infty(X)$-module $\mathcal{T}(X)$ of vector fields on X, i.e., with the structure module of the tangent bundle TX of X. Hence, it is a projective $C^\infty(X)$-module of finite rank. It is the $C^\infty(X)$-dual $\mathcal{T}(X) = \mathcal{O}^1(X)^*$ of the structure module $\mathcal{O}^1(X)$ of the cotangent bundle T^*X of X which is a module of one-forms on X and, conversely, $\mathcal{O}^1(X) = \mathcal{T}(X)^*$.

It follows that the Chevalley–Eilenberg differential calculus over a real ring $C^\infty(X)$ (Definition A.6) is exactly a DGA $(\mathcal{O}^*(X), d)$ of exterior forms on a manifold X, where the Chevalley–Eilenberg coboundary operator d (A.13) coincides with the exterior differential (B.31). Moreover, one can show that $(\mathcal{O}^*(X), d)$ is the minimal differential calculus (Definition A.7), i.e., the $C^\infty(X)$-module $\mathcal{O}^1(X)$ is generated by elements df, $f \in C^\infty(X)$ [58]. Accordingly, the de Rham complex (A.15) of a real ring $C^\infty(X)$ is the *de Rham complex*

$$0 \to \mathbb{R} \longrightarrow C^\infty(X) \xrightarrow{d} \mathcal{O}^1(X) \xrightarrow{d} \cdots \mathcal{O}^k(X) \xrightarrow{d} \cdots \tag{A.18}$$

of exterior forms on a manifold X. Its cohomology is called the *de Rham cohomology* $H^*_{\mathrm{DR}}(X)$ of X. It obeys de Rham Theorem C.10.

Appendix B
Differential Calculus on Fibre Bundles

Unless otherwise stated, fibre bundles throughout the book are smooth manifolds. The differential calculus on fibre bundles and their differential geometry are phrased in terms of jet manifolds [25, 53, 85].

B.1 Geometry of Fibre Bundles

We start with manifolds and fibred manifolds (Definition B.1). Fibre bundles are locally trivial fibred manifolds (Definition B.2).

Remark B.1 We follow the notion of a finite-dimensional real *smooth manifold* without boundary. It conventionally is assumed to be Hausdorff and second-countable (i.e., it has a countable base for topology). Consequently, it is a locally compact space which is a union of a countable number of compact subsets, a separable space (i.e., it has a countable dense subset) and a paracompact space (Remark C.5). Being paracompact, a smooth manifold admits the partition of unity by smooth real functions. One also can show that, given two disjoint closed subsets N and N' of a smooth manifold X, there exists a smooth function f on X such that $f|_N = 0$ and $f|_{N'} = 1$. Unless otherwise stated, manifolds are assumed to be connected and, consequently, arcwise connected.

Let Z be a manifold. We denote by

$$\pi_Z : TZ \to Z, \qquad \pi_Z^* : T^*Z \to Z$$

its tangent and cotangent bundles, respectively. Given coordinates (z^α) on Z, they are equipped with the *holonomic bundle coordinates*

G. Sardanashvily, *Noether's Theorems*, Atlantis Studies
in Variational Geometry 3, DOI 10.2991/978-94-6239-171-0

$$(z^\lambda, \dot{z}^\lambda), \qquad \dot{z}'^\lambda = \frac{\partial z'^\lambda}{\partial z^\mu}\dot{z}^\mu,$$
$$(z^\lambda, \dot{z}_\lambda), \qquad \dot{z}'_\lambda = \frac{\partial z'^\mu}{\partial z^\lambda}\dot{z}_\mu,$$

with respect to the *holonomic frames* $\{\partial_\lambda\}$ and *coframes* $\{dz^\lambda\}$ in the tangent and cotangent spaces to Z, respectively.

Any manifold morphism $f : Z \to Z'$ yields the *tangent morphism*

$$Tf : TZ \to TZ', \qquad \dot{z}'^\lambda \circ Tf = \frac{\partial f^\lambda}{\partial x^\mu}\dot{z}^\mu.$$

Let us consider manifold morphisms of maximal rank. They are immersions (in particular, imbeddings) and submersions. An injective immersion is a submanifold, and a surjective submersion is a fibred manifold (in particular, a fibre bundle).

Given manifolds M and N, by the *rank of a morphism* $f : M \to N$ at a point $p \in M$ is meant the rank of the linear morphism

$$T_p f : T_p M \to T_{f(p)} N.$$

For instance, if f is of maximal rank at $p \in M$, then $T_p f$ is injective when $\dim M \leq \dim N$ and surjective when $\dim N \leq \dim M$. In this case, f is called the *immersion* and the *submersion* at a point $p \in M$, respectively.

Since $p \to \mathrm{rank}_p f$ is a lower semicontinuous function, then the morphism $T_p f$ is of maximal rank on an open neighborhood of p, too. It follows from the inverse function theorem that, if f is an immersion (resp. submersion) at p, then it is locally injective (resp. surjective) around p. If f is both an immersion and a submersion, it is termed a *local diffeomorphism* at p. In this case, there exists an open neighborhood U of p such that $f : U \to f(U)$ is a diffeomorphism onto an open set $f(U) \subset N$.

A manifold morphism f is called the immersion (resp. submersion) if it is an immersion (resp. submersion) at all points of M. A submersion is necessarily an *open map*, i.e., it sends open subsets of M onto open subsets of N.

If an immersion f is open (i.e., f is a homeomorphism onto $f(M)$ equipped with the relative topology from N), it is termed the *imbedding*.

A pair (M, f) is called a *submanifold* of N if f is an injective immersion. A submanifold (M, f) is an *imbedded submanifold* if f is an imbedding. For the sake of simplicity, we usually identify (M, f) with $f(M)$. If $M \subset N$, its natural injection is denoted by $i_M : M \to N$.

There are the following criteria for a submanifold to be imbedded.

Theorem B.1 *Let (M, f) be a submanifold of N.*

(i) The map f is an imbedding if and only if, for each point $p \in M$, there exists a (cubic) coordinate chart (V, ψ) of N centered at $f(p)$ so that $f(M) \cap V$ consists of all points of V with coordinates $(x^1, \ldots, x^m, 0, \ldots, 0)$.

(ii) *Suppose that $f : M \to N$ is a proper map, i.e., the pre-images of compact sets are compact. Then (M, f) is a closed imbedded submanifold of N. In particular, this occurs if M is a compact manifold.*
(iii) *If* $\dim M = \dim N$, *then (M, f) is an open imbedded submanifold of N.*

Fibred Manifolds and Fibre Bundles

Definition B.1 A triple

$$\pi : Y \to X, \qquad \dim X = n > 0, \tag{B.1}$$

is called a *fibred manifold* if a manifold morphism π is a surjective submersion, i.e., the tangent morphism $T\pi : TY \to TX$ is a surjection.

One says that Y is a *total space* of a fibred manifold (B.1), X is its *base*, π is a *fibration*, and $Y_x = \pi^{-1}(x)$ is a *fibre* over $x \in X$.

Any fibre is an imbedded submanifold of Y of dimension $\dim Y - \dim X$. Unless otherwise stated, we assume that $\dim Y \neq \dim X$, i.e., fibred manifolds with discrete fibres are not considered.

Theorem B.2 *A surjection (B.1) is a fibred manifold if and only if a manifold Y admits an atlas of coordinate charts $(U_Y; x^\lambda, y^i)$ such that (x^λ) are coordinates on $\pi(U_Y) \subset X$ and coordinate transition functions read $x'^\lambda = f^\lambda(x^\mu)$, $y'^i = f^i(x^\mu, y^j)$. These coordinates are termed fibred coordinates compatible with a fibration π.*

By a *local section* of the surjection (B.1) is meant an injection $s : U \to Y$ of an open subset $U \subset X$ such that $\pi \circ s = \mathrm{Id}\, U$, i.e., a section sends any point $x \in X$ into a fibre Y_x over this point. A local section also is defined over any subset $N \in X$ as the restriction to N of a local section over an open set containing N. If $U = X$, one calls s the *global section*. Hereafter, by a *section* is meant both a global section and a local section (over an open subset).

Theorem B.3 *The surjection π (B.1) is a fibred manifold if and only if, for each point $y \in Y$, there exists a local section s of $\pi : Y \to X$ passing through y.*

The range $s(U)$ of a local section $s : U \to Y$ of a fibred manifold $Y \to X$ is an imbedded submanifold of Y. It also is a *closed map*, which sends closed subsets of U onto closed subsets of Y. If s is a global section, then $s(X)$ is a closed imbedded submanifold of Y. Global sections of a fibred manifold need not exist.

Theorem B.4 *Let $Y \to X$ be a fibred manifold whose fibres are diffeomorphic to $\mathbb{R}^m$. Any its section over a closed imbedded submanifold (e.g., a point) of X is extended to a global section [141]. In particular, such a fibred manifold always has a global section.*

Given fibred coordinates $(U_Y; x^\lambda, y^i)$, a section s of a fibred manifold $Y \to X$ is represented by collections of local functions $\{s^i = y^i \circ s\}$ on $\pi(U_Y)$.

Definition B.2 A fibred manifold $Y \to X$ is called a *fibre bundle* if it admits a fibred coordinate atlas $\{(\pi^{-1}(U_\xi); x^\lambda, y^i)\}$ over a cover $\{\pi^{-1}(U_\iota)\}$ of Y which is the inverse image of a cover $\mathfrak{U} = \{U_\xi\}$ of X.

In this case, there exists a manifold V, called the *typical fibre*, such that Y is locally diffeomorphic to the splittings

$$\psi_\xi : \pi^{-1}(U_\xi) \to U_\xi \times V, \tag{B.2}$$

glued together by means of *transition functions*

$$\varrho_{\xi\zeta} = \psi_\xi \circ \psi_\zeta^{-1} : U_\xi \cap U_\zeta \times V \to U_\xi \cap U_\zeta \times V \tag{B.3}$$

on overlaps $U_\xi \cap U_\zeta$. Transition functions $\varrho_{\xi\zeta}$ fulfil the *cocycle condition*

$$\varrho_{\xi\zeta} \circ \varrho_{\zeta\kappa} = \varrho_{\xi\kappa} \tag{B.4}$$

on all overlaps $U_\xi \cap U_\zeta \cap U_\iota$. Restricted to a point $x \in X$, *trivialization morphisms* ψ_ξ (B.2) and transition functions $\varrho_{\xi\zeta}$ (B.3) define diffeomorphisms of fibres

$$\psi_\xi(x) : Y_x \to V, \qquad x \in U_\xi, \tag{B.5}$$

$$\varrho_{\xi\zeta}(x) : V \to V, \qquad x \in U_\xi \cap U_\zeta. \tag{B.6}$$

Trivialization charts (U_ξ, ψ_ξ) together with transition functions $\varrho_{\xi\zeta}$ (B.3) constitute a *bundle atlas*

$$\Psi = \{(U_\xi, \psi_\xi), \varrho_{\xi\zeta}\} \tag{B.7}$$

of a fibre bundle $Y \to X$. Two bundle atlases are said to be *equivalent* if their union also is a bundle atlas, i.e., there exist transition functions between trivialization charts of different atlases.

A fibre bundle $Y \to X$ is uniquely defined by a bundle atlas. Given the atlas Ψ (B.7), there is a unique manifold structure on Y for which $\pi : Y \to X$ is a fibre bundle with a typical fibre V and a bundle atlas Ψ. All atlases of a fibre bundle are equivalent.

Remark B.2 The notion of a fibre bundle (Definition B.2) is the notion of a *smooth locally trivial fibre bundle*. In a general setting, a *continuous fibre bundle* is defined as a continuous surjective submersion of topological spaces $Y \to X$. A continuous map $\pi : Y \to X$ is termed a submersion if, for any point $y \in Y$, there exists an open neighborhood U of the point $\pi(y)$ and a right continuous inverse $\sigma : U \to Y$ of π through y so that $\pi \circ \sigma = \mathrm{Id}\, U$ and $\sigma \circ \pi(y) = y$, i.e., there exists a local section σ of π and π is open. The notion of a locally trivial continuous fibre bundle is a repetition of that of a smooth fibre bundle, where trivialization morphisms ψ_ξ and transition functions $\varrho_{\xi\zeta}$ are continuous.

We have the following useful criteria for a fibred manifold to be a fibre bundle.

Theorem B.5 *If a fibration $\pi : Y \to X$ is a proper map, then $Y \to X$ is a fibre bundle. In particular, a fibred manifold with a compact total space is a fibre bundle.*

Theorem B.6 *A fibred manifold whose fibres are diffeomorphic either to a compact manifold or $\mathbb{R}^r$ is a fibre bundle [105].*

A comprehensive relation between fibred manifolds and fibre bundles is given in Remark B.11. It involves the notion of an Ehresmann connection.

Unless otherwise stated, we restrict our consideration to fibre bundles. Without a loss of generality, we further assume that a cover $\mathfrak{U}$ for a bundle atlas of $Y \to X$ also is a cover for a manifold atlas of the base X. Then, given the bundle atlas Ψ (B.7), a fibre bundle Y is provided with the associated *bundle coordinates*

$$x^\lambda(y) = (x^\lambda \circ \pi)(y), \qquad y^i(y) = (y^i \circ \psi_\xi)(y), \qquad y \in \pi^{-1}(U_\xi),$$

where x^λ are coordinates on $U_\xi \subset X$ and y^i, called the *fibre coordinates*, are coordinates on a typical fibre V.

The forthcoming Theorems B.7 and B.9 describe the particular covers which one can choose for a bundle atlas.

A fibred manifold $Y \to X$ is termed trivial if it is diffeomorphic to a product $X \times V$.

Theorem B.7 *Any fibre bundle over a contractible base is trivial.*

However, a fibred manifold over a contractible base need not be trivial, even its fibres are mutually diffeomorphic [53].

It follows from Theorem B.7 that any cover of a base X consisting of *domains* (i.e., contractible open subsets) is a bundle cover.

Theorem B.8 *Every fibre bundle $Y \to X$ admits a bundle atlas over a countable cover $\mathfrak{U}$ of X where each member U_ξ of $\mathfrak{U}$ is a domain whose closure is compact [67].*

If a base X is compact, there is a bundle atlas of Y over a finite cover of X which obeys the condition of Theorem B.8.

Theorem B.9 *Every fibre bundle $Y \to X$ admits a bundle atlas over a finite cover of X, but its members need not be contractible and connected [61].*

Morphisms of fibre bundles, by definition, are *fibrewise morphisms*, sending a fibre to a fibre. Namely, a *bundle morphism* of a fibre bundle $\pi : Y \to X$ to a fibre bundle $\pi' : Y' \to X'$ is defined as a pair (Φ, f) of manifold morphisms which form a commutative diagram

$$\begin{array}{ccc} Y & \xrightarrow{\Phi} & Y' \\ {\scriptstyle\pi}\downarrow & & \downarrow{\scriptstyle\pi'} \\ X & \xrightarrow{f} & X' \end{array}, \qquad \pi' \circ \Phi = f \circ \pi. \tag{B.8}$$

Bundle injections and surjections are called *bundle monomorphisms and epimorphisms*, respectively. A bundle diffeomorphism is termed the *bundle isomorphism*, or the *bundle automorphism* if it is an isomorphism to itself. For the sake of brevity, a bundle morphism over $f = \mathrm{Id}\, X$ often is said to be the bundle morphism over X, and is denoted by $Y \underset{X}{\longrightarrow} Y'$. In particular, a bundle automorphism over X is called the *vertical automorphism*.

A bundle monomorphism $\Phi : Y \to Y'$ over X is called the *subbundle* of a fibre bundle $Y' \to X$ if $\Phi(Y)$ is a submanifold of Y'. There is the following useful criterion for an image and an inverse image of a bundle morphism to be subbundles.

Theorem B.10 *Let $\Phi : Y \to Y'$ be a bundle morphism over X. Given a global section s of the fibre bundle $Y' \to X$ such that $s(X) \subset \Phi(Y)$, by the kernel of a bundle morphism Φ with respect to a section s is meant the inverse image* $\mathrm{Ker}_s \Phi = \Phi^{-1}(s(X))$ *of $s(X)$ by Φ. If $\Phi : Y \to Y'$ is a bundle morphism of constant rank over X, then $\Phi(Y)$ and* $\mathrm{Ker}_s \Phi$ *are subbundles of Y' and Y, respectively.*

There are the following standard constructions of new fibre bundles from the old ones.

- Given a fibre bundle $\pi : Y \to X$ and a manifold morphism $f : X' \to X$, by the *pull-back* of Y by f is termed the manifold

  $$f^*Y = \{(x', y) \in X' \times Y \, : \, \pi(y) = f(x')\}$$

  together with the natural projection $(x', y) \to x'$. It is a fibre bundle over X' such that a fibre of f^*Y over a point $x' \in X'$ is that of Y over the point $f(x') \in X$. There is the canonical bundle morphism

  $$f_Y : f^*Y \ni (x', y)|_{\pi(y)=f(x')} \underset{f}{\longrightarrow} y \in Y.$$

  Any section s of a fibre bundle $Y \to X$ yields the *pull-back section* $f^*s(x') = (x', s(f(x'))$ of $f^*Y \to X'$.
- If $X' \subset X$ is a submanifold of X and $i_{X'}$ is the corresponding natural injection, then the pull-back bundle $i^*_{X'}Y = Y|_{X'}$ is called the *restriction* of a fibre bundle Y to a submanifold $X' \subset X$. If X' is an imbedded submanifold, any section of the pull-back bundle $Y|_{X'} \to X'$ is the restriction to X' of some section of $Y \to X$.
- Let $\pi : Y \to X$ and $\pi' : Y' \to X$ be fibre bundles over the same base X. Their *bundle product* $Y \times_X Y'$ over X is defined as the pull-back

  $$Y \underset{X}{\times} Y' = \pi^*Y' \quad \text{or} \quad Y \underset{X}{\times} Y' = \pi'^*Y$$

together with its natural surjection onto X. Fibres of the bundle product $Y \times Y'$ are the Cartesian products $Y_x \times Y'_x$ of fibres of fibre bundles Y and Y'.

Let us consider a composition

$$\pi : Y \to \Sigma \to X \tag{B.9}$$

of fibre bundles

$$\pi_{Y\Sigma} : Y \to \Sigma, \tag{B.10}$$

$$\pi_{\Sigma X} : \Sigma \to X. \tag{B.11}$$

One can show that it is a fibre bundle, called the *composite bundle*. It is provided with bundle coordinates $(x^\lambda, \sigma^m, y^i)$, where (x^λ, σ^m) are bundle coordinates on the fibre bundle (B.11), i.e., transition functions of coordinates σ^m are independent of coordinates y^i.

For instance, the tangent bundle TY of a fibre bundle $Y \to X$ is a composite bundle $TY \to Y \to X$.

The following fact makes composite bundles useful for applications.

Theorem B.11 *Given the composite bundle (B.9), let h be a global section of the fibre bundle $\Sigma \to X$ (B.11). Then the restriction*

$$Y^h = h^* Y \tag{B.12}$$

of the fibre bundle $Y \to \Sigma$ (B.10) to $h(X) \subset \Sigma$ is a subbundle of the fibre bundle $Y \to X$.

Vector and Affine Bundles

A *vector bundle* is a fibre bundle $Y \to X$ such that:

- its typical fibre V and all the fibres $Y_x = \pi^{-1}(x), x \in X$, are real finite-dimensional vector spaces;
- there is the bundle atlas Ψ (B.7) of $Y \to X$ whose trivialization morphisms ψ_ξ (B.5) and transition functions $\varrho_{\xi\zeta}$ (B.6) are linear isomorphisms.

Accordingly, a vector bundle is provided with *linear fibre coordinates* (y^i) possessing linear transition functions $y'^i = A^i_j(x)y^j$. We have

$$y = y^i e_i(\pi(y)) = y^i \psi_\xi(\pi(y))^{-1}(e_i), \qquad \pi(y) \in U_\xi,$$

where $\{e_i\}$ is a fixed basis for the typical fibre V of Y, and $\{e_i(x)\}$ are the fibre bases (or the *frames*) for the fibres Y_x of Y associated to the bundle atlas Ψ.

By virtue of Theorem B.4, any vector bundle has a global section, e.g., the canonical global *zero-valued section* $\widehat{0}(x) = 0$. Global sections of a vector bundle $Y \to X$ constitute a projective $C^\infty(X)$-module $Y(X)$ of finite rank. It is called the *structure module of a vector bundle*.

Theorem B.12 *Let a vector bundle $Y \to X$ admit $m = \dim V$ nowhere vanishing global sections s_i which are linearly independent, i.e., $\overset{m}{\wedge} s_i \neq 0$. Then Y is trivial.*

By a morphism of vector bundles is meant a *linear bundle morphism*, which is a linear fibrewise map whose restriction to each fibre is a linear map.

Given a linear bundle morphism $\Phi : Y' \to Y$ of vector bundles over X, its *kernel* $\operatorname{Ker} \Phi$ is defined as the inverse image $\Phi^{-1}(\widehat{0}(X))$ of the canonical zero-valued section $\widehat{0}(X)$ of Y. By virtue of Theorem B.10, if Φ is of constant rank, its kernel and its range are vector subbundles of the vector bundles Y' and Y, respectively. For instance, monomorphisms and epimorphisms of vector bundles fulfil this condition.

There are the following particular constructions of new vector bundles from the old ones.

- Let $Y \to X$ be a vector bundle with a typical fibre V. By $Y^* \to X$ is denoted the *dual vector bundle* with the typical fibre V^* dual of V. An *interior product* of Y and Y^* is defined as a fibred morphism

$$\rfloor : Y \otimes Y^* \underset{X}{\longrightarrow} X \times \mathbb{R}.$$

- Let $Y \to X$ and $Y' \to X$ be vector bundles with typical fibres V and V', respectively. Their *Whitney sum* $Y \oplus_X Y'$ is a vector bundle over X with a typical fibre $V \oplus V'$.
- Let $Y \to X$ and $Y' \to X$ be vector bundles with typical fibres V and V', respectively. Their *tensor product* $Y \otimes_X Y'$ is a vector bundle over X with a typical fibre $V \otimes V'$. Similarly, the *exterior product* of vector bundles $Y \underset{X}{\wedge} Y'$ is defined. The exterior product

$$\wedge Y = X \times \mathbb{R} \underset{X}{\oplus} Y \underset{X}{\oplus} \overset{2}{\wedge} Y \underset{X}{\oplus} \cdots \overset{k}{\wedge} Y, \qquad k = \dim Y - \dim X, \tag{B.13}$$

is called the *exterior bundle*.

Remark B.3 Given vector bundles Y and Y' over the same base X, every linear bundle morphism

$$\Phi : Y_x \ni \{e_i(x)\} \to \{\Phi_i^k(x) e'_k(x)\} \in Y'_x$$

over X defines a global section

$$\Phi : x \to \Phi_i^k(x) e^i(x) \otimes e'_k(x)$$

of the tensor product $Y \otimes Y'^*$, and vice versa.

A sequence $Y' \xrightarrow{i} Y \xrightarrow{j} Y''$ of vector bundles over the same base X is called *exact* at Y if $\operatorname{Ker} j = \operatorname{Im} i$. A sequence of vector bundles

$$0 \to Y' \xrightarrow{i} Y \xrightarrow{j} Y'' \to 0 \tag{B.14}$$

over X is said to be a *short exact sequence* if it is exact at all terms Y', Y, and Y''. This means that i is a bundle monomorphism, j is a bundle epimorphism, and $\operatorname{Ker} j = \operatorname{Im} i$. Then Y'' is the *quotient bundle* Y/Y' whose structure module is the quotient $Y(X)/Y'(X)$ of the structure modules of Y and Y'. Given an exact sequence of vector bundles (B.14), there is the exact sequence of their duals

$$0 \to Y''^* \xrightarrow{j^*} Y^* \xrightarrow{i^*} Y'^* \to 0.$$

One says that an exact sequence (B.14) is *split* if there exists a bundle monomorphism $s : Y'' \to Y$ such that $j \circ s = \operatorname{Id} Y''$ or, equivalently,

$$Y = i(Y') \oplus s(Y'') = Y' \oplus Y''.$$

The following is a corollary of Theorem A.2 and Serre–Swan Theorem A.10.

Theorem B.13 *Every exact sequence of vector bundles (B.14) is split.*

The tangent bundle TZ and the cotangent bundle T^*Z of a manifold Z exemplify vector bundles.

Remark B.4 Given an atlas $\Psi_Z = \{(U_\iota, \phi_\iota)\}$ of a manifold Z, the tangent bundle is provided with the *holonomic bundle atlas*

$$\Psi_T = \{(U_\iota, \psi_\iota = T\phi_\iota)\}, \tag{B.15}$$

where $T\phi_\iota$ is the tangent morphism to ϕ_ι. The associated linear fibre coordinates are holonomic (or *induced*) coordinates $(\dot z^\lambda)$ with respect to the holonomic frames $\{\partial_\lambda\}$ in tangent spaces T_zZ.

A tensor product of tangent and cotangent bundles

$$T = (\overset{m}{\otimes} TZ) \otimes (\overset{k}{\otimes} T^*Z), \qquad m, k \in \mathbb{N}, \tag{B.16}$$

is called the *tensor bundle*, provided with holonomic fibre coordinates $\dot x^{\alpha_1\cdots\alpha_m}_{\beta_1\cdots\beta_k}$ possessing transition functions

$$\dot x'^{\alpha_1\cdots\alpha_m}_{\beta_1\cdots\beta_k} = \frac{\partial x'^{\alpha_1}}{\partial x^{\mu_1}} \cdots \frac{\partial x'^{\alpha_m}}{\partial x^{\mu_m}} \frac{\partial x^{\nu_1}}{\partial x'^{\beta_1}} \cdots \frac{\partial x^{\nu_k}}{\partial x'^{\beta_k}} \dot x^{\mu_1\cdots\mu_m}_{\nu_1\cdots\nu_k}.$$

Its sections are termed *tensor fields* or (m, k)-tensor fields.

Let $\pi_Y : TY \to Y$ be the tangent bundle of a fibre bundle $\pi : Y \to X$. Given bundle coordinates (x^λ, y^i) on Y, it is equipped with the holonomic coordinates $(x^\lambda, y^i, \dot{x}^\lambda, \dot{y}^i)$ with transition functions

$$\dot{x}'^\lambda = \frac{\partial x'^\lambda}{\partial x^\mu}\dot{x}^\mu, \qquad \dot{y}'^i = \frac{\partial y'^i}{\partial y^j}\dot{y}^j + \frac{\partial y'^i}{\partial x^\mu}\dot{x}^\mu.$$

Definition B.3 The *vertical tangent bundle* of a fibre bundle $Y \to X$ is defined to be a subbundle $VY = \mathrm{Ker}\,(T\pi)$ of the tangent bundle $TY \to Y$ which consists of vectors tangent to fibres of Y.

A vertical tangent bundle is provided with the *holonomic bundle coordinates* $(x^\lambda, y^i, \overline{y}^i = \dot{y}^i|_{\dot{x}^\lambda=0})$ with respect to the vertical frames $\{\partial_i\}$. They possess transition functions

$$\overline{y}'^i = \frac{\partial y'^i}{\partial y^j}\overline{y}^j.$$

Every bundle morphism $\Phi : Y \to Y'$ yields a linear bundle morphism

$$V\Phi : VY \underset{\Phi}{\to} VY', \qquad \overline{y}'^i \circ V\Phi = \frac{\partial \Phi^i}{\partial y^j}\overline{y}^j \tag{B.17}$$

of the vertical tangent bundles. It is called the *vertical tangent morphism*.

In many important cases, the vertical tangent bundle $VY \to Y$ of a fibre bundle $Y \to X$ is trivial, and it is isomorphic to the bundle product

$$VY = Y \underset{X}{\times} \overline{Y} \tag{B.18}$$

where $\overline{Y} \to X$ is some vector bundle. It follows that VY can be provided with bundle coordinates $(x^\lambda, y^i, \overline{y}^i)$ such that transition functions of fibre coordinates $\overline{y}^i$ are independent of coordinates y^i. One calls (B.18) the *vertical splitting*. For instance, every vector bundle $Y \to X$ admits the *canonical vertical splitting* $VY = Y \oplus_X Y$.

Let $T^*Y \to Y$ be the cotangent bundle of a fibre bundle $Y \to X$. It is provided with the holonomic coordinates $(x^\lambda, y^i, \dot{x}_\lambda, \dot{y}_i)$ with transition functions

$$\dot{x}'_\lambda = \frac{\partial x^\mu}{\partial x'^\lambda}\dot{x}_\mu + \frac{\partial y^j}{\partial x'^\lambda}\dot{y}_j, \qquad \dot{y}'_i = \frac{\partial y^j}{\partial y'^i}\dot{y}_j.$$

Definition B.4 The *vertical cotangent bundle* $V^*Y \to Y$ of a fibre bundle $Y \to X$ is defined as the dual of the vertical tangent bundle $VY \to Y$ (Definition B.3).

It is not a subbundle of the cotangent bundle T^*Y, but there is the canonical surjection

$$\zeta : T^*Y \ni \dot{x}_\lambda dx^\lambda + \dot{y}_i dy^i \to \dot{y}_i \overline{d} y^i \in V^*Y, \qquad (B.19)$$

where $\{\overline{d}y^i\}$, possessing transition functions

$$\overline{d}y'^i = \frac{\partial y'^i}{\partial y^j}\overline{d}y^j,$$

are the duals of the holonomic frames $\{\partial_i\}$ of the vertical tangent bundle VY. The vertical cotangent bundle is endowed with the *holonomic coordinates* $(x^\lambda, y^i, p_i = \dot{y}_i)$ possessing transition functions

$$p'_i = \frac{\partial y^j}{\partial y'^i}p_j. \qquad (B.20)$$

For any fibre bundle Y, there exist the exact sequences of vector bundles

$$0 \to VY \longrightarrow TY \xrightarrow{\pi_T} Y \underset{X}{\times} TX \to 0, \qquad (B.21)$$

$$0 \to Y \underset{X}{\times} T^*X \to T^*Y \to V^*Y \to 0. \qquad (B.22)$$

Their splitting, by definition, is a connection on $Y \to X$ (Sect. B.3).

Let $\overline{\pi} : \overline{Y} \to X$ be a vector bundle with a typical fibre $\overline{V}$. An *affine bundle* modelled over a vector bundle $\overline{Y} \to X$ is a fibre bundle $\pi : Y \to X$ whose typical fibre V is an affine space modelled over $\overline{V}$ such that the following conditions hold.

- All the fibres Y_x of Y are affine spaces modelled over the corresponding fibres $\overline{Y}_x$ of the vector bundle $\overline{Y}$.
- There is an affine bundle atlas $\Psi = \{(U_\alpha, \psi_\chi), \varrho_{\chi\zeta}\}$ of $Y \to X$ whose local trivializations morphisms ψ_χ (B.5) and transition functions $\varrho_{\chi\zeta}$ (B.6) are affine isomorphisms.

Dealing with affine bundles, we use only *affine fibre coordinates* (y^i) associated to an affine bundle atlas Ψ. There are the bundle morphisms

$$Y \underset{X}{\times} \overline{Y} \underset{X}{\longrightarrow} Y, \qquad (y^i, \overline{y}^i) \to y^i + \overline{y}^i,$$

$$Y \underset{X}{\times} Y \underset{X}{\longrightarrow} \overline{Y}, \qquad (y^i, y'^i) \to y^i - y'^i,$$

where $(\overline{y}^i)$ are linear fibre coordinates on a vector bundle $\overline{Y}$.

By virtue of Theorem B.4, affine bundles admit global sections, but in contrast with vector bundles, there is not their canonical global section. Every global section s of an affine bundle $Y \to X$ modelled over a vector bundle $\overline{Y} \to X$ yields the bundle morphisms

$$Y \ni y \to y - s(\pi(y)) \in \overline{Y}, \tag{B.23}$$
$$\overline{Y} \ni \overline{y} \to s(\pi(y)) + \overline{y} \in Y. \tag{B.24}$$

In particular, every vector bundle Y has a natural structure of an affine bundle owing to the morphisms (B.24) where $s = \widehat{0}$ is the canonical zero-valued section of Y. Due to the morphism (B.23), any affine bundle $Y \to X$ admits fibre coordinates $\overline{y}^i = y^i - s^i(\pi(y))$, possessing linear transition functions.

By a morphism of affine bundles is meant an *affine bundle morphism* $\Phi : Y \to Y'$ whose restriction to each fibre of Y is an affine map. Every affine bundle morphism $\Phi : Y \to Y'$ of an affine bundle Y modelled over a vector bundle $\overline{Y}$ to an affine bundle Y' modelled over a vector bundle $\overline{Y}'$ yields an unique linear bundle morphism

$$\overline{\Phi} : \overline{Y} \to \overline{Y}', \qquad \overline{y}'^i \circ \overline{\Phi} = \frac{\partial \Phi^i}{\partial y^j} \overline{y}^j,$$

called the linear derivative of Φ.

Every affine bundle $Y \to X$ modelled over a vector bundle $\overline{Y} \to X$ admits the *canonical vertical splitting* $VY = Y \times_X \overline{Y}$.

Let us note that Theorems B.8 and B.9 on a particular cover for bundle atlases remain true in the case of linear and affine atlases of vector and affine bundles.

Vector Fields

Vector fields on a manifold Z are global sections of the tangent bundle $TZ \to Z$ of Z.

The set $\mathcal{T}(Z)$ of vector fields on Z is both a $C^\infty(Z)$-module and a real Lie algebra with respect to the *Lie bracket*

$$[v, u] = (v^\lambda \partial_\lambda u^\mu - u^\lambda \partial_\lambda v^\mu)\partial_\mu, \qquad u = u^\lambda \partial_\lambda, \qquad v = v^\lambda \partial_\lambda.$$

Given a vector field u on X, a *curve* $c : \mathbb{R} \supset (,) \to Z$ in Z is said to be the *integral curve* of u if $Tc = u(c)$. Every vector field u on a manifold Z can be seen as an *infinitesimal generator* of a local one-parameter group of local diffeomorphisms (a *flow*), and vice versa [82]. One-dimensional orbits of this group are integral curves of u. A vector field is termed *complete* if its flow is a one-parameter group of diffeomorphisms of Z. For instance, every vector field on a compact manifold is complete.

A vector field u on a fibre bundle $Y \to X$ is called *projectable* if it projects onto a vector field on X, i.e., there exists a vector field τ on X such that $\tau \circ \pi = T\pi \circ u$. A projectable vector field takes a coordinate form

$$u = u^\lambda(x^\mu)\partial_\lambda + u^i(x^\mu, y^j)\partial_i, \qquad \tau = u^\lambda \partial_\lambda. \tag{B.25}$$

Its flow is a local one-parameter group of automorphisms of $Y \to X$ over a local one-parameter group of diffeomorphisms of X whose infinitesimal generator is τ.

A projectable vector field is termed *vertical* if its projection onto X vanishes, i.e., if it lives into the vertical tangent bundle VY.

Let

$$\gamma : \mathcal{T}(X) \to \mathcal{T}(Y) \tag{B.26}$$

be an $\mathbb{R}$-linear module morphism which sends a vector field $\tau = \tau^\lambda \partial_\lambda$ on a base X onto a projectable vector field

$$\gamma\tau = \tau^\lambda \partial_\lambda + (\gamma\tau)^i \partial_i \tag{B.27}$$

on Y over τ.

Definition B.5 The module morphism (B.26) is called the *lift of vector fields* on a base X onto a fibre bundle Y.

In particular, a vector field τ on a base X of a fibre bundle $Y \to X$ gives rise to the horizontal vector field $\Gamma\tau$ (B.55) on Y by means of a connection Γ on this bundle.

Definition B.6 The lift γ (B.26) of vector fields is said to be *functorial* if it is a monomorphism of a real Lie algebra $\mathcal{T}(X)$ of vector fields on X to a real Lie algebra $\mathcal{T}(Y)$ of vector fields on Y, i.e.,

$$[\gamma\tau, \gamma\tau'] = \gamma[\tau, \tau'], \qquad \tau, \tau' \in \mathcal{T}(X). \tag{B.28}$$

For instance, the above mentioned lift $\Gamma\tau$ of vector fields by means of a connection Γ is functorial if and only if this connection is flat (Theorem B.16).

Every tensor bundle (B.16) admits the canonical functorial lift of vector fields

$$\widetilde{\tau} = \tau^\mu \partial_\mu + [\partial_\nu \tau^{\alpha_1} \dot{x}^{\nu\alpha_2\cdots\alpha_m}_{\beta_1\cdots\beta_k} + \cdots - \partial_{\beta_1}\tau^\nu \dot{x}^{\alpha_1\cdots\alpha_m}_{\nu\beta_2\cdots\beta_k} - \cdots]\dot{\partial}^{\beta_1\cdots\beta_k}_{\alpha_1\cdots\alpha_m}, \tag{B.29}$$

where we employ the compact notation $\dot{\partial}_\lambda = \partial/\partial\dot{x}^\lambda$. In particular, we have the canonical functorial lifts

$$\widetilde{\tau} = \tau^\mu \partial_\mu + \partial_\nu \tau^\alpha \dot{x}^\nu \frac{\partial}{\partial \dot{x}^\alpha}, \qquad \widetilde{\tau} = \tau^\mu \partial_\mu - \partial_\beta \tau^\nu \dot{x}_\nu \frac{\partial}{\partial \dot{x}_\beta} \tag{B.30}$$

of a vector field τ on X onto the tangent and cotangent bundles TX and T^*X.

Definition B.7 A fibre bundle admitting functorial lift of vector fields on its base is called the *natural bundle* [83, 145].

Differential Forms

An *exterior r-form* on a manifold Z is defined as a section

$$\phi = \frac{1}{r!}\phi_{\lambda_1\dots\lambda_r} dz^{\lambda_1} \wedge \dots \wedge dz^{\lambda_r}$$

of the exterior product $\overset{r}{\wedge} T^*Z \to Z$, where

$$dz^{\lambda_1} \wedge \dots \wedge dz^{\lambda_r} = \frac{1}{r!}\varepsilon^{\lambda_1\dots\lambda_r}{}_{\mu_1\dots\mu_r} dx^{\mu_1} \otimes \dots \otimes dx^{\mu_r},$$

$$\varepsilon^{\dots\lambda_i\dots\lambda_j\dots}{}_{\dots\mu_p\dots\mu_k\dots} = -\varepsilon^{\dots\lambda_j\dots\lambda_i\dots}{}_{\dots\mu_p\dots\mu_k\dots} = -\varepsilon^{\dots\lambda_i\dots\lambda_j\dots}{}_{\dots\mu_k\dots\mu_p\dots},$$

$$\varepsilon^{\lambda_1\dots\lambda_r}{}_{\lambda_1\dots\lambda_r} = 1.$$

Let $\mathcal{O}^r(Z)$ denote a vector space of exterior r-forms on a manifold Z. By definition, $\mathcal{O}^0(Z) = C^\infty(Z)$ is a ring of smooth real functions on Z. All exterior forms on Z constitute an $\mathbb{N}$-graded commutative algebra $\mathcal{O}^*(Z)$ of global sections of the exterior bundle $\wedge T^*Z$ (B.13). It is endowed with an *exterior product*

$$\phi = \frac{1}{r!}\phi_{\lambda_1\dots\lambda_r} dz^{\lambda_1} \wedge \dots \wedge dz^{\lambda_r}, \qquad \sigma = \frac{1}{s!}\sigma_{\mu_1\dots\mu_s} dz^{\mu_1} \wedge \dots \wedge dz^{\mu_s},$$

$$\phi \wedge \sigma = \frac{1}{r!s!}\phi_{\nu_1\dots\nu_r}\sigma_{\nu_{r+1}\dots\nu_{r+s}} dz^{\nu_1} \wedge \dots \wedge dz^{\nu_{r+s}} =$$

$$\frac{1}{r!s!(r+s)!}\varepsilon^{\nu_1\dots\nu_{r+s}}{}_{\alpha_1\dots\alpha_{r+s}}\phi_{\nu_1\dots\nu_r}\sigma_{\nu_{r+1}\dots\nu_{r+s}} dz^{\alpha_1} \wedge \dots \wedge dz^{\alpha_{r+s}},$$

$$\phi \wedge \sigma = (-1)^{|\phi||\sigma|}\sigma \wedge \phi,$$

where the symbol $|\phi|$ stands for the form degree. A graded algebra $\mathcal{O}^*(Z)$ also is provided with an *exterior differential*

$$d\phi = dz^\mu \wedge \partial_\mu \phi = \frac{1}{r!}\partial_\mu \phi_{\lambda_1\dots\lambda_r} dz^\mu \wedge dz^{\lambda_1} \wedge \dots \wedge dz^{\lambda_r} \tag{B.31}$$

which obeys the relations

$$d \circ d = 0, \qquad d(\phi \wedge \sigma) = d(\phi) \wedge \sigma + (-1)^{|\phi|}\phi \wedge d(\sigma).$$

The exterior differential d (B.31) brings $\mathcal{O}^*(Z)$ into a DGA which is the minimal Chevalley–Eilenberg differential calculus $\mathcal{O}^*\mathcal{A}$ over a real ring $\mathcal{A} = C^\infty(Z)$. Its de Rham complex is (A.18).

Given a manifold morphism $f : Z \to Z'$, any exterior k-form ϕ on Z' yields the *pull-back exterior form* $f^*\phi$ on Z given by a condition

$$f^*\phi(v^1, \dots, v^k)(z) = \phi(Tf(v^1), \dots, Tf(v^k))(f(z))$$

for an arbitrary collection of tangent vectors $v^1, \dots, v^k \in T_z Z$. We have the relations

$$f^*(\phi \wedge \sigma) = f^*\phi \wedge f^*\sigma, \qquad df^*\phi = f^*(d\phi).$$

In particular, given a fibre bundle $\pi : Y \to X$, the pull-back onto Y of exterior forms on X by π provides a monomorphism of graded commutative algebras $\mathcal{O}^*(X) \to \mathcal{O}^*(Y)$. Elements of its range $\pi^*\mathcal{O}^*(X)$ are called *basic forms*. Exterior forms

$$\phi : Y \to \overset{r}{\wedge} T^*X, \qquad \phi = \frac{1}{r!}\phi_{\lambda_1\dots\lambda_r} dx^{\lambda_1} \wedge \dots \wedge dx^{\lambda_r},$$

on Y such that $u\rfloor\phi = 0$ for an arbitrary vertical vector field u on Y are said to be *horizontal forms*. Horizontal forms of degree $n = \dim X$ are termed *densities*. We use for them the compact notation

$$\begin{aligned} &L = \frac{1}{n!} L_{\mu_1\dots\mu_n} dx^{\mu_1} \wedge \dots \wedge dx^{\mu_n} = \mathcal{L}\omega, \qquad \mathcal{L} = L_{1\dots n}, \\ &\omega = dx^1 \wedge \dots \wedge dx^n = \frac{1}{n!}\varepsilon_{\mu_1\dots\mu_n} dx^{\mu_1} \wedge \dots \wedge dx^{\mu_n}, \\ &\omega_\lambda = \partial_\lambda\rfloor\omega, \qquad \omega_{\mu\lambda} = \partial_\mu\rfloor\partial_\lambda\rfloor\omega, \end{aligned} \tag{B.32}$$

where ε is the skew-symmetric *Levi–Civita symbol* with a component $\varepsilon_{\mu_1\dots\mu_n} = 1$.

The *interior product* (or *contraction*) of a vector field u and an exterior r-form ϕ on a manifold Z is given by a coordinate expression

$$u\rfloor\phi = \sum_{k=1}^{r} \frac{(-1)^{k-1}}{r!} u^{\lambda_k}\phi_{\lambda_1\dots\lambda_k\dots\lambda_r} dz^{\lambda_1} \wedge \dots \wedge \widehat{dz}^{\lambda_k} \wedge \dots \wedge dz^{\lambda_r} =$$
$$\frac{1}{(r-1)!} u^\mu \phi_{\mu\alpha_2\dots\alpha_r} dz^{\alpha_2} \wedge \dots \wedge dz^{\alpha_r},$$

where the caret $\widehat{}$ denotes omission. It obeys relations

$$\phi(u_1, \dots, u_r) = u_r\rfloor \cdots u_1\rfloor\phi, \qquad u\rfloor(\phi \wedge \sigma) = u\rfloor\phi \wedge \sigma + (-1)^{|\phi|}\phi \wedge u\rfloor\sigma.$$

The *Lie derivative* of an exterior form ϕ along a vector field u is

$$\mathbf{L}_u\phi = u\rfloor d\phi + d(u\rfloor\phi), \qquad \mathbf{L}_u(\phi \wedge \sigma) = \mathbf{L}_u\phi \wedge \sigma + \phi \wedge \mathbf{L}_u\sigma.$$

It is a derivation of the graded algebra $\mathcal{O}^*(Z)$ such that

$$\mathbf{L}_u \circ \mathbf{L}_{u'} - \mathbf{L}_{u'} \circ \mathbf{L}_u = \mathbf{L}_{[u,u']}.$$

In particular, if f is a function, then $\mathbf{L}_u f = u(f) = u\rfloor df$.

Remark B.5 An exterior form ϕ is invariant under a local one-parameter group of diffeomorphisms $G(t)$ of Z (i.e., $G(t)^*\phi = \phi$) if and only if its Lie derivative along the infinitesimal generator u of this group vanishes, i.e., $\mathbf{L}_u\phi = 0$.

A *tangent-valued r-form* on a manifold Z is defined as a section

$$\phi = \frac{1}{r!}\phi^\mu_{\lambda_1\ldots\lambda_r} dz^{\lambda_1} \wedge \cdots \wedge dz^{\lambda_r} \otimes \partial_\mu$$

of a tensor bundle $\overset{r}{\wedge} T^*Z \otimes TZ \to Z$.

Remark B.6 There is one-to-one correspondence between the tangent-valued one-forms ϕ on a manifold Z and the linear bundle endomorphisms

$$\widehat{\phi} : TZ \to TZ, \quad \widehat{\phi} : T_zZ \ni v \to v\rfloor\phi(z) \in T_zZ, \tag{B.33}$$
$$\widehat{\phi}^* : T^*Z \to T^*Z, \quad \widehat{\phi}^* : T^*_zZ \ni v^* \to \phi(z)\rfloor v^* \in T^*_zZ, \tag{B.34}$$

over Z (Remark B.3). For instance, the *canonical tangent-valued one-form*

$$\theta_Z = dz^\lambda \otimes \partial_\lambda \tag{B.35}$$

on Z corresponds to the identity morphisms (B.33) and (B.34).

Remark B.7 Let $Z = TX$, and let TTX be the tangent bundle of TX. There is a bundle endomorphism

$$J(\partial_\lambda) = \dot\partial_\lambda, \qquad J(\dot\partial_\lambda) = 0$$

of TTX over X. It corresponds to the canonical tangent-valued form

$$\theta_J = dx^\lambda \otimes \dot\partial_\lambda \tag{B.36}$$

on the tangent bundle TX. It is readily observed that $J \circ J = 0$.

A space $\mathcal{O}^*(Z) \otimes \mathcal{T}(Z)$ of tangent-valued forms is provided with the *Frölicher–Nijenhuis bracket*

$$\begin{aligned}
&[\,,\,]_{\mathrm{FN}} : \mathcal{O}^r(Z) \otimes \mathcal{T}(Z) \times \mathcal{O}^s(Z) \otimes \mathcal{T}(Z) \to \mathcal{O}^{r+s}(Z) \otimes \mathcal{T}(Z),\\
&[\alpha \otimes u, \beta \otimes v]_{\mathrm{FN}} = (\alpha \wedge \beta) \otimes [u, v] + (\alpha \wedge \mathbf{L}_u\beta) \otimes v - \\
&\qquad (\mathbf{L}_v\alpha \wedge \beta) \otimes u + (-1)^r(d\alpha \wedge u\rfloor\beta) \otimes v + (-1)^r(v\rfloor\alpha \wedge d\beta) \otimes u,\\
&\alpha \in \mathcal{O}^r(Z), \qquad \beta \in \mathcal{O}^s(Z), \qquad u, v \in \mathcal{T}(Z).
\end{aligned} \tag{B.37}$$

Its coordinate expression is

$$[\phi,\sigma]_{\rm FN} = \frac{1}{r!s!}(\phi^\nu_{\lambda_1\dots\lambda_r}\partial_\nu\sigma^\mu_{\lambda_{r+1}\dots\lambda_{r+s}} - \sigma^\nu_{\lambda_{r+1}\dots\lambda_{r+s}}\partial_\nu\phi^\mu_{\lambda_1\dots\lambda_r} -$$
$$r\phi^\mu_{\lambda_1\dots\lambda_{r-1}\nu}\partial_{\lambda_r}\sigma^\nu_{\lambda_{r+1}\dots\lambda_{r+s}} + s\sigma^\mu_{\nu\lambda_{r+2}\dots\lambda_{r+s}}\partial_{\lambda_{r+1}}\phi^\nu_{\lambda_1\dots\lambda_r})dz^{\lambda_1}\wedge\cdots\wedge dz^{\lambda_{r+s}}\otimes\partial_\mu,$$
$$\phi\in\mathcal{O}^r(Z)\otimes\mathcal{T}(Z), \qquad \sigma\in\mathcal{O}^s(Z)\otimes\mathcal{T}(Z).$$

There are the relations

$$[\phi,\sigma]_{\rm FN} = (-1)^{|\phi||\psi|+1}[\sigma,\phi]_{\rm FN},$$
$$[\phi,[\sigma,\theta]_{\rm FN}]_{\rm FN} = [[\phi,\sigma]_{\rm FN},\theta]_{\rm FN} + (-1)^{|\phi||\sigma|}[\sigma,[\phi,\theta]_{\rm FN}]_{\rm FN}. \qquad (B.38)$$

Given a tangent-valued form θ, the *Nijenhuis differential* on $\mathcal{O}^*(Z)\otimes\mathcal{T}(Z)$ is defined as a morphism

$$d_\theta : \psi \to d_\theta\psi = [\theta,\psi]_{\rm FN}, \qquad \psi\in\mathcal{O}^*(Z)\otimes\mathcal{T}(Z). \qquad (B.39)$$

By virtue of the relation (B.38), it has a property

$$d_\phi[\psi,\theta]_{\rm FN} = [d_\phi\psi,\theta]_{\rm FN} + (-1)^{|\phi||\psi|}[\psi,d_\phi\theta]_{\rm FN}.$$

In particular, if $\theta = u$ is a vector field, the Nijenhuis differential is a *Lie derivative of tangent-valued forms*

$$\mathbf{L}_u\sigma = d_u\sigma = [u,\sigma]_{\rm FN} = \frac{1}{s!}(u^\nu\partial_\nu\sigma^\mu_{\lambda_1\dots\lambda_s} - \sigma^\nu_{\lambda_1\dots\lambda_s}\partial_\nu u^\mu +$$
$$s\sigma^\mu_{\nu\lambda_2\dots\lambda_s}\partial_{\lambda_1}u^\nu)dx^{\lambda_1}\wedge\cdots\wedge dx^{\lambda_s}\otimes\partial_\mu, \qquad \sigma\in\mathcal{O}^s(Z)\otimes\mathcal{T}(Z).$$

Let $Y\to X$ be a fibre bundle. In the book, we deal with the following particular subspaces of the space $\mathcal{O}^*(Y)\otimes\mathcal{T}(Y)$ of tangent-valued forms on Y:

- *horizontal tangent-valued forms*

$$\phi : Y \to \overset{r}{\wedge} T^*X \underset{Y}{\otimes} TY, \quad \phi = dx^{\lambda_1}\wedge\cdots\wedge dx^{\lambda_r}\otimes\frac{1}{r!}[\phi^\mu_{\lambda_1\dots\lambda_r}(y)\partial_\mu + \phi^i_{\lambda_1\dots\lambda_r}(y)\partial_i],$$

- *vertical-valued forms*

$$\phi : Y \to \overset{r}{\wedge} T^*X \underset{Y}{\otimes} VY, \quad \phi = \frac{1}{r!}\phi^i_{\lambda_1\dots\lambda_r}(y)dx^{\lambda_1}\wedge\cdots\wedge dx^{\lambda_r}\otimes\partial_i,$$

- vertical-valued one-forms, called the *soldering forms*, $\sigma = \sigma^i_\lambda(y)dx^\lambda\otimes\partial_i$,
- *basic soldering forms* $\sigma = \sigma^i_\lambda(x)dx^\lambda\otimes\partial_i$.

Remark B.8 The tangent bundle TX is provided with the canonical basic soldering form θ_J (B.36). Due to the canonical vertical splitting

$$VTX = TX \underset{X}{\times} TX, \tag{B.40}$$

the canonical soldering form (B.36) on TX defines the canonical tangent-valued form θ_X (B.35) on X. By this reason, tangent-valued one-forms on a manifold X also are called the soldering forms.

Let us also mention TX-valued forms

$$\phi : Y \to \overset{r}{\wedge} T^*X \underset{Y}{\otimes} TX, \quad \phi = \frac{1}{r!}\phi^{\mu}_{\lambda_1 \ldots \lambda_r} dx^{\lambda_1} \wedge \cdots \wedge dx^{\lambda_r} \otimes \partial_\mu, \tag{B.41}$$

and V^*Y-valued forms

$$\phi : Y \to \overset{r}{\wedge} T^*X \underset{Y}{\otimes} V^*Y, \quad \phi = \frac{1}{r!}\phi_{\lambda_1 \ldots \lambda_r i} dx^{\lambda_1} \wedge \cdots \wedge dx^{\lambda_r} \otimes \overline{d}y^i. \tag{B.42}$$

Let us note that (B.41) are not tangent-valued forms, while (B.42) are not exterior forms. They exemplify *vector-valued forms*. Given a vector bundle $E \to X$, by a E-valued k-form on X is meant a section of a fibre bundle $(\overset{k}{\wedge} T^*X) \otimes_X E^* \to X$.

B.2 Jet Manifolds

This section addresses first and second order jet manifolds of sections of fibre bundles (see Sect. B.4) for higher order jets) [53, 83, 89, 133, 138].

Given a fibre bundle $Y \to X$ with bundle coordinates (x^λ, y^i), let us consider the equivalence classes $j^1_x s$ of its sections s, which are identified by their values $s^i(x)$ and the values of their partial derivatives $\partial_\mu s^i(x)$ at a point $x \in X$. They are called the *first order jets* of sections at x. Let us note that the definition of jets is coordinate-independent. A key point is that the set J^1Y of first order jets $j^1_x s$, $x \in X$, is a smooth manifold with respect to the adapted coordinates $(x^\lambda, y^i, y^i_\lambda)$ such that

$$y^i_\lambda(j^1_x s) = \partial_\lambda s^i(x), \qquad y'^i_\lambda = \frac{\partial x^\mu}{\partial x'^\lambda}(\partial_\mu + y^j_\mu \partial_j)y'^i. \tag{B.43}$$

Remark B.9 Let us note that there are different notions of jets. Jets of sections are the particular jets of maps [83, 110] and the jets of submanifolds [53, 85]. Let us also mention the jets of modules over a commutative ring [85, 131] which are representative objects of differential operators on modules. In particular, given a manifold X, the jets of a projective $C^\infty(X)$-module P of finite rank are exactly the jets of sections of the vector bundle over X whose module of sections is P in accordance with the Serre–Swan Theorem A.10. The notion of jets is extended to modules over graded commutative rings, though the jets of modules over a noncommutative ring fails to be defined [58, 131].

Definition B.8 Endowed with a smooth manifold structure, a set of first order jets J^1Y is termed the first order *jet manifold* of a fibre bundle $Y \to X$.

A jet manifold J^1Y admits the natural fibrations

$$\pi^1 : J^1Y \ni j^1_x s \to x \in X, \tag{B.44}$$
$$\pi^1_0 : J^1Y \ni j^1_x s \to s(x) \in Y. \tag{B.45}$$

A glance at the transformation law (B.43) shows that π^1_0 is an affine bundle modelled over a vector bundle

$$T^*X \underset{Y}{\otimes} VY \to Y. \tag{B.46}$$

It is convenient to call π^1 (B.44) the *jet bundle*, while π^1_0 (B.45) is said to be the *affine jet bundle*. Let us note that, if $Y \to X$ is a vector or an affine bundle, the jet bundle π^1 (B.44) is so.

Jets can be expressed in terms of familiar tangent-valued forms owing to the canonical imbeddings

$$\lambda_{(1)} : J^1Y \underset{Y}{\to} T^*X \underset{Y}{\otimes} TY, \qquad \lambda_{(1)} = dx^\lambda \otimes (\partial_\lambda + y^i_\lambda \partial_i) = dx^\lambda \otimes d_\lambda, \tag{B.47}$$
$$\theta_{(1)} : J^1Y \underset{Y}{\to} T^*Y \underset{Y}{\otimes} VY, \qquad \theta_{(1)} = (dy^i - y^i_\lambda dx^\lambda) \otimes \partial_i = \theta^i \otimes \partial_i, \tag{B.48}$$

where d_λ are said to be *total derivatives*, and θ^i are called *local contact forms*.

Remark B.10 We usually identify a jet manifold J^1Y with its images under the canonical morphisms (B.47) and (B.48), and represent the jets $j^1_x s = (x^\lambda, y^i, y^i_\mu)$ by the tangent-valued forms $\lambda_{(1)}$ (B.47) and $\theta_{(1)}$ (B.48).

Sections and morphisms of fibre bundles admit prolongations to jet manifolds as follows.

Any section s of a fibre bundle $Y \to X$ has the *jet prolongation* to a section

$$(J^1 s)(x) = j^1_x s, \qquad y^i_\lambda \circ J^1 s = \partial_\lambda s^i(x), \tag{B.49}$$

of a jet bundle $J^1Y \to X$. A *section of a jet bundle* $J^1Y \to X$ is termed *integrable* if it is the jet prolongation (B.49) of some section of a fibre bundle $Y \to X$.

Any bundle morphism $\Phi : Y \to Y'$ (B.8) over a diffeomorphism f admits the *jet prolongation* to a bundle morphism of affine jet bundles

$$J^1\Phi : J^1Y \underset{\Phi}{\longrightarrow} J^1Y', \qquad y'^i_\lambda \circ J^1\Phi = \frac{\partial (f^{-1})^\mu}{\partial x'^\lambda} d_\mu \Phi^i. \tag{B.50}$$

Any projectable vector field u (B.25) on a fibre bundle $Y \to X$ has the *jet prolongation* to a projectable vector field

$$J^1u = u^\lambda\partial_\lambda + u^i\partial_i + (d_\lambda u^i - y^i_\mu\partial_\lambda u^\mu)\partial_i^\lambda, \tag{B.51}$$

on a jet manifold J^1Y. In order to obtain (B.51), the canonical bundle morphism

$$r_1 : J^1TY \to TJ^1Y, \qquad \dot{y}^i_\lambda \circ r_1 = (\dot{y}^i)_\lambda - y^i_\mu \dot{x}^\mu_\lambda$$

is used. In particular, there is the canonical isomorphism

$$VJ^1Y = J^1VY, \qquad \dot{y}^i_\lambda = (\dot{y}^i)_\lambda.$$

Taking the first order jet manifold of the jet bundle $J^1Y \to X$, we obtain the *repeated jet manifold* J^1J^1Y provided with the adapted coordinates $(x^\lambda, y^i, y^i_\lambda, \widehat{y}^i_\mu, y^i_{\mu\lambda})$ possessing transition functions

$$y'^i_\lambda = \frac{\partial x^\alpha}{\partial x'^\lambda} d_\alpha y'^i, \qquad \widehat{y}'^i_\lambda = \frac{\partial x^\alpha}{\partial x'^\lambda}\widehat{d}_\alpha y'^i, \qquad y'^i_{\mu\lambda} = \frac{\partial x^\alpha}{\partial x'^\mu}\widehat{d}_\alpha y'^i_\lambda,$$

$$d_\alpha = \partial_\alpha + y^j_\alpha\partial_j + y^j_{\nu\alpha}\partial^\nu_j, \qquad \widehat{d}_\alpha = \partial_\alpha + \widehat{y}^j_\alpha\partial_j + y^j_{\nu\alpha}\partial^\nu_j.$$

There exist the two following different affine fibrations of J^1J^1Y over J^1Y:

- the familiar affine jet bundle (B.45):

$$\pi_{11} : J^1J^1Y \to J^1Y, \qquad y^i_\lambda \circ \pi_{11} = y^i_\lambda,$$

- the affine bundle

$$J^1\pi^1_0 : J^1J^1Y \to J^1Y, \qquad y^i_\lambda \circ J^1\pi^1_0 = \widehat{y}^i_\lambda.$$

The points $q \in J^1J^1Y$, where $\pi_{11}(q) = J^1\pi^1_0(q)$, form an affine subbundle $\widehat{J}^2Y \to J^1Y$ of J^1J^1Y called the *sesquiholonomic jet manifold*. It is given by coordinate conditions $\widehat{y}^i_\lambda = y^i_\lambda$, and is coordinated by $(x^\lambda, y^i, y^i_\lambda, y^i_{\mu\lambda})$. A *second order jet manifold* J^2Y of a fibre bundle $Y \to X$ can be defined as the affine subbundle of the fibre bundle $\widehat{J}^2Y \to J^1Y$ by means of coordinate conditions $y^i_{\lambda\mu} = y^i_{\mu\lambda}$. It is endowed with adapted coordinates $(x^\lambda, y^i, y^i_\lambda, y^i_{\lambda\mu} = y^i_{\mu\lambda})$, possessing transition functions

$$y'^i_\lambda = \frac{\partial x^\alpha}{\partial x'^\lambda} d_\alpha y'^i, \qquad y'^i_{\mu\lambda} = \frac{\partial x^\alpha}{\partial x'^\mu} d_\alpha y'^i_\lambda.$$

A second order jet manifold J^2Y also can be introduced as a set of equivalence classes j^2_xs of sections s of a fibre bundle $Y \to X$, which are identified by their values and the values of their first and second order partial derivatives at points $x \in X$. The equivalence classes j^2_xs are called the *second order jets* of sections.

Let s be a section of a fibre bundle $Y \to X$, and let J^1s be its jet prolongation to a section of a jet bundle $J^1Y \to X$. The latter gives rise to a section J^1J^1s of a repeated jet bundle $J^1J^1Y \to X$. This section lives in a second order jet manifold J^2Y. It is termed the second order jet prolongation of a section s, and is denoted by J^2s.

B.3 Connections on Fibre Bundles

There are several equivalent definitions of a connection on a fibre bundle. We start with the traditional notion of a connection as a splitting of the exact sequences (B.21) and (B.22) (Definition B.9), but then follow its Definition B.10 as a global section of the affine jet bundle (B.45) [83, 100, 133, 138].

Connections as Tangent-Valued Forms

Let $Y \to X$ be a fibre bundle coordinated by (x^λ, y^i) (Definition B.2).

Definition B.9 A *connection* on a fibre bundle $Y \to X$ is defined as a linear bundle monomorphism

$$\Gamma : Y \underset{X}{\times} TX \to TY, \qquad \Gamma : \dot{x}^\lambda \partial_\lambda \to \dot{x}^\lambda(\partial_\lambda + \Gamma^i_\lambda \partial_i), \tag{B.52}$$

over Y which splits the exact sequence (B.21), i.e., $\pi_T \circ \Gamma = \mathrm{Id}\,(Y \times_X TX)$.

This also is a definition of connections on fibred manifolds (Remark B.11).

By virtue of Theorem B.13, a connection always exists. Local functions $\Gamma^i_\lambda(y)$ in (B.52) are called the *components* of a connection Γ with respect to bundle coordinates (x^λ, y^i) on $Y \to X$.

The image of $Y \times_X TX$ by a connection Γ defines the *horizontal subbundle* $HY \subset TY$ which splits the tangent bundle TY as follows:

$$TY = HY \underset{Y}{\oplus} VY, \qquad \dot{x}^\lambda \partial_\lambda + \dot{y}^i \partial_i = \dot{x}^\lambda(\partial_\lambda + \Gamma^i_\lambda \partial_i) + (\dot{y}^i - \dot{x}^\lambda \Gamma^i_\lambda)\partial_i. \tag{B.53}$$

Its annihilator is locally generated by one-forms $dy^i - \Gamma^i_\lambda dx^\lambda$.

The morphism Γ (B.52) uniquely yields a horizontal tangent-valued one-form

$$\Gamma = dx^\lambda \otimes (\partial_\lambda + \Gamma^i_\lambda \partial_i) \tag{B.54}$$

on Y which projects onto the canonical tangent-valued form θ_X (B.35) on X. With this form, called the *connection form*, the morphism (B.52) reads

$$\Gamma : \partial_\lambda \to \partial_\lambda \rfloor \Gamma = \partial_\lambda + \Gamma^i_\lambda \partial_i.$$

Given a connection Γ and the corresponding horizontal subbundle (B.53), a vector field u on a fibre bundle $Y \to X$ is termed *horizontal* if it lives in HY. A horizontal vector field takes a form

$$u = u^\lambda(y)(\partial_\lambda + \Gamma^i_\lambda \partial_i).$$

In particular, let τ be a vector field on X. By means of the connection form Γ (B.54), we obtain a projectable horizontal vector field

$$\Gamma\tau = \tau \rfloor \Gamma = \tau^\lambda(\partial_\lambda + \Gamma^i_\lambda \partial_i) \tag{B.55}$$

on Y, called the *horizontal lift* of τ. Conversely, any projectable horizontal vector field u on Y is the horizontal lift $\Gamma\tau$ of its projection τ on X. Moreover, the horizontal distribution HY is generated by the horizontal lifts $\Gamma\tau$ (B.55) of vector fields τ on X. The horizontal lift

$$\mathcal{T}(X) \ni \tau \to \Gamma\tau \in \mathcal{T}(Y)$$

is a $C^\infty(X)$-linear module morphism. Given the splitting (B.52), the dual splitting of the exact sequence (B.22) is

$$\Gamma : V^*Y \to T^*Y, \qquad \Gamma : \overline{d}y^i \to dy^i - \Gamma^i_\lambda dx^\lambda. \tag{B.56}$$

Hence, a connection Γ on $Y \to X$ also is represented by a vertical-valued form

$$\Gamma = (dy^i - \Gamma^i_\lambda dx^\lambda) \otimes \partial_i.$$

Remark B.11 Let $\pi : Y \to X$ be a fibred manifold. Any connection Γ on $Y \to X$ yields a horizontal lift of a vector field on X onto Y, but need not defines the similar lift of a *path* in X into Y. Let

$$\mathbb{R} \supset [,] \ni t \to x(t) \in X, \qquad \mathbb{R} \ni t \to y(t) \in Y,$$

be paths in X and Y, respectively. Then $t \to y(t)$ is called a *horizontal lift* of $x(t)$ if

$$\pi(y(t)) = x(t), \qquad \dot{y}(t) \in H_{y(t)}Y, \qquad t \in \mathbb{R},$$

where $HY \subset TY$ is a horizontal subbundle associated to Γ. If, for each path $x(t)$ $(t_0 \leq t \leq t_1)$ and any $y_0 \in \pi^{-1}(x(t_0))$, there exists a horizontal lift $y(t)$ $(t_0 \leq t \leq t_1)$ so that $y(t_0) = y_0$, then Γ is termed the *Ehresmann connection*. A fibred manifold is a fibre bundle if and only if it admits an Ehresmann connection [67].

Connections as Jet Bundle Sections

Throughout the book, we follow the equivalent Definition B.10 of connections on a fibre bundle $Y \to X$ as sections of the affine jet bundle $J^1Y \to Y$ (B.45).

Let $Y \to X$ be a fibre bundle and J^1Y its first order jet manifold. Given the canonical morphisms (B.47) and (B.48), we have the corresponding morphisms

$$\widehat{\lambda}_{(1)} : J^1Y \underset{X}{\times} TX \ni \partial_\lambda \to d_\lambda = \partial_\lambda \rfloor \lambda_{(1)} \in J^1Y \underset{Y}{\times} TY,$$
$$\widehat{\theta}_{(1)} : J^1Y \underset{Y}{\times} V^*Y \ni \overline{d}y^i \to \theta^i = \theta_{(1)} \rfloor dy^i \in J^1Y \underset{Y}{\times} T^*Y$$

(Remark B.3). They yield the *canonical horizontal splittings* of pull-back bundles

$$J^1Y \underset{Y}{\times} TY = \widehat{\lambda}_{(1)}(TX) \underset{J^1Y}{\oplus} VY, \tag{B.57}$$
$$\dot{x}^\lambda \partial_\lambda + \dot{y}^i \partial_i = \dot{x}^\lambda(\partial_\lambda + y^i_\lambda \partial_i) + (\dot{y}^i - \dot{x}^\lambda y^i_\lambda)\partial_i,$$
$$J^1Y \underset{Y}{\times} T^*Y = T^*X \underset{J^1Y}{\oplus} \widehat{\theta}_{(1)}(V^*Y), \tag{B.58}$$
$$\dot{x}_\lambda dx^\lambda + \dot{y}_i dy^i = (\dot{x}_\lambda + \dot{y}_i y^i_\lambda) dx^\lambda + \dot{y}_i (dy^i - y^i_\lambda dx^\lambda).$$

Let Γ be a global section of $J^1Y \to Y$. Substituting the tangent-valued form

$$\lambda_{(1)} \circ \Gamma = dx^\lambda \otimes (\partial_\lambda + \Gamma^i_\lambda \partial_i)$$

in the canonical splitting (B.57), we obtain the familiar horizontal splitting (B.53) of TY by means of a connection Γ on $Y \to X$. Accordingly, substitution of the tangent-valued form

$$\theta_{(1)} \circ \Gamma = (dy^i - \Gamma^i_\lambda dx^\lambda) \otimes \partial_i$$

in the canonical splitting (B.58) leads to the dual splitting (B.56) of T^*Y by means of a connection Γ.

Definition B.10 A connection Γ on a fibre bundle $Y \to X$ equivalently is defined as a global section

$$\Gamma : Y \to J^1Y, \qquad (x^\lambda, y^i, y^i_\lambda) \circ \Gamma = (x^\lambda, y^i, \Gamma^i_\lambda), \tag{B.59}$$

of the affine jet bundle $J^1Y \to Y$.

The following are corollaries of this definition.

- Since $J^1Y \to Y$ is affine, a connection on a fibre bundle $Y \to X$ exists in accordance with Theorem B.4.
- Connections on a fibre bundle $Y \to X$ make up an affine space modelled over a vector space of soldering forms on Y, i.e., sections of the vector bundle (B.46).

- Connection components possess transition functions

$$\Gamma'^i_\lambda = \frac{\partial x^\mu}{\partial x'^\lambda}(\partial_\mu + \Gamma^j_\mu \partial_j)y'^i.$$

- Every connection Γ (B.59) on a fibre bundle $Y \to X$ yields a first order differential operator

$$D_\Gamma : J^1Y \underset{Y}{\to} T^*X \underset{Y}{\otimes} VY, \qquad D_\Gamma = \lambda_{(1)} - \Gamma \circ \pi^1_0 = (y^i_\lambda - \Gamma^i_\lambda)dx^\lambda \otimes \partial_i, \quad \text{(B.60)}$$

  on Y called the *covariant differential* relative to a connection Γ. If $s : X \to Y$ is a section, from (B.60) we obtain its covariant differential

$$\nabla^\Gamma s = D_\Gamma \circ J^1 s : X \to T^*X \otimes VY, \qquad \nabla^\Gamma s = (\partial_\lambda s^i - \Gamma^i_\lambda \circ s)dx^\lambda \otimes \partial_i, \quad \text{(B.61)}$$

  and the *covariant derivative* $\nabla^\Gamma_\tau = \tau \rfloor \nabla^\Gamma$ along a vector field τ on X.

A section s is said to be an *integral section* for a connection Γ if it belongs to the kernel of the covariant differential D_Γ (B.60), i.e.,

$$\nabla^\Gamma s = 0 \quad \text{or} \quad J^1 s = \Gamma \circ s.$$

Theorem B.14 *For any global section $s : X \to Y$, there always exists a connection Γ such that s is an integral section for Γ.*

Proof This connection Γ is an extension of the local section $s(x) \to J^1 s(x)$ of the affine jet bundle $J^1 Y \to Y$ over a closed imbedded submanifold $s(X) \subset Y$ in accordance with Theorem B.4. □

Curvature and Torsion

Let Γ be a connection on a fibre bundle $Y \to X$. Its *curvature* is defined as the Nijenhuis differential

$$R = \frac{1}{2} d_\Gamma \Gamma = \frac{1}{2}[\Gamma, \Gamma]_{\mathrm{FN}} : Y \to \overset{2}{\wedge} T^*X \otimes VY, \quad \text{(B.62)}$$

$$R = \frac{1}{2} R^i_{\lambda\mu} dx^\lambda \wedge dx^\mu \otimes \partial_i, \quad R^i_{\lambda\mu} = \partial_\lambda \Gamma^i_\mu - \partial_\mu \Gamma^i_\lambda + \Gamma^j_\lambda \partial_j \Gamma^i_\mu - \Gamma^j_\mu \partial_j \Gamma^i_\lambda.$$

This is a VY-valued horizontal two-form on Y.

Given a connection Γ and a soldering form σ, the *torsion form* of Γ with respect to σ is defined as

$$T = d_\Gamma \sigma = d_\sigma \Gamma : Y \to \overset{2}{\wedge} T^*X \otimes VY,$$
$$T = (\partial_\lambda \sigma^i_\mu + \Gamma^j_\lambda \partial_j \sigma^i_\mu - \partial_j \Gamma^i_\lambda \sigma^j_\mu) dx^\lambda \wedge dx^\mu \otimes \partial_i. \tag{B.63}$$

Flat Connections

A connection g on a fibre bundle $Y \to X$ is called the *flat* connection if its curvature R (B.62) vanishes. The following fact is important for our consideration [100].

Theorem B.15 *Let Γ be a flat connection on a fibre bundle $Y \to X$. Then there exists a bundle coordinate atlas (x^λ, y^i) on Y where transition functions of fibre coordinates $y^i \to y'^i(y^j)$ are independent of base coordinates x^λ (it is termed the atlas of constant local trivializations) such that, relative to these coordinates, the connection form (B.54) reads*

$$\Gamma = dx^\lambda \otimes \partial_\lambda. \tag{B.64}$$

Conversely, if a fibre bundle $Y \to X$ admits an atlas of constant local trivializations, there exists a flat connection on $Y \to X$ which takes the form (B.64) with respect to this atlas.

In particular, if $Y \to X$ is a trivial bundle, one can associate the flat connection (B.64) to each its trivialization.

Theorem B.16 *A connection Γ on a fibre bundle $Y \to X$ is flat if and only if the horizontal lift (B.55) of vector fields on X is functorial, i.e., the relation (B.28) holds.*

Proof Given vector fields τ, τ' on X and their horizontal lifts $\Gamma\tau$ and $\Gamma\tau'$ (B.55) on Y, we have the relation

$$R(\tau, \tau') = -\Gamma[\tau, \tau'] + [\Gamma\tau, \Gamma\tau'] = \tau^\lambda \tau'^\mu R^i_{\lambda\mu} \partial_i. \qquad \square$$

Theorem B.17 *A connection Γ on a fibre bundle $Y \to X$ is flat if and only if there exists its local integral section through any point $y \in Y$.*

Proof The proof follows from the expression (B.64). $\square$

Linear Connections

A connection Γ on a vector bundle $Y \to X$ is called the *linear connection* if

$$\Gamma : Y \to J^1Y, \qquad \Gamma = dx^\lambda \otimes (\partial_\lambda + \Gamma_\lambda{}^i{}_j(x) y^j \partial_i), \tag{B.65}$$

is a linear bundle morphism over X. Let us note that linear connections are principal connections (Sect. 8.1), and they always exist (Theorem 8.1).

The curvature R (B.62) of a linear connection Γ (B.65) reads

$$R = \frac{1}{2} R_{\lambda\mu}{}^i{}_j(x) y^j dx^\lambda \wedge dx^\mu \otimes \partial_i,$$
$$R_{\lambda\mu}{}^i{}_j = \partial_\lambda \Gamma_\mu{}^i{}_j - \partial_\mu \Gamma_\lambda{}^i{}_j + \Gamma_\lambda{}^h{}_j \Gamma_\mu{}^i{}_h - \Gamma_\mu{}^h{}_j \Gamma_\lambda{}^i{}_h. \tag{B.66}$$

The following are constructions of new linear connections from the old ones.

- Let $Y \to X$ be a vector bundle, coordinated by (x^λ, y^i), and $Y^* \to X$ its dual, coordinated by (x^λ, y_i). Any linear connection Γ (B.65) on a vector bundle $Y \to X$ defines the *dual linear connection*

$$\Gamma^* = dx^\lambda \otimes (\partial_\lambda - \Gamma_\lambda{}^j{}_i(x) y_j \partial^i) \tag{B.67}$$

  on $Y^* \to X$.
- Let Y coordinated by (x^λ, y^i) and Y' coordinated by (x^λ, y^a) be vector bundles over the same base X. Their tensor product $Y \otimes Y'$ is endowed with the bundle coordinates (x^λ, y^{ia}). Linear connections Γ and Γ' on $Y \to X$ and $Y' \to X$ define the linear *tensor product connection*

$$\Gamma \otimes \Gamma' = dx^\lambda \otimes \left[\partial_\lambda + (\Gamma_\lambda{}^i{}_j y^{ja} + \Gamma'_\lambda{}^a{}_b y^{ib}) \frac{\partial}{\partial y^{ia}} \right] \tag{B.68}$$

  on $Y \otimes_X Y' \to X$.

An important example of linear connections is a linear connection

$$\Gamma = dx^\lambda \otimes (\partial_\lambda + \Gamma_\lambda{}^\mu{}_\nu \dot{x}^\nu \dot{\partial}_\mu) \tag{B.69}$$

on the tangent bundle TX of a manifold X. We agree to call it the *world connection* on X. The *dual world connection* (B.67) on the cotangent bundle T^*X is

$$\Gamma^* = dx^\lambda \otimes (\partial_\lambda - \Gamma_\lambda{}^\mu{}_\nu \dot{x}_\mu \dot{\partial}^\nu). \tag{B.70}$$

Then, using the construction of the tensor product connection (B.68), one can introduce the corresponding linear connection on an arbitrary tensor bundle T (B.16).

A *curvature of a world connection* is defined as the curvature R (B.66) of the connection Γ (B.69) on the tangent bundle TX. It reads

$$R = \frac{1}{2} R_{\lambda\mu}{}^\alpha{}_\beta \dot{x}^\beta dx^\lambda \wedge dx^\mu \otimes \dot{\partial}_\alpha, \tag{B.71}$$
$$R_{\lambda\mu}{}^\alpha{}_\beta = \partial_\lambda \Gamma_\mu{}^\alpha{}_\beta - \partial_\mu \Gamma_\lambda{}^\alpha{}_\beta + \Gamma_\lambda{}^\gamma{}_\beta \Gamma_\mu{}^\alpha{}_\gamma - \Gamma_\mu{}^\gamma{}_\beta \Gamma_\lambda{}^\alpha{}_\gamma.$$

By a *torsion of a world connection* is meant the torsion (B.63) of the connection Γ (B.69) on TX relative to the canonical soldering form θ_J (B.36):

$$T = \frac{1}{2} T_\mu{}^\nu{}_\lambda dx^\lambda \wedge dx^\mu \otimes \dot{\partial}_\nu, \qquad T_\mu{}^\nu{}_\lambda = \Gamma_\mu{}^\nu{}_\lambda - \Gamma_\lambda{}^\nu{}_\mu. \tag{B.72}$$

A world connection is called *symmetric* if its torsion (B.72) vanishes.

Remark B.12 Every manifold X can be provided with a nondegenerate fibre metric

$$g \in \overset{2}{\vee} \mathcal{O}^1(X), \qquad g = g_{\lambda\mu} dx^\lambda \otimes dx^\mu,$$
$$g \in \overset{2}{\vee} \mathcal{T}(X), \qquad g = g^{\lambda\mu} \partial_\lambda \otimes \partial_\mu,$$

in the tangent and cotangent bundles. We call it the *world metric* on X. For any world metric g, there exists a unique symmetric world connection Γ (B.69):

$$\Gamma_\lambda{}^\nu{}_\mu = \{_\lambda{}^\nu{}_\mu\} = -\frac{1}{2} g^{\nu\rho}(\partial_\lambda g_{\rho\mu} + \partial_\mu g_{\rho\lambda} - \partial_\rho g_{\lambda\mu}), \tag{B.73}$$

termed the *Christoffel symbols*, such that $\nabla^\Gamma g = 0$. It is called the *Levi–Civita connection* associated to g.

B.4 Higher Order Jet Manifolds

The notion of first and second order jets of sections of a fibre bundle (Sect. B.2) is naturally extended to the higher order ones [53, 83, 133, 138].

Remark B.13 Let $Y \to X$ be a fibre bundle. Given its bundle coordinates (x^λ, y^i), a *multi-index* Λ of the length $|\Lambda| = k$ throughout denotes a collection of indices $(\lambda_1 ... \lambda_k)$ modulo permutations. By $\Lambda + \Sigma$ is meant a multi-index $(\lambda_1 \dots \lambda_k \sigma_1 \dots \sigma_r)$. For instance $\lambda + \Lambda = (\lambda\lambda_1 ... \lambda_r)$. By $\Lambda\Sigma$ is denoted the union of collections $(\lambda_1 \dots \lambda_k; \sigma_1 \dots \sigma_r)$ where the indices λ_i and σ_j are not permitted. Summation over a multi-index Λ means separate summation over each its index λ_i. We use the compact notation $\partial_\Lambda = \partial_{\lambda_k} \circ \cdots \circ \partial_{\lambda_1}$, $\Lambda = (\lambda_1 ... \lambda_k)$.

Definition B.11 An *r-order jet manifold* $J^r Y$ of sections of a fibre bundle $Y \to X$ is defined as the disjoint union of equivalence classes $j^r_x s$ of sections s of $Y \to X$ such that sections s and s' belong to the same equivalence class $j^r_x s$ if and only if

$$s^i(x) = s'^i(x), \qquad \partial_\Lambda s^i(x) = \partial_\Lambda s'^i(x), \qquad 0 < |\Lambda| \leq r.$$

In brief, one can say that sections of $Y \to X$ are identified by the $r + 1$ terms of their Taylor series at points of X. The particular choice of coordinates does not matter for this definition. The equivalence classes $j^r_x s$ are called the *r-order jets* of sections. Their set $J^r Y$ is endowed with an atlas of the adapted coordinates

$$(x^\lambda, y^i_\Lambda), \qquad y^i_\Lambda \circ s = \partial_\Lambda s^i(x), \qquad 0 \leq |\Lambda| \leq r, \tag{B.74}$$

possessing transition functions

$$y'^i_{\lambda+\Lambda} = \frac{\partial x^\mu}{\partial' x^\lambda} d_\mu y'^i_\Lambda, \tag{B.75}$$

where the symbol d_λ stands for the *higher order total derivative*

$$d_\lambda = \partial_\lambda + \sum_{0\leq|\Lambda|\leq r-1} y^i_{\Lambda+\lambda}\partial^\Lambda_i, \qquad d'_\lambda = \frac{\partial x^\mu}{\partial x'^\lambda} d_\mu.$$

These derivatives act on exterior forms on J^rY and obey the relations

$$\begin{aligned}
&[d_\lambda, d_\mu] = 0, \qquad d_\lambda \circ d = d \circ d_\lambda, \\
&d_\lambda(\phi \wedge \sigma) = d_\lambda(\phi) \wedge \sigma + \phi \wedge d_\lambda(\sigma), \\
&d_\lambda(dx^\mu) = 0, \qquad d_\lambda(dy^i_\Lambda) = dy^i_{\lambda+\Lambda}.
\end{aligned}$$

We use the compact notation $d_\Lambda = d_{\lambda_r} \circ \cdots \circ d_{\lambda_1}$, $\Lambda = (\lambda_r ... \lambda_1)$.

The coordinates (B.74) brings a set J^rY into a manifold. They are compatible with the natural surjections $\pi^r_k : J^rY \to J^kY$, $r > k$, which form a composite bundle

$$\begin{aligned}
&\pi^r : J^rY \xrightarrow{\pi^r_{r-1}} J^{r-1}Y \xrightarrow{\pi^{r-1}_{r-2}} \cdots \xrightarrow{\pi^1_0} Y \xrightarrow{\pi} X, \\
&\pi^k_s \circ \pi^r_k = \pi^r_s, \qquad \pi^s \circ \pi^r_s = \pi^r.
\end{aligned}$$

A glance at the transition functions (B.75), when $|\Lambda| = r$, shows that the fibration

$$\pi^r_{r-1} : J^rY \to J^{r-1}Y \tag{B.76}$$

is an affine bundle modelled over a vector bundle

$$\overset{r}{\vee} T^*X \underset{J^{r-1}Y}{\otimes} VY \to J^{r-1}Y. \tag{B.77}$$

In particular, $J^rY \to Y$ is an affine bundle with fibre coordinates (y^i_Λ), $|\Lambda| = 1, \ldots, r$, which we agree to call the *jet coordinates*. For the sake of convenience, let us also denote $J^0Y = Y$.

Remark B.14 Let us recall that a base of any affine bundle is a strong deformation retract of its total space. Consequently, Y is a strong deformation retract of J^1Y, which in turn is a strong deformation retract of J^2Y, and so on. It follows that a fibre bundle Y is a strong deformation retract of any finite order jet manifold J^rY. Therefore, by virtue of the Vietoris–Begle theorem [23], there is an isomorphism

$$H^*(J^rY; \mathbb{R}) = H^*(Y; \mathbb{R}) \tag{B.78}$$

of cohomology groups of J^rY and Y with coefficients in the constant sheaf $\mathbb{R}$.

In the calculus in higher order jets, we have the r-order *jet prolongation functor* such that, given fibre bundles Y and Y' over X, every bundle morphism $\Phi : Y \to Y'$ (B.8) over a diffeomorphism f of X admits the r-order *jet prolongation* to a morphism of r-order jet manifolds

$$J^r\Phi : J^rY \ni j^r_x s \to j^r_{f(x)}(\Phi \circ s \circ f^{-1}) \in J^rY'.$$

The jet prolongation functor is exact. If Φ is an injection or a surjection, so is $J^r\Phi$. It also preserves an algebraic structure. In particular, if $Y \to X$ is a vector bundle, $J^rY \to X$ is well. If $Y \to X$ is an affine bundle modelled over a vector bundle $\overline{Y} \to X$, then $J^rY \to X$ is an affine bundle modelled over a vector bundle $J^r\overline{Y} \to X$.

Every section s of a fibre bundle $Y \to X$ admits the r-order *jet prolongation* to a section $(J^rs)(x) = j^r_x s$ of a jet bundle $J^rY \to X$.

Let $\mathcal{O}^*_k = \mathcal{O}^*(J^kY)$ be a DGA of exterior forms on a jet manifold J^kY. Every exterior form ϕ on a jet manifold J^kY gives rise to the pull-back form $\pi^{k+i}_k{}^*\phi$ on a jet manifold $J^{k+i}Y$. We have a direct sequence of $C^\infty(X)$-algebras

$$\mathcal{O}^*(X) \xrightarrow{\pi^*} \mathcal{O}^*(Y) \xrightarrow{\pi^1_0{}^*} \mathcal{O}^*_1 \xrightarrow{\pi^2_1{}^*} \cdots \xrightarrow{\pi^r_{r-1}{}^*} \mathcal{O}^*_r.$$

Remark B.15 By virtue of de Rham Theorem C.10, the cohomology of the de Rham complex of $\mathcal{O}^*_k$ equals the cohomology $H^*(J^kY;\mathbb{R})$ of J^kY with coefficients in the constant sheaf $\mathbb{R}$. The latter in turn coincides with the sheaf cohomology $H^*(Y;\mathbb{R})$ of Y (Remark B.14) and, thus, it equals the de Rham cohomology $H^*_{\rm DR}(Y)$ of Y.

Given a k-order jet manifold J^kY of $Y \to X$, there exists the canonical bundle morphism $r_{(k)} : J^kTY \to TJ^kY$ over a surjection

$$J^kY \underset{X}{\times} J^kTX \to J^kY \underset{X}{\times} TX$$

whose coordinate expression is

$$\dot{y}^i_\Lambda \circ r_{(k)} = (\dot{y}^i)_\Lambda - \sum (\dot{y}^i)_{\mu+\Sigma}(\dot{x}^\mu)_\Xi, \qquad 0 \leq |\Lambda| \leq k,$$

where the sum is taken over all partitions $\Sigma + \Xi = \Lambda$ and $0 < |\Xi|$. In particular, we have the canonical isomorphism

$$r_{(k)} : J^kVY \to VJ^kY, \qquad (\dot{y}^i)_\Lambda = \dot{y}^i_\Lambda \circ r_{(k)}, \tag{B.79}$$

over J^kY. As a consequence, every projectable vector field u (B.25) on a fibre bundle $Y \to X$ admits the k-order *jet prolongation* to a vector field

$$J^k u = r_{(k)} \circ J^k u : J^k Y \to T J^k Y,$$
$$J^k u = u^\lambda \partial_\lambda + u^i \partial_i + \sum_{0<|\Lambda|\leq k} (d_\Lambda(u^i - y^i_\mu u^\mu) + y^i_{\mu+\Lambda} u^\mu)\partial_i^\Lambda, \qquad \text{(B.80)}$$

on $J^k Y$ (cf. (B.51) for $k = 1$). In particular, the k-order jet prolongation (B.80) of a vertical vector field $u = u^i \partial_i$ on $Y \to X$ is a vertical vector field

$$J^k u = u^i \partial_i + \sum_{0<|\Lambda|\leq k} d_\Lambda u^i \partial_i^\Lambda$$

on $J^k Y \to X$ owing to the isomorphism (B.79).

Similarly to the canonical monomorphisms (B.47) and (B.48), there are the canonical bundle monomorphisms over $J^k Y$:

$$\lambda_{(k)} : J^{k+1} Y \longrightarrow T^* X \underset{J^k Y}{\otimes} T J^k Y, \qquad \lambda_{(k)} = dx^\lambda \otimes d_\lambda,$$
$$\theta_{(k)} : J^{k+1} Y \longrightarrow T^* J^k Y \underset{J^k Y}{\otimes} V J^k Y, \qquad \theta_{(k)} = \sum_{|\Lambda|\leq k} (dy^i_\Lambda - y^i_{\lambda+\Lambda} dx^\lambda) \otimes \partial_i^\Lambda.$$

The one-forms

$$\theta^i_\Lambda = dy^i_\Lambda - y^i_{\lambda+\Lambda} dx^\lambda$$

are called the *local contact forms*. They generate an ideal of a DGA $\mathcal{O}^*_k$ of exterior forms on $J^k Y$ which is termed the *ideal of contact forms*.

B.5 Differential Operators and Equations

Jet manifolds provides the conventional language of theory of differential equations and differential operators if they need not be linear [25, 53, 85].

Definition B.12 A system of k-order partial differential equations on a fibre bundle $Y \to X$ is defined as a closed subbundle $\mathfrak{E}$ of a jet bundle $J^k Y \to X$. For the sake of brevity, we agree to call $\mathfrak{E}$ the *differential equation*.

Let $J^k Y$ be provided with the adapted coordinates (x^λ, y^i_Λ). There exists a local coordinate system (z^A), $A = 1, \ldots, \mathrm{codim}\mathfrak{E}$, on $J^k Y$ such that $\mathfrak{E}$ is locally given (in the sense of item (i) of Theorem B.1) by equations

$$\mathcal{E}^A(x^\lambda, y^i_\Lambda) = 0, \qquad A = 1, \ldots, \mathrm{codim}\mathfrak{E}. \qquad \text{(B.81)}$$

Definition B.13 A *solution* of the k-order differential equation $\mathfrak{E}$ (B.81) on $Y \to X$ is defined as a section $\overline{s}$ of the k-order jet bundle $J^k Y \to X$ which lives in $\mathfrak{E}$. By a *classical solution* of a differential equation $\mathfrak{E}$ on $Y \to X$ is meant a section s of $Y \to X$ such that its k-order jet prolongation $J^k s$ takes its values into $\mathfrak{E}$, i.e., s obeys the differential equations

$$\mathcal{E}^A(x^\lambda, \partial_\Lambda s^i(x)) = 0.$$

In applications, differential equations are mostly associated to differential operators. There are several equivalent definitions of (nonlinear) differential operators. We start with the following.

Definition B.14 Let $Y \to X$ and $E \to X$ be fibre bundles, which are assumed to have global sections. A *k-order E-valued differential operator* on a fibre bundle $Y \to X$ is defined as a section $\mathcal{E}$ of the pull-back bundle

$$\mathrm{pr}_1 : E_Y^k = J^k Y \underset{X}{\times} E \to J^k Y. \tag{B.82}$$

Given bundle coordinates (x^λ, y^i) on Y and (x^λ, χ^a) on E, the pull-back (B.82) is provided with coordinates $(x^\lambda, y^j_\Sigma, \chi^a)$, $0 \leq |\Sigma| \leq k$. With respect to these coordinates, a differential operator $\mathcal{E}$ seen as a closed imbedded submanifold $\mathcal{E} \subset E_Y^k$ is given by the equalities

$$\chi^a = \mathcal{E}^a(x^\lambda, y^j_\Sigma). \tag{B.83}$$

There is obvious one-to-one correspondence between the sections $\mathcal{E}$ (B.83) of the fibre bundle (B.82) and the bundle morphisms

$$\Phi : J^k Y \underset{X}{\longrightarrow} E, \qquad \Phi = \mathrm{pr}_2 \circ \mathcal{E} \Longleftrightarrow \mathcal{E} = (\mathrm{Id}\, J^k Y, \Phi). \tag{B.84}$$

Therefore, there is the following equivalent definition of differential operators on Y.

Definition B.15 Let $Y \to X$ and $E \to X$ be fibre bundles. A bundle morphism $J^k Y \to E$ over X is called the *E-valued k-order differential operator* on $Y \to X$.

It is readily observed that the differential operator Φ (B.84) sends each section s of $Y \to X$ onto a section $\Phi \circ J^k s$ of $E \to X$. The mapping

$$\Delta_\Phi : s \to \Phi \circ J^k s, \qquad \chi^a(x) = \mathcal{E}^a(x^\lambda, \partial_\Sigma s^j(x)),$$

is termed the *standard form of a differential operator.*

Let e be a global section of a fibre bundle $E \to X$, a *kernel of a E-valued differential operator* Φ is defined as a kernel

$$\mathrm{Ker}_e \Phi = \Phi^{-1}(e(X)) \tag{B.85}$$

of the bundle morphism Φ (B.84). If it is a closed subbundle of a jet bundle $J^k Y \to X$, one says that $\mathrm{Ker}_e \Phi$ (B.85) is a *differential equation associated to the differential operator* Φ. By virtue of Theorem B.10, this condition holds if Φ is a bundle morphism of constant rank.

If $E \to X$ is a vector bundle, by a kernel of a E-valued differential operator usually is meant its kernel with respect to the canonical zero-valued section $\widehat{0}$ of $E \to X$.

If $Y \to X$ and $E \to X$ are vector bundles, a k-order E-valued differential operator is called the *linear differential operator* if it is a linear morphism on fibres of $J^kY \to X$. Then Serre–Swan Theorem A.10 states one-to-one correspondence between linear differential operators on vector bundles in accordance with Definition B.15 and differential operators (Definition A.3) on projective $C^\infty(X)$-modules of finite rank (Sect. A.2).

Appendix C
Calculus on Sheaves

In the book, we address algebraic systems on topological spaces. They are described in terms of sheaves. In this Appendix, the relevant material on sheaves of modules over a commutative ring is summarized. We follow the terminology of [23, 75].

C.1 Sheaf Cohomology

A *sheaf* on a topological space X is a continuous fibre bundle $\pi : S \to X$ (Remark B.2) in modules over a commutative ring $\mathcal{K}$, where the surjection π is a local homeomorphism and fibres S_x, $x \in X$, called the *stalks*, are provided with the discrete topology. Global sections of a sheaf S make up a $\mathcal{K}$-module $S(X)$, termed the *structure module* of S.

Any sheaf is generated by a presheaf. A *presheaf* $S_{\{U\}}$ on a topological space X is defined if a module S_U over a commutative ring $\mathcal{K}$ is assigned to every open subset $U \subset X$ ($S_\emptyset = 0$) and if, for any pair of open subsets $V \subset U$, there exists a *restriction morphism* $r_V^U : S_U \to S_V$ such that

$$r_U^U = \mathrm{Id}\, S_U, \qquad r_W^U = r_W^V r_V^U, \qquad W \subset V \subset U.$$

Every presheaf $S_{\{U\}}$ on a topological space X yields a sheaf on X whose stalk S_x at a point $x \in X$ is the direct limit of the modules S_U, $x \in U$ (Definition A.1), with respect to a restriction morphisms r_V^U. It means that, for each open neighborhood U of a point x, every element $s \in S_U$ determines an element $s_x \in S_x$, called the *germ* of s at x. Two elements $s \in S_U$ and $s' \in S_V$ belong to the same germ at x if and only if there exists an open neighborhood $W \subset U \cap V$ of x such that $r_W^U s = r_W^V s'$.

Remark C.1 Let $C^0_{\{U\}}$ be a presheaf of local continuous real functions on a topological space X. Two such functions s and s' define the same germ s_x if they coincide on an open neighborhood of x. Hence, we obtain a *sheaf C^0_X of germs of local continuous real functions* on X. Similarly, a *sheaf C^∞_X of germs of local smooth*

G. Sardanashvily, *Noether's Theorems*, Atlantis Studies
in Variational Geometry 3, DOI 10.2991/978-94-6239-171-0

real functions on a manifold X is defined. Let us also mention a presheaf of local real functions which are constant on connected open subsets of X. It generates the *constant sheaf* on X denoted by $\mathbb{R}$.

Two different presheaves may generate the same sheaf. Conversely, every sheaf S defines a presheaf $S(\{U\})$ of modules $S(U)$ of its local sections. It is termed the *canonical presheaf* of a sheaf S. If a sheaf S is constructed from a presheaf $S_{\{U\}}$, there are natural module morphisms

$$S_U \ni s \to s(U) \in S(U), \qquad s(x) = s_x, \quad x \in U,$$

which are neither monomorphisms nor epimorphisms in general. For instance, it may happen that a nonzero presheaf defines a zero sheaf. A sheaf generated by the canonical presheaf of a sheaf S coincides with S.

A direct sum and a tensor product of presheaves (as families of modules) and sheaves (as fibre bundles in modules) are naturally defined. By virtue of Theorem A.3, a direct sum (resp. a tensor product) of presheaves generates a direct sum (resp. a tensor product) of the corresponding sheaves.

Remark C.2 In the terminology of [144], a sheaf is introduced as a presheaf which satisfies the following additional axioms.

(S1) Suppose that $U \subset X$ is an open subset and $\{U_\alpha\}$ is its open cover. If $s, s' \in S_U$ obey the condition $r^U_{U_\alpha}(s) = r^U_{U_\alpha}(s')$ for all U_α, then $s = s'$.

(S2) Let U and $\{U_\alpha\}$ be as in previous item. Suppose that we are given a family of presheaf elements $\{s_\alpha \in S_{U_\alpha}\}$ such that

$$r^{U_\alpha}_{U_\alpha \cap U_\lambda}(s_\alpha) = r^{U_\lambda}_{U_\alpha \cap U_\lambda}(s_\lambda)$$

for all U_α, U_λ. Then there exists a presheaf element $s \in S_U$ such that $s_\alpha = r^U_{U_\alpha}(s)$.

Canonical presheaves are in one-to-one correspondence with presheaves obeying these axioms. For instance, the presheaves of continuous, smooth and locally constant functions in Remark C.1 satisfy the axioms (S1) and (S2).

Remark C.3 The notion of a sheaf can be extended to sets, but not to noncommutative groups. One can consider a presheaf of such groups, but it generates a sheaf of sets because a direct limit of noncommutative groups need not be a group.

There is a useful construction of a sheaf on a topological space X from local sheaves on open subsets which make up a cover of X.

Theorem C.1 *Let $\{U_\zeta\}$ be an open cover of a topological space X and S_ζ a sheaf on U_ζ for every U_ζ. Let us suppose that, if $U_\zeta \cap U_\xi \neq \emptyset$, there is a sheaf isomorphism*

$$\rho_{\zeta\xi} : S_\xi \mid_{U_\zeta \cap U_\xi} \to S_\zeta \mid_{U_\zeta \cap U_\xi}$$

and, for every triple $(U_\zeta, U_\xi, U_\iota)$*, these isomorphisms fulfil the cocycle condition*

$$\rho_{\xi\zeta} \circ \rho_{\zeta\iota}(S_\iota|_{U_\zeta \cap U_\xi \cap U_\iota}) = \rho_{\xi\iota}(S_\iota|_{U_\zeta \cap U_\xi \cap U_\iota}).$$

Then there exists a sheaf S *together with sheaf isomorphisms* $\phi_\zeta : S|_{U_\zeta} \to S_\zeta$ *so that*

$$\phi_\zeta|_{U_\zeta \cap U_\xi} = \rho_{\zeta\xi} \circ \phi_\xi|_{U_\zeta \cap U_\xi}.$$

A *morphism of a presheaf* $S_{\{U\}}$ to a presheaf $S'_{\{U\}}$ on the same topological space X is defined as a set of module morphisms $\gamma_U : S_U \to S'_U$ which commute with restriction morphisms. A morphism of presheaves yields a *morphism of sheaves* generated by these presheaves. This is a bundle morphism over X such that $\gamma_x : S_x \to S'_x$ is the direct limit of morphisms γ_U, $x \in U$. Conversely, any morphism of sheaves $S \to S'$ on a topological space X yields a morphism of canonical presheaves of local sections of these sheaves. Let Hom $(S|_U, S'|_U)$ be the commutative group of sheaf morphisms $S|_U \to S'|_U$ for any open subset $U \subset X$. These groups are assembled into a presheaf, and define a sheaf Hom (S, S') on X. There is a monomorphism

$$\text{Hom}\,(S, S')(U) \to \text{Hom}\,(S(U), S'(U)), \tag{C.1}$$

which need not be an isomorphism.

By virtue of Theorem A.5, if a presheaf morphism is a monomorphism or an epimorphism, so is the corresponding sheaf morphism. Furthermore, the following holds.

Theorem C.2 *A short exact sequence*

$$0 \to S'_{\{U\}} \to S_{\{U\}} \to S''_{\{U\}} \to 0 \tag{C.2}$$

of presheaves on the same topological space yields a short exact sequence of sheaves generated by these presheaves

$$0 \to S' \to S \to S'' \to 0, \tag{C.3}$$

where the quotient sheaf $S'' = S/S'$ *is isomorphic to that generated by the quotient presheaf* $S''_{\{U\}} = S_{\{U\}}/S'_{\{U\}}$. *If the exact sequence of presheaves (C.2) is split, i.e.,*

$$S_{\{U\}} \cong S'_{\{U\}} \oplus S''_{\{U\}},$$

the corresponding splitting $S \cong S' \oplus S''$ *of the exact sequence of sheaves (C.3) holds.*

The converse is more intricate. A sheaf morphism induces a morphism of the corresponding canonical presheaves. If $S \to S'$ is a monomorphism, $S(\{U\}) \to S'(\{U\})$ also is a monomorphism. However, if $S \to S'$ is an epimorphism, $S(\{U\}) \to$

$S'(\{U\})$ need not be so. Therefore, the short exact sequence (C.3) of sheaves yields an exact sequence of canonical presheaves

$$0 \to S'(\{U\}) \to S(\{U\}) \to S''(\{U\}), \tag{C.4}$$

where $S(\{U\}) \to S''(\{U\})$ is not necessarily an epimorphism. At the same time, there is a short exact sequence of presheaves

$$0 \to S'(\{U\}) \to S(\{U\}) \to S''_{\{U\}} \to 0, \tag{C.5}$$

where the quotient presheaf $S''_{\{U\}} = S(\{U\})/S'(\{U\})$ generates a quotient sheaf $S'' = S/S'$, but need not be its canonical presheaf.

Theorem C.3 *Let the exact sequence of sheaves (C.3) be split. Then*

$$S(\{U\}) \cong S'(\{U\}) \oplus S''(\{U\}),$$

and the canonical presheaves make up the short exact sequence

$$0 \to S'(\{U\}) \to S(\{U\}) \to S''(\{U\}) \to 0.$$

Let us turn now to sheaf cohomology. We follow its definition in [75].

Let $S_{\{U\}}$ be a presheaf of modules on a topological space X, and let $\mathfrak{U} = \{U_i\}_{i\in I}$ be an open cover of X. One constructs a cochain complex where a p-cochain is defined as a function s^p which associates an element $s^p(i_0, \dots, i_p) \in S_{U_{i_0}\cap\cdots\cap U_{i_p}}$ to each $(p+1)$-tuple $(i_0, \dots, i_p)$ of indices in I. These p-cochains are assembled into a module $C^p(\mathfrak{U}, S_{\{U\}})$. Let us introduce a coboundary operator

$$\delta^p : C^p(\mathfrak{U}, S_{\{U\}}) \to C^{p+1}(\mathfrak{U}, S_{\{U\}}),$$
$$\delta^p s^p(i_0, \dots, i_{p+1}) = \sum_{k=0}^{p+1} (-1)^k r_W^{W_k} s^p(i_0, \dots, \widehat{i}_k, \dots, i_{p+1}), \tag{C.6}$$
$$W = U_{i_0} \cap \cdots \cap U_{i_{p+1}}, \qquad W_k = U_{i_0} \cap \cdots \cap \widehat{U}_{i_k} \cap \cdots \cap U_{i_{p+1}}.$$

One can verify that $\delta^{p+1} \circ \delta^p = 0$. Thus, we obtain a cochain complex of modules

$$0 \to C^0(\mathfrak{U}, S_{\{U\}}) \xrightarrow{\delta^0} \cdots C^p(\mathfrak{U}, S_{\{U\}}) \xrightarrow{\delta^p} C^{p+1}(\mathfrak{U}, S_{\{U\}}) \longrightarrow \cdots. \tag{C.7}$$

Its cohomology groups

$$H^p(\mathfrak{U}; S_{\{U\}}) = \operatorname{Ker} \delta^p / \operatorname{Im} \delta^{p-1}$$

are modules. Of course, they depend on an open cover $\mathfrak{U}$ of the topological space X.

Remark C.4 Throughout the book, only *proper covers* are considered, i.e., $U_i \neq U_j$ if $i \neq j$. A cover $\mathfrak{U}'$ is said to be a *refinement* of a cover $\mathfrak{U}$ if, for each $U' \in \mathfrak{U}'$, there exists $U \in \mathfrak{U}$ such that $U' \subset U$.

Let $\mathfrak{U}'$ be a refinement of a cover $\mathfrak{U}$. Then there is a morphism of cohomology groups

$$H^*(\mathfrak{U}; S_{\{U\}}) \to H^*(\mathfrak{U}'; S_{\{U\}}).$$

Let us take the direct limit of cohomology groups $H^*(\mathfrak{U}; S_{\{U\}})$ with respect to these morphisms, where $\mathfrak{U}$ runs through all open covers of X. This limit $H^*(X; S_{\{U\}})$ is called the *cohomology of X with coefficients in a presheaf $S_{\{U\}}$*.

Let S be a sheaf on a topological space X. *Cohomology of X with coefficients in S* or, shortly, *sheaf cohomology* of X is defined as cohomology

$$H^*(X; S) = H^*(X; S(\{U\}))$$

with coefficients in the canonical presheaf $S(\{U\})$ of a sheaf S.

In this case, a p-cochain $s^p \in C^p(\mathfrak{U}, S(\{U\}))$ is a collection $s^p = \{s^p(i_0, \ldots, i_p)\}$ of local sections $s^p(i_0, \ldots, i_p)$ of a sheaf S over $U_{i_0} \cap \cdots \cap U_{i_p}$ for each $(p+1)$-tuple $(U_{i_0}, \ldots, U_{i_p})$ of elements of the cover $\mathfrak{U}$. The coboundary operator (C.6) reads

$$\delta^p s^p(i_0, \ldots, i_{p+1}) = \sum_{k=0}^{p+1} (-1)^k s^p(i_0, \ldots, \widehat{i_k}, \ldots, i_{p+1})|_{U_{i_0} \cap \cdots \cap U_{i_{p+1}}}.$$

For instance,

$$\begin{aligned} \delta^0 s^0(i, j) &= [s^0(j) - s^0(i)]|_{U_i \cap U_j}, \\ \delta^1 s^1(i, j, k) &= [s^1(j, k) - s^1(i, k) + s^1(i, j)]|_{U_i \cap U_j \cap U_k}. \end{aligned} \quad \text{(C.8)}$$

A glance at the expression (C.8) shows that a zero-cocycle is a collection $s = \{s(i)\}_I$ of local sections of a sheaf S over $U_i \in \mathfrak{U}$ such that $s(i) = s(j)$ on $U_i \cap U_j$. It follows from the axiom (S2) in Remark C.2 that s is a global section of a sheaf S, while each $s(i)$ is its restriction $s|_{U_i}$ to U_i. Consequently, the cohomology group $H^0(\mathfrak{U}, S(\{U\}))$ is isomorphic to the structure module $S(X)$ of global sections of a sheaf S. A one-cocycle is a collection $\{s(i, j)\}$ of local sections of a sheaf S over overlaps $U_i \cap U_j$ which satisfy the *cocycle condition*

$$[s(j, k) - s(i, k) + s(i, j)]|_{U_i \cap U_j \cap U_k} = 0.$$

C.2 Abstract de Rham Theorem

If X is a paracompact space, the study of its sheaf cohomology is essentially simplified owing to the following fact [75].

Theorem C.4 *Cohomology of a paracompact space X with coefficients in a sheaf S equals cohomology of X with coefficients in any presheaf generating a sheaf S.*

Remark C.5 We follow the definition of a *paracompact topological space* in [75] as a Hausdorff space such that any its open cover admits a *locally finite* open refinement, i.e., any point has an open neighborhood which intersects only a finite number of elements of this refinement. A topological space X is paracompact if and only if any cover $\{U_\xi\}$ of X admits the subordinate *partition of unity* $\{f_\xi\}$, i.e.:

(i) f_ξ are real positive continuous functions on X;

(ii) supp $f_\xi \subset U_\xi$;

(iii) each point $x \in X$ has an open neighborhood which intersects only a finite number of the sets supp f_ξ;

(iv) $\sum\limits_\xi f_\xi(x) = 1$ for all $x \in X$.

A key point of the analysis of sheaf cohomology is that short exact sequences of presheaves and sheaves yield long exact sequences of sheaf cohomology groups.

Let $S_{\{U\}}$ and $S'_{\{U\}}$ be presheaves on the same topological space X. It is readily observed that, given an open cover $\mathfrak{U}$ of X, any morphism $S_{\{U\}} \to S'_{\{U\}}$ yields a cochain morphism of complexes

$$C^*(\mathfrak{U}, S_{\{U\}}) \to C^*(\mathfrak{U}, S'_{\{U\}})$$

and the corresponding morphism

$$H^*(\mathfrak{U}, S_{\{U\}}) \to H^*(\mathfrak{U}, S'_{\{U\}})$$

of cohomology groups of these complexes. Passing to the direct limit through all refinements of $\mathfrak{U}$, we come to a morphism of the cohomology groups

$$H^*(X, S_{\{U\}}) \to H^*(X, S'_{\{U\}})$$

of X with coefficients in the presheaves $S_{\{U\}}$ and $S'_{\{U\}}$. In particular, any sheaf morphism $S \to S'$ yields a morphism of canonical presheaves $S(\{U\}) \to S'(\{U\})$ and the corresponding cohomology morphism $H^*(X, S) \to H^*(X, S')$.

By virtue of Theorems A.7 and A.8, every short exact sequence

$$0 \to S'_{\{U\}} \longrightarrow S_{\{U\}} \longrightarrow S''_{\{U\}} \to 0$$

of presheaves on the same topological space X and the corresponding exact sequence of complexes (C.7) yield the long exact sequence

$$\begin{aligned}0 \to H^0(X; S'_{\{U\}}) \longrightarrow H^0(X; S_{\{U\}}) \longrightarrow H^0(X; S''_{\{U\}}) \longrightarrow \\ H^1(X; S'_{\{U\}}) \longrightarrow \cdots H^p(X; S'_{\{U\}}) \longrightarrow H^p(X; S_{\{U\}}) \longrightarrow \\ H^p(X; S''_{\{U\}}) \longrightarrow H^{p+1}(X; S'_{\{U\}}) \longrightarrow \cdots\end{aligned}$$

of the cohomology groups of X with coefficients in these presheaves.

This result however is not extended to an exact sequence of sheaves, unless X is a paracompact space. Let

$$0 \to S' \longrightarrow S \longrightarrow S'' \to 0$$

be a short exact sequence of sheaves on X. It yields the short exact sequence of presheaves (C.5) where a presheaf $S''_{\{U\}}$ generates a sheaf S''. If X is paracompact,

$$H^*(X; S''_{\{U\}}) = H^*(X; S'')$$

by virtue of Theorem C.4, and we have the exact sequence of sheaf cohomology

$$\begin{aligned}0 \to H^0(X; S') \longrightarrow H^0(X; S) \longrightarrow H^0(X; S'') \longrightarrow \\ H^1(X; S') \longrightarrow \cdots H^p(X; S') \longrightarrow H^p(X; S) \longrightarrow \\ H^p(X; S'') \longrightarrow H^{p+1}(X; S') \longrightarrow \cdots .\end{aligned}$$

Let us turn now to the abstract de Rham theorem which provides a powerful tool of studying algebraic systems on paracompact spaces.

Let us consider an exact sequence of sheaves

$$0 \to S \xrightarrow{h} S_0 \xrightarrow{h^0} S_1 \xrightarrow{h^1} \cdots S_p \xrightarrow{h^p} \cdots . \tag{C.9}$$

It is said to be a *resolution of a sheaf* S if each sheaf $S_{p\geq 0}$ is *acyclic*, i.e., its cohomology groups $H^{k>0}(X; S_p)$ vanish.

Any exact sequence of sheaves (C.9) yields the sequence of their structure modules

$$0 \to S(X) \xrightarrow{h_*} S_0(X) \xrightarrow{h^0_*} S_1(X) \xrightarrow{h^1_*} \cdots S_p(X) \xrightarrow{h^p_*} \cdots \tag{C.10}$$

which is always exact at terms $S(X)$ and $S_0(X)$ (see the exact sequence (C.4)). The sequence (C.10) is a cochain complex because $h^{p+1}_* \circ h^p_* = 0$. If X is a paracompact space and the exact sequence (C.9) is a resolution of S, the forthcoming *abstract de Rham theorem* establishes an isomorphism of cohomology of the complex (C.10) to cohomology of X with coefficients in a sheaf S [75].

Theorem C.5 *Given the resolution (C.9) of a sheaf S on a paracompact topological space X and the induced complex (C.10), there are isomorphisms*

$$H^0(X; S) = \operatorname{Ker} h^0_*, \qquad H^q(X; S) = \operatorname{Ker} h^q_*/\operatorname{Im} h^{q-1}_*, \qquad q > 0. \tag{C.11}$$

We also refer to the following minor modification of Theorem C.5 [56, 143].

Theorem C.6 *Let*

$$0 \to S \xrightarrow{h} S_0 \xrightarrow{h^0} S_1 \xrightarrow{h^1} \cdots \xrightarrow{h^{p-1}} S_p \xrightarrow{h^p} S_{p+1}, \qquad p > 1,$$

be an exact sequence of sheaves on a paracompact topological space X, where the sheaves S_q, $0 \leq q < p$, are acyclic, and let

$$0 \to S(X) \xrightarrow{h_*} S_0(X) \xrightarrow{h_*^0} S_1(X) \xrightarrow{h_*^1} \cdots \xrightarrow{h_*^{p-1}} S_p(X) \xrightarrow{h_*^p} S_{p+1}(X)$$

be the corresponding cochain complex of structure modules of these sheaves. Then the isomorphisms (C.11) hold for $0 \leq q \leq p$.

In our book, we appeal to a *fine resolution* of sheaves, i.e., a resolution by fine sheaves.

A sheaf S on a paracompact space X is called *fine* if, for each locally finite open cover $\mathfrak{U} = \{U_i\}_{i\in I}$ of X, there exists a system $\{h_i\}$ of endomorphisms $h_i : S \to S$ such that:

(i) there is a closed subset $V_i \subset U_i$ and $h_i(S_x) = 0$ if $x \notin V_i$,

(ii) $\sum\limits_{i\in I} h_i$ is the identity map of S.

Theorem C.7 *A fine sheaf on a paracompact space is acyclic.*

There are the following important examples of fine sheaves.

Theorem C.8 *Let X be a paracompact topological space which admits the partition of unity performed by elements of the structure module $\mathfrak{A}(X)$ of some sheaf $\mathfrak{A}$ of germs of local real functions on X. Then any sheaf S of $\mathfrak{A}(X)$-modules on X, including $\mathfrak{A}$ itself, is fine.*

Remark C.6 In particular, a sheaf C^0_X of germs of local continuous real functions on a paracompact space is fine, and so is any sheaf of $C^0(X)$-modules. A smooth manifold X admits the partition of unity performed by smooth real functions (Remark B.1). It follows that the sheaf C^∞_X of germs of local smooth real functions on X is fine, and so is any sheaf of $C^\infty(X)$-modules, e.g., the sheaves of sections of smooth vector bundles over X.

C.3 Local-Ringed Spaces

Local-ringed spaces are sheaves of local rings. In our book, graded manifolds are defined as local-ringed spaces (Sect. 6.3).

A sheaf $\mathfrak{A}$ on a topological space X is said to be the *ringed space* if its stalk $\mathfrak{A}_x$ at each point $x \in X$ is a commutative ring [144]. A ringed space often is denoted by a pair $(X, \mathfrak{A})$ of a topological space X and a sheaf $\mathfrak{A}$ of rings on X which are called the *body* and the *structure sheaf* of a ringed space, respectively.

Definition C.1 A ringed space is said to be the *local-ringed space* (the geometric space in the terminology of [144]) if it is a sheaf of local rings.

Remark C.7 A sheaf C^0_X of germs of local continuous real functions on a topological space X is a local-ringed space. Its stalk C^0_x, $x \in X$, contains an unique maximal ideal of germs of functions vanishing at x. A sheaf C^0_X of germs of local smooth real functions on a manifold X also is so.

Morphisms of local-ringed spaces are defined to be particular morphisms of sheaves on different topological spaces as follows.

Let $\varphi : X \to X'$ be a continuous map. Given a sheaf S on X, its *direct image* $\varphi_* S$ on X' is generated by the presheaf of assignments

$$X' \supset U' \to S(\varphi^{-1}(U'))$$

for any open subset $U' \subset X'$. Conversely, given a sheaf S' on X', its *inverse image* $\varphi^* S'$ on X is defined as the pull-back onto X of a continuous fibre bundle S' over X', i.e., $\varphi^* S'_x = S_{\varphi(x)}$. This sheaf is generated by the presheaf which associates to any open $V \subset X$ the direct limit of modules $S'(U)$ over all open subsets $U \subset X'$ such that $V \subset f^{-1}(U)$.

Remark C.8 Let $i : X \to X'$ be a closed subspace of X'. Then $i_* S$ is a unique sheaf on X' such that

$$i_* S|_X = S, \qquad i_* S|_{X' \setminus X} = 0.$$

Indeed, if $x' \in X \subset X'$, then $i_* S(U') = S(U' \cap X)$ for any open neighborhood U of this point. If $x' \notin X$, there exists its neighborhood U' such that $U' \cap X$ is empty, i.e., $i_* S(U') = 0$. A sheaf $i_* S$ is termed the *trivial extension* of a sheaf S.

By a *morphism of ringed spaces* $(X, \mathfrak{A}) \to (X', \mathfrak{A}')$ is meant a pair (φ, Φ) of a continuous map $\varphi : X \to X'$ and a sheaf morphism $\Phi : \mathfrak{A}' \to \varphi_* \mathfrak{A}$ or, equivalently, a sheaf morphism $\varphi^* \mathfrak{A}' \to \mathfrak{A}$ [144]. Restricted to each stalk, a sheaf morphism Φ is assumed to be a ring homomorphism. A morphism of ringed spaces is said to be:

- a monomorphism if φ is an injection and Φ is an epimorphism,
- an epimorphism if φ is a surjection, while Φ is a monomorphism.

Let $(X, \mathfrak{A})$ be a local-ringed space. By a *sheaf* $\mathfrak{d}\mathfrak{A}$ *of derivations* of the sheaf $\mathfrak{A}$ is meant a subsheaf of endomorphisms of $\mathfrak{A}$ such that any section u of $\mathfrak{d}\mathfrak{A}$ over an open subset $U \subset X$ is a derivation of a ring $\mathfrak{A}(U)$. It should be emphasized that, since (C.1) is not necessarily an isomorphism, a derivation of a ring $\mathfrak{A}(U)$ need not be a section of a sheaf $\mathfrak{d}\mathfrak{A}|_U$. Namely, it may happen that, given open sets $U' \subset U$, there is no restriction morphism $\mathfrak{d}(\mathfrak{A}(U)) \to \mathfrak{d}(\mathfrak{A}(U'))$.

Given a local-ringed space $(X, \mathfrak{A})$, a sheaf P on X is called the *sheaf of $\mathfrak{A}$-modules* if every stalk P_x, $x \in X$, is an $\mathfrak{A}_x$-module or, equivalently, if $P(U)$ is an $\mathfrak{A}(U)$-module for any open subset $U \subset X$. A sheaf of $\mathfrak{A}$-modules P is said to be *locally free* if there exists an open neighborhood U of every point $x \in X$ such that $P(U)$ is a free $\mathfrak{A}(U)$-module. If all these free modules are of finite rank (resp. of the same finite rank), one says that P is of *finite type* (resp. of constant rank). The structure module of a locally free sheaf is called the *locally free module*.

The following is a generalization of Theorem C.8 [75].

Theorem C.9 *Let X be a paracompact space which admits the partition of unity by elements of the structure module $S(X)$ of some sheaf S of real functions on X. Let P be a sheaf of S-modules, i.e., the stalks P_x, $x \in X$, of P are S_x-modules. Then P is fine and, consequently, acyclic.*

In particular, all sheaves $\mathcal{O}^k_X$, $k \in \mathbb{N}_+$, of germs of exterior forms on X are fine. These sheaves constitute the *de Rham complex*

$$0 \to \mathbb{R} \longrightarrow C^\infty_X \xrightarrow{d} \mathcal{O}^1_X \xrightarrow{d} \cdots \mathcal{O}^k_X \xrightarrow{d} \cdots . \tag{C.12}$$

The corresponding complex of structure modules of these sheaves is the de Rham complex (A.18) of exterior forms on a manifold X.

Due to the Poincaré lemma, the complex (C.12) is exact and, thereby, is a fine resolution of the constant sheaf $\mathbb{R}$ on a manifold. Then a corollary of Theorem C.5 is the classical *de Rham theorem*.

Theorem C.10 *There is an isomorphism*

$$H^k_{\mathrm{DR}}(X) = H^k(X; \mathbb{R}) \tag{C.13}$$

of the de Rham cohomology $H^*_{\mathrm{DR}}(X)$ of a manifold X to the cohomology $H^*(X; \mathbb{R})$ of X with coefficients in the constant sheaf $\mathbb{R}$.

The sheaf cohomology $H^k(X; \mathbb{R})$ in turn is related to cohomology of other types as follows.

Let us consider a short exact sequence of constant sheaves

$$0 \to \mathbb{Z} \longrightarrow \mathbb{R} \longrightarrow U(1) \to 0,$$

where $U(1) = \mathbb{R}/\mathbb{Z}$ is a circle group of complex numbers of unit module. This exact sequence yields a long exact sequence of sheaf cohomology groups

$$\begin{gathered} 0 \to \mathbb{Z} \longrightarrow \mathbb{R} \longrightarrow U(1) \longrightarrow H^1(X; \mathbb{Z}) \longrightarrow H^1(X; \mathbb{R}) \longrightarrow \cdots \\ H^p(X; \mathbb{Z}) \longrightarrow H^p(X; \mathbb{R}) \longrightarrow H^p(X; U(1)) \longrightarrow H^{p+1}(X; \mathbb{Z}) \longrightarrow \cdots , \\ H^0(X; \mathbb{Z}) = \mathbb{Z}, \qquad H^0(X; \mathbb{R}) = \mathbb{R}, \qquad H^0(X; U(1)) = U(1). \end{gathered}$$

This exact sequence defines a homomorphism

$$H^*(X; \mathbb{Z}) \to H^*(X; \mathbb{R}) \tag{C.14}$$

of cohomology with coefficients in the constant sheaf $\mathbb{Z}$ to that with coefficients in $\mathbb{R}$. Combining the isomorphism (C.13) and the homomorphism (C.14) leads to a cohomology homomorphism $H^*(X; \mathbb{Z}) \to H^*_{\mathrm{DR}}(X)$. Its kernel contains all cyclic elements of cohomology groups $H^k(X; \mathbb{Z})$.

Since manifolds are assumed to be paracompact, there is the following isomorphism between their sheaf cohomology and singular cohomology.

Theorem C.11 *The sheaf cohomology $H^*(X; \mathbb{Z})$ (resp. $H^*(X; \mathbb{Q})$, $H^*(X; \mathbb{R})$) of a paracompact topological space X with coefficients in the constant sheaf $\mathbb{Z}$ (resp. $\mathbb{Q}$, $\mathbb{R}$) is isomorphic to the singular cohomology of X with coefficients in a ring $\mathbb{Z}$ (resp. $\mathbb{Q}$, $\mathbb{R}$) [23, 139].*

Let us note that singular cohomology of paracompact topological spaces coincide with the Čhech and Alexandery ones. Since singular cohomology is a *topological invariant* (i.e., homotopic topological spaces have the same singular cohomology) [139], the sheaf cohomology $H^*(X; \mathbb{Z})$, $H^*(X; \mathbb{Q})$, $H^*(X; \mathbb{R})$ and, consequently, the de Rham cohomology of smooth manifolds are topological invariants.

Appendix D
Noether Identities of Differential Operators

Noether and higher-stage Noether identities of a Lagrangian system in Sect. 7.1 are particular Noether identities of differential operators which are described in homology terms as follows [61, 123].

Let $E \to X$ be a vector bundle, and let $\mathcal{E}$ be a E-valued k-order differential operator on a fibre bundle $Y \to X$ in accordance with Definition B.14. It is represented by a section $\mathcal{E}^a$ (B.83) of the pull-back bundle (B.82) endowed with bundle coordinates $(x^\lambda, y^j_\Sigma, \chi^a)$, $0 \leq |\Sigma| \leq k$.

Definition D.1 One says that a differential operator $\mathcal{E}$ obeys *Noether identities* (NI) if there exist an r-order differential operator Φ on the pull-back bundle

$$E_Y = Y \underset{X}{\times} E \to X \tag{D.1}$$

so that its projection to E is a linear differential operator and its kernel contains $\mathcal{E}$:

$$\Phi = \sum_{0\leq|\Lambda|} \Phi^\Lambda_a \chi^a_\Lambda, \qquad \sum_{0\leq|\Lambda|} \Phi^\Lambda_a \mathcal{E}^a_\Lambda = 0.$$

Any differential operator admits NI, e.g.,

$$\Phi = \sum_{0\leq|\Lambda|,|\Sigma|} T^{\Lambda\Sigma}_{ab} d_\Sigma \mathcal{E}^b \chi^a_\Lambda, \qquad T^{\Lambda\Sigma}_{ab} = -T^{\Sigma\Lambda}_{ba}. \tag{D.2}$$

Therefore, NI must be separated into the trivial and nontrivial ones. For this purpose, we describe them as boundaries and cycles of some chain complex [61, 123].

Lemma D.1 *One can associate to a differential operator $\mathcal{E}$ the chain complex (D.3) whose boundaries vanish on* Ker $\mathcal{E}$.

Proof Let us consider the composite graded manifold $(Y, \mathfrak{A}_{E_Y})$ modelled over the vector bundle $E_Y \to Y$. Let $\mathcal{S}^0_\infty[E_Y, Y]$ be the ring (6.39) of graded functions on

G. Sardanashvily, *Noether's Theorems*, Atlantis Studies
in Variational Geometry 3, DOI 10.2991/978-94-6239-171-0

an infinite order jet manifold $J^\infty Y$ possessing the local generating basis (y^i, ε^a) of Grassmann parity $[y^i] = 0$, $[\varepsilon^a] = 1$ and, in accordance with the terminology of Chap. 7, antifield number $\mathrm{Ant}[y^i] = 0$, $\mathrm{Ant}[\varepsilon^a] = 1$. It is provided with a nilpotent right odd derivation $\overline{\delta} = \partial_a \mathcal{E}^a$ (Remark 6.10). Then we have a chain complex

$$0 \leftarrow \mathrm{Im}\,\delta \xleftarrow{\overline{\delta}} \mathcal{S}^0_\infty[E_Y, Y]_1 \xleftarrow{\overline{\delta}} \mathcal{S}^0_\infty[E_Y, Y]_2 \tag{D.3}$$

of graded functions of antifield number $k \leq 2$. By very definition, its one-boundaries $\overline{\delta}\Phi$, $\Phi \in \mathcal{S}^0_\infty[E_Y, Y]_2$ vanish on Ker $\mathcal{E}$. □

Every one-cycle

$$\Phi = \sum_{0\leq|\Lambda|} \Phi^\Lambda_a \varepsilon^a_\Lambda \in \mathcal{S}^0_\infty[E_Y, Y]_1 \tag{D.4}$$

of the complex (D.3) defines a linear differential operator on the pull-back bundle E_Y (D.1) such that it is linear on E and its kernel contains $\mathcal{E}$, i.e.,

$$\delta\Phi = 0, \qquad \sum_{0\leq|\Lambda|} \Phi^\Lambda_a d_\Lambda \mathcal{E}^a = 0. \tag{D.5}$$

In accordance with Definition D.1, the one-cycles (D.4) define the NI (D.5) of a differential operator $\mathcal{E}$. These NI are trivial if a cycle is a boundary, i.e., it takes the form (D.2). Accordingly, nontrivial NI modulo the trivial ones are associated to elements of the homology $H_1(\delta)$ of the complex (D.4).

A differential operator is called *degenerate* if it obeys nontrivial NI.

One can say something more if the $\mathcal{O}^0_\infty Y$-module $H_1(\delta)$ is finitely generated, i.e., it possesses the following particular structure. There are elements $\Delta \in H_1(\delta)$ making up a projective $C^\infty(X)$-module $\mathcal{C}_{(0)}$ of finite rank which, by virtue of Serre–Swan Theorem A.10, is isomorphic to the structure module of sections of some vector bundle $E_0 \to X$. Let $\{\Delta^r\}$:

$$\Delta^r = \sum_{0\leq|\Lambda|} \Delta^{\Lambda r}_a \varepsilon^a_\Lambda, \qquad \Delta^{\Lambda r}_a \in \mathcal{O}^0_\infty Y,$$

be local bases for this $C^\infty(X)$-module. Then every element $\Phi \in H_1(\delta)$ factorizes as

$$\Phi = \sum_{0\leq|\Xi|} G^\Xi_r d_\Xi \Delta^r, \qquad G^\Xi_r \in \mathcal{O}_\infty Y, \tag{D.6}$$

through elements of $\mathcal{C}_{(0)}$, i.e., any NI (D.5) is a corollary of the NI

$$\sum_{0\leq|\Lambda|} \Delta^{\Lambda r}_a d_\Lambda \mathcal{E}^a = 0, \tag{D.7}$$

called complete NI.

Given an integer $N = 2k$ or $N = 2k + 1$, $k = 0, 1, \ldots$, let us denote

$$P\{N\} = \mathcal{S}^0_\infty[E_N \underset{Y}{\oplus} \cdots \underset{Y}{\oplus} E_1 \underset{Y}{\oplus} E_0 \underset{Y}{\oplus} E_Y, Y \underset{X}{\times} (E_0 \underset{X}{\oplus} E_2 \underset{X}{\oplus} \cdots \underset{X}{\oplus} E_k)], \tag{D.8}$$

where $\mathcal{S}^0_\infty[,]$ is the graded commutative ring (6.39).

Lemma D.2 *If the homology $H_1(\delta)$ of the complex (D.3) is finitely generated, this complex can be extended to the one-exact complex (D.9) with a boundary operator whose nilpotency conditions are equivalent to complete NI.*

Proof Let us consider the graded commutative ring $P\{0\}$ (D.8). It possesses the local generating basis $\{y^i, \varepsilon^a, \varepsilon^r\}$ of Grassmann parity $[\varepsilon^r] = 0$ and antifield number $\mathrm{Ant}[\varepsilon^r] = 2$. This ring is provided with the nilpotent graded derivation

$$\delta_0 = \delta + \partial_r \Delta^r.$$

Its nilpotency conditions are equivalent to the complete NI (D.7). Then the module $P\{0\}_{\leq 3}$ of graded functions of antifield number ≤ 3 is decomposed into a chain complex

$$0 \leftarrow \mathrm{Im}\,\delta \xleftarrow{\delta} \mathcal{S}^0_\infty[E_Y, Y]_1 \xleftarrow{\delta_0} P\{0\}_2 \xleftarrow{\delta_0} P\{0\}_3. \tag{D.9}$$

Let $H_*(\delta_0)$ denote its homology. We have $H_0(\delta_0) = H_0(\delta) = 0$. Furthermore, any one-cycle Φ up to a boundary takes the form (D.6) and, therefore, it is a δ_0-boundary

$$\Phi = \sum_{0\leq|\Sigma|} G_r^\Xi d_\Xi \Delta^r = \delta_0 \left(\sum_{0\leq|\Sigma|} G_r^\Xi \varepsilon^r_\Xi \right).$$

Hence, $H_1(\delta_0) = 0$, i.e., the complex (D.9) is one-exact. □

Let us consider the second homology $H_2(\delta_0)$ of the complex (D.9). Its two-chains read

$$\Phi = G + H = \sum_{0\leq|\Lambda|} G_r^\Lambda \varepsilon^r_\Lambda + \sum_{0\leq|\Lambda|,|\Sigma|} H_{ab}^{\Lambda\Sigma} \varepsilon^a_\Lambda \varepsilon^b_\Sigma. \tag{D.10}$$

Its two-cycles define the first-stage NI

$$\delta_0 \Phi = 0, \qquad \sum_{0\leq|\Lambda|} G_r^\Lambda d_\Lambda \Delta^r + \delta H = 0. \tag{D.11}$$

Conversely, let the equality (D.11) hold. Then it is a cycle condition of the two-chain (D.10). The first-stage NI (D.11) are trivial either if the two-cycle Φ (D.10) is a boundary or its summand G vanishes on $\mathrm{Ker}\,\mathcal{E}$.

Lemma D.3 *First-stage NI can be identified with nontrivial elements of the homology $H_2(\delta_0)$ if and only if any δ-cycle $\Phi \in \mathcal{S}^0_\infty[E_Y, Y]_2$ is a δ_0-boundary.*

Proof The proof is similar to that of Theorem 7.3 [61, 123]. □

If the condition of Lemma D.3 is satisfied, let us assume that nontrivial first-stage NI are finitely generated as follows. There exists a graded projective $C^\infty(X)$-module $\mathscr{C}_{(1)} \subset H_2(\delta_0)$ of finite rank possessing a local basis $\Delta_{(1)}$:

$$\Delta^{r_1} = \sum_{0\leq|\Lambda|} \Delta_r^{\Lambda r_1} \varepsilon_\Lambda^r + h^{r_1},$$

such that any element $\Phi \in H_2(\delta_0)$ factorizes as

$$\Phi = \sum_{0\leq|\Xi|} \Phi_{r_1}^{\Xi} d_\Xi \Delta^{r_1} \tag{D.12}$$

through elements of $\mathscr{C}_{(1)}$. Thus, all nontrivial first-stage NI (D.11) result from the equalities

$$\sum_{0\leq|\Lambda|} \Delta_r^{r_1\Lambda} d_\Lambda \Delta^r + \delta h^{r_1} = 0, \tag{D.13}$$

called the complete first-stage NI.

Lemma D.4 *If nontrivial first-stage NI are finitely generated, the one-exact complex (D.9) is extended to the two-exact one (D.14) with a boundary operator whose nilpotency conditions are equivalent to complete Noether and first-stage NI.*

Proof By virtue of Serre–Swan Theorem A.10, a module $\mathscr{C}_{(1)}$ is isomorphic to a module of sections of some vector bundle $E_1 \to X$. Let us consider the graded commutative ring $P\{1\}$ (D.8) of graded functions on $J^\infty Y$ possessing the local generating bases $\{y^i, \varepsilon^a, \varepsilon^r, \varepsilon^{r_1}\}$ of Grassmann parity $[\varepsilon^{r_1}] = 1$ and antifield number $\mathrm{Ant}[\varepsilon^{r_1}] = 3$. It can be provided with the nilpotent graded derivation

$$\delta_1 = \delta_0 + \partial_{r_1} \Delta^{r_1}.$$

Its nilpotency conditions are equivalent to the complete NI (D.7) and the complete first-stage NI (D.13). Then a module $P\{1\}_{\leq 4}$ of graded functions of antifield number ≤ 4 is decomposed into a chain complex

$$0 \leftarrow \mathrm{Im}\,\delta \overset{\delta}{\longleftarrow} \mathscr{S}_\infty[E_Y, Y]_1 \overset{\delta_0}{\longleftarrow} P\{0\}_2 \overset{\delta_1}{\longleftarrow} P\{1\}_3 \overset{\delta_1}{\longleftarrow} P^0\{1\}_4. \tag{D.14}$$

Let $H_*(\delta_1)$ denote its homology. It is readily observed that

$$H_0(\delta_1) = H_0(\delta) = 0, \qquad H_1(\delta_1) = H_1(\delta_0) = 0.$$

By virtue of the equality (D.12), any two-cycle of the complex (D.14) is a boundary

$$\Phi = \sum_{0\leq|\Xi|} \Phi^{\Xi}_{r_1} d_\Xi \Delta^{r_1} = \delta_1 \left(\sum_{0\leq|\Xi|} \Phi^{\Xi}_{r_1} \varepsilon^{r_1}_\Xi \right).$$

It follows that $H_2(\delta_1) = 0$, i.e., the complex (D.14) is two-exact. □

If the third homology $H_3(\delta_1)$ of the complex (D.14) is not trivial, its elements correspond to second-stage NI, and so on. Iterating the arguments, we come to the following.

A degenerate differential operator $\mathcal{E}$ is called N-stage reducible if it admits finitely generated nontrivial N-stage NI, but no nontrivial $(N+1)$-stage ones. It is characterized as follows [61, 123].

- There are vector bundles $E_0, \ldots, E_N$ over X, and a graded commutative ring $\mathcal{S}^0_\infty[E_Y, Y]$ is enlarged to a graded commutative ring $P\{N\}$ (D.8) with the local generating basis $(y^i, \varepsilon^a, \varepsilon^r, \varepsilon^{r_1}, \ldots, \varepsilon^{r_N})$ of Grassmann parity $[\varepsilon^{r_k}] = k \operatorname{mod} 2$ and antifield number $\mathrm{Ant}[\varepsilon^{r_k}_\Lambda] = k+2$.
- A graded commutative ring $P\{N\}$ is provided with the nilpotent right graded derivation

$$\delta_{\mathrm{KT}} = \delta_N = \delta_0 + \sum_{1\leq k\leq N} \partial_{r_k} \Delta^{r_k}, \tag{D.15}$$
$$\Delta^{r_k} = \sum_{0\leq|\Lambda|} \Delta^{\Lambda r_k}_{r_{k-1}} \varepsilon^{r_{k-1}}_\Lambda + \sum_{0\leq\Sigma, 0\leq\Xi} (h^{\Xi\Sigma r_k}_{a r_{k-2}} \varepsilon^a_\Xi \varepsilon^{r_{k-2}}_\Sigma + \ldots),$$

  of antifield number -1.
- With this graded derivation, a module $P\{N\}_{\leq N+3}$ of graded functions of antifield number $\leq (N+3)$ is decomposed into the exact *KT complex*

$$0 \leftarrow \mathrm{Im}\,\delta \xleftarrow{\delta} \mathcal{S}^0_\infty[E_Y, Y]_1 \xleftarrow{\delta_0} P\{0\}_2 \xleftarrow{\delta_1} P\{1\}_3 \cdots \tag{D.16}$$
$$\xleftarrow{\delta_{N-1}} P\{N-1\}_{N+1} \xleftarrow{\delta_{\mathrm{KT}}} P^0\{N\}_{N+2} \xleftarrow{\delta_{\mathrm{KT}}} P\{N\}_{N+3},$$

  which satisfies the homology regularity condition that any $\delta_{k<N-1}$-cycle

$$\Phi \in P\{k\}_{k+3} \subset P\{k+1\}_{k+3}$$

  is a δ_{k+1}-boundary (cf. Definition 7.2).
- The nilpotentness $\delta^2_{\mathrm{KT}} = 0$ of the KT operator (D.15) is equivalent to the complete nontrivial NI (D.7) and the complete nontrivial $(k \leq N)$-stage NI

$$\sum_{0\leq|\Lambda|} \Delta^{\Lambda r_k}_{r_{k-1}} d_\Lambda \left(\sum_{0\leq|\Sigma|} \Delta^{\Sigma r_{k-1}}_{r_{k-2}} \varepsilon^{r_{k-2}}_\Sigma \right) + \delta \left(\sum_{0\leq\Sigma,\Xi} h^{\Xi\Sigma r_k}_{a r_{k-2}} \varepsilon^a_\Xi \varepsilon^{r_{k-2}}_\Sigma \right) = 0. \tag{D.17}$$

Example

Let us study the following example of reducible NI of a differential operator which is relevant to topological BF theory (Chap. 12).

Let us consider the fibre bundles

$$Y = X \times \mathbb{R}, \qquad E = \overset{n-1}{\wedge} TX, \qquad 2 < n, \tag{D.18}$$

coordinated by (x^λ, y) and $(x^\lambda, \chi^{\mu_1\dots\mu_{n-1}})$, respectively. We study the E-valued differential operator

$$\mathcal{E}^{\mu_1\dots\mu_{n-1}} = -\varepsilon^{\mu\mu_1\dots\mu_{n-1}} y_\mu, \tag{D.19}$$

where ε is the Levi–Civita symbol. It defines a first order differential equation $d_H y = 0$ on the fibre bundle Y (D.18). Putting

$$E_Y = \mathbb{R} \underset{X}{\times} \overset{n-1}{\wedge} TX,$$

let us consider the graded commutative ring $\mathcal{S}^*_\infty[E_Y, Y]$ of graded functions on $J^\infty Y$. It possesses the local generating basis $(y, \varepsilon^{\mu_1\dots\mu_{n-1}})$ of Grassmann parity $[\varepsilon^{\mu_1\dots\mu_{n-1}}] = 1$ and antifield number $\mathrm{Ant}[\varepsilon^{\mu_1\dots\mu_{n-1}}] = 1$. With the nilpotent derivation

$$\overline{\delta} = \frac{\partial}{\partial \varepsilon^{\mu_1\dots\mu_{n-1}}} \mathcal{E}^{\mu_1\dots\mu_{n-1}},$$

we have the complex (D.3). Its one-chains read

$$\Phi = \sum_{0\le|\Lambda|} \Phi^\Lambda_{\mu_1\dots\mu_{n-1}} \varepsilon^{\mu_1\dots\mu_{n-1}}_\Lambda,$$

and the cycle condition $\overline{\delta}\Phi = 0$ takes a form

$$\Phi^\Lambda_{\mu_1\dots\mu_{n-1}} \mathcal{E}^{\mu_1\dots\mu_{n-1}}_\Lambda = 0. \tag{D.20}$$

This equality is satisfied if and only if

$$\Phi^{\lambda_1\dots\lambda_k}_{\mu_1\dots\mu_{n-1}} \varepsilon^{\mu\mu_1\dots\mu_{n-1}} = -\Phi^{\mu\lambda_2\dots\lambda_k}_{\mu_1\dots\mu_{n-1}} \varepsilon^{\lambda_1\mu_1\dots\mu_{n-1}}.$$

It follows that Φ factorizes as

$$\Phi = \sum_{0\le|\Xi|} G^\Xi_{\nu_2\dots\nu_{n-1}} d_\Xi \Delta^{\nu_2\dots\nu_{n-1}} \omega$$

through graded functions

$$\Delta^{\nu_2\ldots\nu_{n-1}} = \Delta^{\lambda,\nu_2\ldots\nu_{n-1}}_{\alpha_1\ldots\alpha_{n-1}}\varepsilon^{\alpha_1\ldots\alpha_{n-1}}_{\lambda} = \delta^{\lambda}_{\alpha_1}\delta^{\nu_2}_{\alpha_2}\cdots\delta^{\nu_{n-1}}_{\alpha_{n-1}}\varepsilon^{\alpha_1\ldots\alpha_{n-1}}_{\lambda} = d_{\nu_1}\varepsilon^{\nu_1\nu_2\ldots\nu_{n-1}}, \tag{D.21}$$

which provide the complete NI

$$d_{\nu_1}\mathscr{E}^{\nu_1\nu_2\ldots\nu_{n-1}} = 0. \tag{D.22}$$

They can be written in a form

$$d_H d_H y = 0. \tag{D.23}$$

The graded functions (D.21) form a basis for a projective $C^\infty(X)$-module of finite rank which is isomorphic to the structure module of sections of a vector bundle $E_0 = \wedge^{n-2}TX$. Therefore, let us extend the graded commutative ring $\mathscr{S}^0_\infty[E_Y, Y]$ to that $P\{0\}$ (D.8) possessing the local generating basis $(y, \varepsilon^{\mu_1\ldots\mu_{n-1}}, \varepsilon^{\mu_2\ldots\mu_{n-1}}\}$, where $\varepsilon^{\mu_2\ldots\mu_{n-1}}$ are even antifields of antifield number 2. We have the nilpotent graded derivation

$$\delta_0 = \overline{\delta} + \frac{\partial}{\partial\varepsilon^{\mu_2\ldots\mu_{n-1}}}\Delta^{\mu_2\ldots\mu_{n-1}}$$

of $\mathscr{P}^0_\infty\{0\}$. Its nilpotency is equivalent to the complete NI (D.22). Then we obtain the one-exact complex (D.9).

Iterating the arguments, let us consider the vector bundles

$$E_{N=n-2} = X\times\mathbb{R}, \qquad E_k = \overset{n-k-2}{\wedge} TX, \qquad k = 1,\ldots,n-3,$$

and the graded commutative ring $P\{N\}$, possessing the local generating basis

$$(y, \varepsilon^{\mu_1\ldots\mu_{n-1}}, \varepsilon^{\mu_2\ldots\mu_{n-1}}, \ldots, \varepsilon^{\mu_{n-1}}, \varepsilon)$$

of Grassmann parity $[\varepsilon] = n$, $[\varepsilon^{\mu_{k+2}\ldots\mu_{n-1}}] = k \bmod 2$ and of antifield number $\mathrm{Ant}[\varepsilon] = n$, $\mathrm{Ant}[\varepsilon^{\mu_{k+2}\ldots\mu_{n-1}}] = k+2$. It is provided with the nilpotent graded derivation

$$\delta_{\mathrm{KT}} = \delta_0 + \sum_{1\leq k\leq n-3}\frac{\partial}{\partial\varepsilon^{\mu_{k+2}\ldots\mu_{n-1}}} + \frac{\partial}{\partial\varepsilon}d_{\mu_{n-1}}\varepsilon^{\mu_{n-1}}, \tag{D.24}$$
$$\Delta^{\mu_{k+2}\ldots\mu_{n-1}} = d_{\mu_{k+1}}\varepsilon^{\mu_{k+1}\mu_{k+2}\ldots\mu_{n-1}},$$

of antifield number -1. Its nilpotency results from the complete NI (D.22) and the equalities

$$d_{\mu_{k+2}}\Delta^{\mu_{k+2}\ldots\mu_{n-1}} = 0, \qquad k = 0,\ldots,n-3, \tag{D.25}$$

which are the $(k+1)$-stage NI (D.17). Then the KT complex (D.16) reads

$$0 \leftarrow \mathrm{Im}\,\overline{\delta} \xleftarrow{\overline{\delta}} \mathcal{S}^0_\infty[E_Y, Y]_1 \xleftarrow{\delta_0} P\{0\}_2 \xleftarrow{\delta_1} P\{1\}_3 \cdots \tag{D.26}$$
$$\xleftarrow{\delta_{n-3}} P\{n-3\}_{n-1} \xleftarrow{\delta_{\mathrm{KT}}} P\{n-2\}_n \xleftarrow{\delta_{\mathrm{KT}}} P\{n-2\}_{n+1}.$$

It obeys the above mentioned homology regularity condition as follows.

Lemma D.5 *Any δ_k-cycle $\Phi \in \mathcal{P}^0_\infty\{k\}_{k+3}$ up to a δ_k-boundary takes a form*

$$\Phi = \sum_{(k_1+\cdots+k_i+3i=k+3)} \sum_{(0\leq|\Lambda_1|,\ldots,|\Lambda_i|)} G^{\Lambda_1\cdots\Lambda_i}_{\mu^1_{k_1+2}\cdots\mu^1_{n-1};\ldots;\mu^i_{k_i+2}\cdots\mu^i_{n-1}} \tag{D.27}$$
$$d_{\Lambda_1}\Delta^{\mu^1_{k_1+2}\cdots\mu^1_{n-1}}\cdots d_{\Lambda_i}\Delta^{\mu^i_{k_i+2}\cdots\mu^i_{n-1}}, \qquad k_j = -1, 0, 1, \ldots, n-3,$$

where $k_j = -1$ stands for $\varepsilon^{\mu_1\cdots\mu_{n-1}}$ and $\Delta^{\mu_1\cdots\mu_{n-1}} = \mathcal{E}^{\mu_1\cdots\mu_{n-1}}$. It follows that Φ is a δ_{k+1}-boundary.

Proof Let us choose some basis element $\varepsilon^{\mu_{k+2}\cdots\mu_{n-1}}$ and denote it, simply, by ε. Let Φ contain a summand $\phi_1\varepsilon$, linear in ε. Then the cycle condition reads

$$\delta_k\Phi = \delta_k(\Phi - \phi_1\varepsilon) + (-1)^{[\varepsilon]}\delta_k(\phi_1)\varepsilon + \phi\Delta = 0, \qquad \Delta = \delta_k\varepsilon.$$

It follows that Φ contains a summand $\psi\Delta$ such that

$$(-1)^{[\varepsilon]+1}\delta_k(\psi)\Delta + \phi\Delta = 0.$$

This equality implies a relation

$$\phi_1 = (-1)^{[\varepsilon]+1}\delta_k(\psi) \tag{D.28}$$

because the reduction conditions (D.25) involve total derivatives of Δ, but not Δ. Hence, $\Phi = \Phi' + \delta_k(\psi\varepsilon)$, where Φ' contains no term linear in ε. Furthermore, let ε be even and Φ have a summand $\sum \phi_r\varepsilon^r$ polynomial in ε. Then the cycle condition leads to the equalities $\phi_r\Delta = -\delta_k\phi_{r-1}$, $r \geq 2$. Since ϕ_1 (D.28) is δ_k-exact, then $\phi_2 = 0$ and, consequently, $\phi_{r>2} = 0$. Thus, a cycle Φ up to a δ_k-boundary contains no term polynomial in c. It reads

$$\Phi = \sum_{(k_1+\cdots+k_i+3i=k+3)} \sum_{(0<|\Lambda_1|,\ldots,|\Lambda_i|)} G^{\Lambda_1\cdots\Lambda_i}_{\mu^1_{k_1+2}\cdots\mu^1_{n-1};\ldots;\mu^i_{k_i+2}\cdots\mu^i_{n-1}}$$
$$\varepsilon_{\Lambda_1}^{\mu^1_{k_1+2}\cdots\mu^1_{n-1}}\cdots\varepsilon_{\Lambda_i}^{\mu^i_{k_i+2}\cdots\mu^i_{n-1}}. \tag{D.29}$$

However, the terms polynomial in ε may appear under general coordinate transformations

$$\varepsilon'^{\nu_{k+2}\cdots\nu_{n-1}} = \det\left(\frac{\partial x^\alpha}{\partial x'^\beta}\right)\frac{\partial x'^{\nu_{k+2}}}{\partial x^{\mu_{k+2}}}\cdots\frac{\partial x'^{\nu_{n-1}}}{\partial x^{\mu_{n-1}}}\varepsilon^{\mu_{k+2}\cdots\mu_{n-1}}$$

of a chain Φ (D.29). In particular, Φ contains the summand

$$\sum_{k_1+\cdots+k_i+3i=k+3} F_{\nu^1_{k_1+2}\cdots\nu^1_{n-1};\ldots;\nu^i_{k_i+2}\cdots\nu^i_{n-1}}\varepsilon'^{\nu^1_{k_1+2}\cdots\nu^1_{n-1}}\cdots\varepsilon'^{\nu^i_{k_i+2}\cdots\nu^i_{n-1}},$$

which must vanish if Φ is a cycle. This takes place only if Φ factorizes through the graded densities $\Delta^{\mu_{k+2}\cdots\mu_{n-1}}$ (D.24) in accordance with the expression (D.27). □

Following the proof of Lemma D.5, one also can show that any δ_k-cycle $\Phi \in \mathcal{P}^0_\infty\{k\}_{k+2}$ up to a boundary takes a form

$$\Phi = \sum_{0\leq|\Lambda|} G^\Lambda_{\mu_{k+2}\cdots\mu_{n-1}} d_\Lambda \Delta^{\mu_{k+2}\cdots\mu_{n-1}},$$

i.e., the homology $H_{k+2}(\delta_k)$ of the complex (D.26) is finitely generated by the cycles $\Delta^{\mu_{k+2}\cdots\mu_{n-1}}$.

References

1. Almorox, A.: Supergauge theories in graded manifolds. Differential Geometric Methods in Mathematical Physics. Lecture Notes in Mathematics, vol. 1251, pp. 114–136. Springer, Berlin (1987)
2. Anderson, I., Duchamp, T.: On the existence of global variational principles. Am. J. Math. **102**, 781–867 (1980)
3. Anderson, I.: Introduction to the variational bicomplex. Contemp. Math. **132**, 51–73 (1992)
4. Atiyah, M., Macdonald, I.: Introduction to Commutative Algebra. Addison-Wesley, London (1969)
5. Bak, D., Cangemi, D., Jackiw, R.: Energy-momentum conservation law in gravity theories. Phys. Rev. D **49**, 5173–5181 (1994)
6. Barnich, G., Brandt, F., Henneaux, M.: Local BRST cohomology in gauge theories. Phys. Rep. **338**, 439–569 (2000)
7. Bartocci, C., Bruzzo, U., Hernández Ruipérez, D.: The Geometry of Supermanifolds. Kluwer, Boston (1991)
8. Bashkirov, D., Sardanashvily, G.: On the BV quantization of gauge gravitation theory. Int. J. Geom. Methods Mod. Phys. **2**, 203–226 (2005)
9. Bashkirov, D., Giachetta, G., Mangiarotti, L., Sardanashvily, G.: Noether's second theorem for BRST symmetries. J. Math. Phys. 46, 053517 (23 pp) (2005)
10. Bashkirov, D., Giachetta, G., Mangiarotti, L., Sardanashvily, G.: Noether's second theorem in a general setting. Reducible gauge theories. J. Phys. A **38**, 5329–5344 (2005)
11. Bashkirov, D., Giachetta, G., Mangiarotti, L., Sardanashvily, G.: The antifield Koszul-Tate complex of reducible Noether identities. J. Math. Phys. 46, 103513 (19 pp) (2005)
12. Bashkirov, D., Giachetta, G., Mangiarotti, L., Sardanashvily, G.: The KT-BRST complex of degenerate Lagrangian systems. Lett. Math. Phys. **83**, 237–252 (2008)
13. Batalin, I., Vilkovisky, G.: Closure of the gauge algebra, generalized Lie algebra equations and Feynman rules. Nucl. Phys. B **234**, 106–124 (1984)
14. Batchelor, M.: The structure of supermanifolds. Trans. Am. Math. Soc. **253**, 329–338 (1979)
15. Bauderon, M.: Differential geometry and Lagrangian formalism in the calculus of variations. In: Differential Geometry, Calculus of Variations, and their Applications. Lecture Notes in Pure and Applied Mathematics, vol. 100, pp. 67–82. Dekker, New York (1985)
16. Birmingham, D., Blau, M.: Topological field theory. Phys. Rep. **209**, 129–340 (1991)
17. Blagojević, M., Hehl, F. (eds.): Gauge Theories of Gravitation. A Reader with Commentaries. Imperial College Press, London (2013)
18. Bolsinov, A., Jovanović, B.: Noncommutative integrability, moment map and geodesic flows. Ann. Glob. Anal. Geom. **23**, 305–322 (2003)

G. Sardanashvily, *Noether's Theorems*, Atlantis Studies in Variational Geometry 3, DOI 10.2991/978-94-6239-171-0

19. Borowiec, A., Ferraris, M., Francaviglia, M., Volovich, I.: Energy-momentum complex for nonlinear gravitational Lagrangians in the first-order formalism. Gen. Rel. Grav. **26**, 637–645 (1994)
20. Borowoiec, A., Fatibene, L., Ferraris, M., Mercadante, S.: Covariant Lagrangian formulation of Chern-Simons theories. Int. J. Geom. Methods Mod. Phys. **4**, 277–284 (2007)
21. Boyer, C., Sánchez Valenzuela, O.: Lie supergroup action on supermanifolds. Trans. Am. Math. Soc. **323**, 151–175 (1991)
22. Brandt, F.: Jet coordinates for local BRST cohomology. Lett. Math. Phys. **55**, 149–159 (2001)
23. Bredon, G.: Sheaf Theory. McGraw-Hill, New York (1967)
24. Bruzzo, U.: The global Utiyama theorem in Einstein-Cartan theory. J. Math. Phys. **28**, 2074–2077 (1987)
25. Bryant, R., Chern, S., Gardner, R., Goldschmidt, H., Griffiths, P.: Exterior Differential Systems. Springer, New York (1991)
26. Cantrijn, A., Ibort, A., De León, M.: On the geometry of multisymplectic manifolds. J. Aust. Math. Soc. Ser. A. **66**, 303–330 (1999)
27. Cariñena, J., Crampin, M., Ibort, I.: On the multisymplectic formalism for first order field theories. Diff. Geom. Appl. **1**, 345–374 (1991)
28. Cariñena, J., Figueroa, H.: Singular Lagrangian in supermechanics. Diff. Geom. Appl. **18**, 33–46 (2003)
29. Cianci, R.: Introduction to Supermanifolds. Bibliopolis, Naples (1990)
30. Cianci, R., Francaviglia, M., Volovich, I.: Variational calculus and Poincaré-Cartan formalism in supermanifolds. J. Phys. A. **28**, 723–734 (1995)
31. Dazord, P., Delzant, T.: Le probleme general des variables actions-angles. J. Diff. Geom. **26**, 223–251 (1987)
32. Dewisme, A., Bouquet, S.: First integrals and symmetries of time-dependent Hamiltonian systems. J. Math. Phys. **34**, 997–1006 (1993)
33. Dodson, C.: Categories, Bundles and Spacetime Topology. Shiva Publishing Limited, Orpington (1980)
34. Dustermaat, J.: On global action-angle coordinates. Commun. Pure Appl. Math. **33**, 687–706 (1980)
35. Echeverría Enríquez, A., Muñoz Lecanda, M., Román Roy, N.: Geometrical setting of time-dependent regular systems. Alternative models. Rev. Math. Phys. **3**, 301–330 (1991)
36. Echeverría Enríquez, A., Muñoz Lecanda, M., Roman-Roy, N.: Geometry of multisymplectic Hamiltonian first-order field theories. J. Math. Phys. **41**, 7402–7444 (2000)
37. Eguchi, T., Gilkey, P., Hanson, A.: Gravitation, gauge theories and differential geometry. Phys. Rep. **66**, 213–393 (1980)
38. Fassó, F.: Superintegrable Hamiltonian systems: geometry and applications. Acta Appl. Math. **87**, 93–121 (2005)
39. Fatibene, L., Ferraris, M., Francaviglia, M.: Nöther formalism for conserved quantities in classical gauge field theories. J. Math. Phys. **35**, 1644–1657 (1994)
40. Fatibene, L., Ferraris, M., Francaviglia, M., McLenaghan, R.: Generalized symmetries in mechanics and field theories. J. Math. Phys. **43**, 3147–3161 (2002)
41. Fatibene, L., Francaviglia, M., Mercadante, S.: Covariant formulation of Chern-Simons theories. J. Geom. Methods Mod. Phys. **2**, 993–1008 (2005)
42. Ferraris, M., Francaviglia, M.: Energy-momentum tensors and stress tensors in geometric field theories. J. Math. Phys. **26**, 1243–1252 (1985)
43. Fiorani, E., Sardanashvily, G.: Noncommutative integrability on noncompact invariant manifold. J. Phys. A **39**, 14035–14042 (2006)
44. Fisch, J., Henneaux, M.: Homological perturbation theory and algebraic structure of the antifield-antibracket formalism for gauge theories. Commun. Math. Phys. **128**, 627–640 (1990)
45. Forger, M., Romero, S.: Covariant Poisson brackets in geometric field theory. Commun. Math. Phys. **256**, 375–410 (2005)

46. Franco, D., Polito, C.: Supersymmetric field-theoretic models on a supermanifold. J. Math. Phys. **45**, 1447–1473 (2004)
47. Fuks, D.: Cohomology of Infinite-Dimensional Lie Algebras. Consultants Bureau, New York (1986)
48. Fulp, R., Lada, T., Stasheff, J.: Sh-Lie algebras induced by gauge transformations. Commun. Math. Phys. **231**, 25–43 (2002)
49. Fulp, R., Lada, T., Stasheff, J.: Noether variational Theorem II and the BV formalism. Rend. Circ. Mat. Palermo **2**(Suppl. 71), 115–126 (2003)
50. Fuster, A., Henneaux, M., Maas, A.: BRST quantization: a short review. Int. J. Geom. Methods Mod. Phys. **2**, 939–964 (2005)
51. García, P.: Gauge algebras, curvature and symplectic structure. J. Diff. Geom. **12**, 209–227 (1977)
52. Giachetta, G., Sardanashvily, G.: Stress-energy-momentum of affine-metric gravity. Generalized Komar superportential. Class. Quant. Grav. **13**, L67–L71 (1996)
53. Giachetta, G., Mangiarotti, L., Sardanashvily, G.: New Lagrangian and Hamiltonian Methods in Field Theory. World Scientific, Singapore (1997)
54. Giachetta, G., Mangiarotti, L., Sardanashvily, G.: Covariant Hamiltonian equations for field theory. J. Phys. A **32**, 6629–6642 (1999)
55. Giachetta, G., Mangiarotti, L., Sardanashvily, G.: Iterated BRST cohomology. Lett. Math. Phys. **53**, 143–156 (2000)
56. Giachetta, G., Mangiarotti, L., Sardanashvily, G.: Cohomology of the infinite-order jet space and the inverse problem. J. Math. Phys. **42**, 4272–4282 (2001)
57. Giachetta, G., Mangiarotti, L., Sardanashvily, G.: Noether conservation laws in higher-dimensional Chern-Simons theory. Mod. Phys. Lett. A. **18**, 2645–2651 (2003)
58. Giachetta, G., Mangiarotti, L., Sardanashvily, G.: Geometric and Algebraic Topological Methods in Quantum Mechanics. World Scientific, Singapore (2005)
59. Giachetta, G., Mangiarotti, L., Sardanashvily, G.: Lagrangian supersymmetries depending on derivatives. Global analysis and cohomology. Commun. Math. Phys. **259**, 103–128 (2005)
60. Giachetta, G., Mangiarotti, L., Sardanashvily, G.: On the notion of gauge symmetries of generic Lagrangian field theory. J. Math. Phys 50, 012903 (19 pp) (2009)
61. Giachetta, G., Mangiarotti, L., Sardanashvily, G.: Advanced Classical Field Theory. World Scientific, Singapore (2009)
62. Giachetta, G., Mangiarotti, L., Sardanashvily, G.: Geometric Formulation of Classical and Quantum Mechanics. World Scientific, Singapore (2010)
63. Gomis, J., París, J., Samuel, S.: Antibracket, antifields and gauge theory quantization. Phys. Rep. **295**, 1–145 (1995)
64. Gotay, M.: A multisymplectic framework for classical field theory and the calculus of variations. In: Francaviglia, M. (ed.) Mechanics, Analysis and Geometry: 200 Years after Lagrange, pp. 203–235. North Holland, Amsterdam (1991)
65. Gotay, M.: A multisymplectic framework for classical field theory and the calculus of variations. II. Space + time decomposition. Diff. Geom. Appl. 1, 375–390 (1991)
66. Gotay, M., Marsden, J.: Stress-energy-momentum tensors and the Belinfante-Rosenfeld formula. Contemp. Math. **132**, 367–392 (1992)
67. Greub, W., Halperin, S., Vanstone, R.: Connections, Curvature and Cohomology. Academic Press, New York (1972)
68. Guillemin, V., Sternberg, S.: Symplectic Techniques in Physics. Cambridge University Press, Cambridge (1984)
69. Günther, G.: The polysymplectic Hamiltonian formalism in field theory and calculus of variations, I: The local case. J. Diff. Geom. **25**, 23–53 (1987)
70. Hamoui, A., Lichnerowicz, A.: Geometry of dynamical systems with time-dependent constraints and time-dependent Hamiltonians: an approach towards quantization. J. Math. Phys. **25**, 923–931 (1984)
71. Hehl, F., McCrea, J., Mielke, E., Ne'eman, Y.: Metric-affine gauge theory of gravity: field equations, Noether identities, world spinors, and breaking of dilaton invariance. Phys. Rep. **258**, 1–171 (1995)

72. Helein, F., Kouneiher, J.: Covariant Hamiltonian formalism for the calculus of variations with several variables. Adv. Theor. Math. Phys. **8**, 565–601 (2004)
73. Hernández Ruipérez, D., Muñoz Masqué. J.: Global variational calculus on graded manifolds. J. Math. Pures Appl. **63**, 283–309 (1984)
74. Van der Heuvel, B.: Energy-momentum conservation in gauge theories. J. Math. Phys. **35**, 1668–1687 (1994)
75. Hirzebruch, F.: Topological Methods in Algebraic Geometry. Springer, New York (1966)
76. Ibragimov, N.: Transformation Groups Applied to Mathematical Physics. Riedel, Boston (1985)
77. Ivanenko, D., Sardanashvily, G.: The gauge treatment of gravity. Phys. Rep. **94**, 1–45 (1983)
78. Jadczyk, A., Pilch, K.: Superspaces and supersymmetries. Commun. Math. Phys. **78**, 373–390 (1981)
79. Julia, B., Silva, S.: Currents and superpotentials in classical gauge invariant theories. Local results with applications to perfect fluids and General Relativity. Class. Quant. Grav. 15, 2137–2215 (1998)
80. Keyl, M.: About the geometric structure of symmetry breaking. J. Math. Phys. **32**, 1065–1071 (1991)
81. Kijowski, J., Tulczyjew, W.: A Symplectic Framework for Field Theories. Springer, Berlin (1979)
82. Kobayashi, S., Nomizu, K.: Foundations of Differential Geometry, vols. 1 and 2. Wiley, New York (1963)
83. Kolář, I., Michor, P., Slovák, J.: Natural Operations in Differential Geometry. Springer, New York (1993)
84. Kosmann-Schwarzbach, Y.: The Noether Theorems. Invariance and the Conservation Laws in the Twentieth Century. Springer, New York (2011)
85. Krasil'shchik, I., Lychagin, V., Vinogradov, A.: Geometry of Jet Spaces and Nonlinear Partial Differential Equations. Gordon and Breach, Glasgow (1985)
86. Krupka, D.: Variational sequences on finite order jet spaces. In: Janyska, J., Krupka, D. (eds.) Differential Geometry and its Applications, pp. 236–254. World Scientific, Singapore (1990)
87. Krupka, D., Musilova, J.: Trivial Lagrangians in field theory. Diff. Geom. Appl. **9**, 293–505 (1998)
88. Krupka, D.: Global variational theory in fibred spaces. In: Krupka, D., Saunders, D. (eds.) Handbook of Global Analysis, pp. 773–836. Elsevier, Amsterdam (2007)
89. Krupka, D.: Introduction to Global Variational Geometry. Springer, Berlin (2015)
90. Krupka, D., Moreno, G., Urban, Z., Volná J.: On a bicomplex induced by the variational sequence. Int. J. Geom. Methods Mod. Phys. 12, 1550057 (15 pp) (2015)
91. Krupkova, O., Smetanova, D.: Legendre transformations for regularizable Lagrangian in field theory. Lett. Math. Phys. **58**, 189–204 (2001)
92. Krupkova, O.: Hamiltonian field theory. J. Geom. Phys. **43**, 93–132 (2002)
93. Lang, S.: Algebra. Addison-Wisley, New York (1993)
94. De León, M., Rodrigues, P.: Methods of Differential Geometry in Analytical Mechanics. North-Holland, Amsterdam (1989)
95. De León, M., de Diego, M.D., Santamaría-Merini, A.: Symmetries in classical field theory. Int. J. Geom. Methods Mod. Phys. **1**, 651–710 (2004)
96. De Leon, M., Saldago, M., Vilariaño, S.: Methods of Differential Geometryin Classical Field Theories. k-Symplectic and k-Cosymplectic Approaches. World Scientific, Singapore (2016)
97. Mac Lane, S.: Homology. Springer, New York (1967)
98. Mangiarotti, L., Sardanashvily, G.: Gauge Mechanics. World Scientific, Singapore (1998)
99. Mangiarotti, L., Sardanashvily, G.: Constraints in Hamiltonian time-dependent mechanics. J. Math. Phys. **41**, 2858–2876 (2000)
100. Mangiarotti, L., Sardanashvily, G.: Connections in Classical and Quantum Field Theory. World Scientific, Singapore (2000)
101. Mangiarotti, L., Sardanashvily, G.: Quantum mechanics with respect to different reference frames. J. Math. Phys. 48, 082104 (12 pp) (2007)

102. Marathe, K., Martucci, G.: The Mathematical Foundations of Gauge Theories. North-Holland, Amsterdam (1992)
103. Marsden, J., Patrick, G., Shkoller, S.: Multisymplectic geometry, variational integrators and nonlinear PDEs. Commun. Math. Phys. **199**, 351–395 (1998)
104. Massey, W.: Homology and Cohomology Theory. Marcel Dekker Inc, New York (1978)
105. Meigniez, G.: Submersions, fibrations and bundles. Trans. Am. Math. Soc. **354**, 3771–3787 (2002)
106. Mishchenko, A., Fomenko, A.: Generalized Liouville method of integration of Hamiltonian systems. Funct. Anal. Appl. **12**, 113–121 (1978)
107. Monterde, J., Munos Masque, J., Vallejo. J.: The Poincaré-Cartan form in superfield theory. Int. J. Geom. Methods Mod. Phys. **3**, 775–822 (2006)
108. Olver, P.: Applications of Lie Groups to Differential Equations. Springer, New York (1986)
109. Palais, R.: A global formulation of Lie theory of transformation groups. Mem. Am. Math. Soc. **22**, 1 (1957)
110. Reinhart, B.: Differential Geometry and Foliations. Springer, New York (1983)
111. Rennie, A.: Smoothness and locality for nonunital spectral triples. K-Theory **28**, 127–165 (2003)
112. Rey, A., Roman-Roy, N., Saldago, M.: Gunther's formalism (k-symplectic formalism) in classical field theory: Skinner-Rusk approach and the evolution operator. J. Math. Phys. **46**, 052901 (2005)
113. Riewe, F.: Nonconservative Lagrangian and Hamiltonian mechanics. Phys. Rev. E **53**, 1890–1899 (1996)
114. Roman-Roy, N.: Multisymplectic Lagrangian and Hamiltonian formalisms of classical field theories. SIGMA 5, 100 (25 pp) (2009)
115. Sardanashvily, G.: On the geometry of spontaneous symmetry breaking. J. Math. Phys. **33**, 1546–1549 (1992)
116. G.Sardanashvily, G., Zakharov, O.: Gauge Gravitation Theory. World Scientific, Singapore (1992)
117. Sardanashvily, G., Zakharov, O.: On application of the Hamilton formalism in fibred manifolds to field theory. Diff. Geom. Appl. **3**, 245–263 (1993)
118. Sardanashvily, G.: Generalized Hamiltonian Formalism for Field Theory. World Scientific, Singapore (1995)
119. Sardanashvily, G.: Stress-energy-momentum tensors in constraint field theories. J. Math. Phys. **38**, 847–866 (1997)
120. Sardanashvily, G.: Stress-energy-momentum conservation law in gauge gravitation theory. Class. Quant. Grav. **14**, 1371–1386 (1997)
121. Sardanashvily, G.: Hamiltonian time-dependent mechanics. J. Math. Phys. **39**, 2714–2729 (1998)
122. Sardanashvily, G.: Cohomology of the variational complex in the class of exterior forms of finite jet order. Int. J. Math. Math. Sci. **30**, 39–48 (2002)
123. Sardanashvily, G.: Noether identities of a differential operator. The Koszul-Tate complex. Int. J. Geom. Methods Mod. Phys. **2**, 873–886 (2005)
124. Sardanashvily, G.: Preface. Gauge gravitation theory from geometric viewpoint. Int. J. Geom. Methods Mod. Phys. 3(1), v–xx (2006)
125. Sardanashvily, G.: Geometry of classical Higgs fields. Int. J. Geom. Methods Mod. Phys. **3**, 139–148 (2006)
126. Sardanashvily, G.: Graded infinite order jet manifolds. Int. J. Geom. Methods Mod. Phys. **4**, 1335–1362 (2007)
127. Sardanashvily, G.: Classical field theory. Advanced mathematical formulation. Int. J. Geom. Methods Mod. Phys. 5, 1163–1189 (2008)
128. Sardanashvily, G.: Gauge conservation laws in a general setting. Superpotential. Int. J. Geom. Methods Mod. Phys. 6, 1047–1056 (2009)
129. Sardanashvily, G.: Superintegrable Hamiltonian systems with noncompact invariant submanifolds. Kepler system. Int. J. Geom. Methods Mod. Phys. 6, 1391–1420 (2009)

130. Sardanashvily, G.: Classical gauge gravitation theory. Int. J. Geom. Methods Mod. Phys. **8**, 1869–1895 (2011)
131. Sardanashvily, G.: Lectures on Differential Geometry of Modules and Rings. Application to Quantum Theory. Lambert Academic Publishing, Saarbrucken (2012)
132. Sardanashvily, G.: Time-dependent superintegrable Hamiltonian systems. Int. J. Geom. Methods Mod. Phys. 9(N8), 1220016 (10 pp) (2012)
133. Sardanashvily, G.: Advanced Differential Geometry for Theoreticians. Fiber Bundles, Jet Manifolds and Lagrangian Theory. Lambert Academic Publishing, Saarbrucken (2013)
134. Sardanashvily, G.: Graded Lagrangian formalism. Int. J. Geom. Methods. Mod. Phys. 10, 1350016 (37 pp) (2013)
135. Sardanashvily, G.: Geometric formulation of non-autonomous mechanics. Int. J. Geom. Methods Mod. Phys. 10, 1350061 (12 pp) (2013)
136. Sardanashvily, G.: Handbook of Integrable Hamiltonian Systems. URSS Publication Group, Moscow (2015)
137. Sardanashvily, G.: Higher-stage Noether identities and second Noether theorems. Adv. Math. Phys. 2015, 127481 (19 pp) (2015)
138. Saunders, D.: The Geometry of Jet Bundles. Cambridge University Press, Cambridge (1989)
139. Spanier, E.: Algebraic Topology. McGraw-Hill, New York (1966)
140. Stavracou, T.: Theory of connections on graded principal bundles. Rev. Math. Phys. **10**, 47–79 (1998)
141. Steenrod, N.: The Topology of Fibre Bundles. Princeton University Press, Princeton (1972)
142. Sussmann, H.: Orbits of families of vector fields and integrability of distributions. Trans. Am. Math. Soc. **180**, 171–188 (1973)
143. Takens, P.: A global version of the inverse problem of the calculus of variations. J. Diff. Geom. **14**, 543–562 (1979)
144. Tennison, B.: Sheaf Theory. Cambridge University Press, Cambridge (1975)
145. Terng, C.: Natural vector bundles and natural differential operators. Am. J. Math. **100**, 775–828 (1978)
146. Trautman, A.: Differential Geometry for Physicists. Bibliopolis, Naples (1984)
147. Tulczyiew, W.: The Euler-Lagrange resolution. In: Differential Geometric Methods in Mathematical Physics (Proc. Conf., Aix-en-Provence/Salamanca, 1979) Lecture Notes in Mathematics, vol. 836, pp. 22–48. Springer, Berlin (1980)
148. Vaisman, I.: Lectures on the Geometry of Poisson Manifolds. Birkhäuser, Basel (1994)
149. Vinogradov, A.: The $\mathcal{C}$-spectral sequence, Lagrangian formalism, and conservation laws. II. The nonlinear theory. J. Math. Anal. Appl. **100**, 41–129 (1984)
150. Vitolo, R.: Finite order variational bicomplex. Math. Proc. Camb. Philos. Soc. **125**, 321–333 (1999)

Index

G. Sardanashvily, *Noether's Theorems*, Atlantis Studies
in Variational Geometry 3, DOI 10.2991/978-94-6239-171-0

D

O

P

Q

R

S

Printed in the United States
By Bookmasters